# Welding Codes, Standards, and Specifications

# Welding Codes, Standards, and Specifications

Jeffrey D. Mouser

**McGraw-Hill**

New York   San Francisco   Washington, D.C.   Auckland   Bogotá
Caracas   Lisbon   London   Madrid   Mexico City   Milan
Montreal   New Delhi   San Juan   Singapore
Sydney   Tokyo   Toronto

**Library of Congress Cataloging-in-Publication Data**

Mouser, Jeffrey D.
    Welding codes, standards, and specifications / Jeffrey D. Mouser.
        p.    cm.
    Includes index.
    ISBN 0-07-043550-2 (alk. paper)
    1. Welding—Standards.   I. Title.
TS227.2.M69    1997
671.5′2′0218—dc21
                                                  97-37604
                                                     CIP

## McGraw-Hill

*A Division of The **McGraw·Hill** Companies*

1 2 3 4 5 6 7 8 9 0   DOC/DOC  9 0 2 1 0 9 8 7

ISBN 0-07-043550-2

*The sponsoring editor for this book was Harold B. Crawford and the
production supervisor was Tina Cameron. It was set in Century
Schoolbook by Progressive Information Technologies.*

*Printed and bound by R.R. Donnelley & Sons Company.*

This book is printed on recycled, acid-free paper containing a minimum of
50% recycled de-inked fiber.

McGraw-Hill books are available at special quantity discounts to use as premiums and
sales promotions, or for use in corporate training programs. For more information,
please write to the Director of Special Sales, McGraw-Hill, 11 West 19th Street, New
York, NY 10011. Or contact your local bookstore.

To my wife, Debbie Mouser

# Contents

## viii    Contents

## Section D.  High-Strength Bolts                                  309

xx    Contents

# Preface

This manuscript was originally designed and compiled for use by inspection and quality assurance personnel. It has however, proven to be a very useful tool to engineers, architects, general and subcontractors, structural steel fabricators, welders, project managers, building officials, instructors and students.

All references to codes and/or standards included in this publication are listed in alphabetical order, rather than in the traditional numerical order, and have been thoroughly cross-referenced. By alphabetizing the codes the user can quickly reference any subject he or she chooses. This book is divided into six sections outlined in the following order:

## A. GENERAL INFORMATION

This section begins with a fairly comprehensive collection of Abbreviations and their respective meanings, followed by the Legend for the Prequalified Full Penetration and Prequalified Partial Penetration Joint configurations. Next is a table of Structural Shape Size Groupings listing available Wide Flange sizes for Groups 1-4, along with Size Grouping for M Shapes, S Shapes, HP Shapes, Standard Channel Iron, Miscellaneous Channel Iron, and Angle Iron.

Continuing, you will discover tables on Wide Flange Dimensions, M Shape Dimensions, S Shape Dimensions, HP Shape Dimensions, Channel Iron Dimensions, Miscellaneous Channel Iron Dimensions, Pipe Dimensions and finally, Structural Tubing Dimensions, including round, square and rectangular shapes. Following these can be found a table on Sheet Steel Thicknesses.

There are actually two tables to be found here, one on Hot and Cold Rolled Sheets and the other covering Galvanized Sheets. Both tables offer thickness equivalents in both inches and millimeters beginning with 3 Gage and ending with 32 Gage.

Last in this Section are the tables on Welding Symbols, from which the user will find information on Typical Welding Symbols, Basic

Welding Symbols, Supplementary Symbols, Basic Joints defining the
Arrow Side, Other Side significance and information on the location
of, and definitions of the different elements that can make up a Welding Symbol and where on, above, or below the Reference Line the elements are located.

## B.   STRUCTURAL STEEL

This Section contains a vast amount of information on the subjects of
Structural Steel and Welding and it to is in alphabetical order. It begins with the subject, Allowable Roughness in Oxygen Cut Surfaces
and ends with Weld Washers, Wind Velocity and Workmanship. The
Maximum and Minimum subject headings are of particular interest
when, for example, the user is searching for Maximum Effective Weld
Sizes, or Maximum Electrode Diameters allowed for each welding
process and each position. By simply looking up the Maximum heading, these and other subjects become easily accessed.

Minimum Edge Distances, Hole Sizes and Preheat and Interpass
Temperatures are only a few of the subjects that are readily available
and easily accessible under the Minimum Heading.

## C.   TABLES

Here the user will find a very complete compilation of over eighty Tables and Figures, all listed in alphabetical order. When searching for
information on Oversized Holes for Anchor bolts for instance, simply
look up the word Anchor Bolts and the table for Oversized Holes for
anchor bolts in base plates can be found, along with its corresponding
code number location.

Tables and Figures on the subjects of Boxing, Effective Throats,
Filler Metal Requirements, Joists, Hole Dimensions, Nut Rotation
from Snug Tight, Transitions and Weld Washers are only a few exam-
ples of the information that is to be found in this section.

## D.   HIGH STRENGTH BOLTS

As its name suggests, this section is devoted almost entirely to High
Strength Bolts. An example of the subjects that can be found in this
section are: Anchor Bolts, A325, A490, and A449 bolts, Slopes of connected parts, Calibrated Wrench and Turn-of-Nut methods of tightening, information on Faying Surfaces, Galvanized Bolts, Installation
and Inspection, just to name a few.

The format of this book remains the same in its entirety. For example, when searching for information on the proper usage of A449 bolts,
simply look up the heading A449, directly below the heading the user
will find the applicable information.

### E.   REINFORCING STEEL

Section E devotes itself entirely to Reinforcing Steel and the welding thereof. Subjects range from Allowable Gaps in different types of joints, to Carbon Equivalents, Contractor Obligations, Filler Metal Requirements, Effective Weld Lengths, Areas, Throats and Sizes, to Undercut and Wind Velocity.

### F.   SHEET STEEL (DECKING)

This section concerns itself only with the welding of Decking. An example of the subjects covered are: Approval Needed, Spot and Puddle Weld Diameters, Cracks, Melting Rates, Qualification Procedures, Studs and Welding Processes.

The simplicity and straightforwardness of this publication is unsurpassed as can be demonstrated by turning to page B-1 and finding the Anchor Bolts entry located in the left hand column. Opposite the words Anchor Bolts and in the right hand column can be found the following three entries:

1. UBC, 2-358, 2210

2. AISC, 5-79, J10

3. AISC, 4-4, Table 1-C

These entries are explained as follows:

1. Uniform Building Code, Volume 2, page 358, Section 2210.

2. American Institute of Steel Construction, Part 5, page 79, Section J10.

3. American Institute of Steel Construction, Part 4, page 4, Table 1-C.

The text found directly beneath Anchor Bolts begins: UBC, . . . "The Protrusions of". This is the information that will be found at UBC, 2-358, 2210. Additional Anchor Bolt information can be found at AISC, 5-79, J10 and at AISC, 4-4, Table 1-C. This format is consistent throughout the compilation. The text found directly beneath the subject in the left hand column is representative of the first item found in the right hand column.

Many of the codes, standards and specifications used in this compilation have been printed in part, while others have been printed in their entirety, therefore, for complete and detailed information regarding any code, standard, or specification, it is necessary that the users of this book have in their possession all required codes and standards.

For added convenience, a comprehensive Table of Contents has been included.

# Acknowledgments

Portions of this book have been reproduced from the 1994 edition of the *Uniform Building Code*™, copyright 1994, with the permission of the publisher, the International Conference of Building Officials. Complete copies of the *Uniform Building Code* can be purchased by calling the ICBO order department at (800) 284-4406.

Additional portions of the document are copied from the BOCA National Building Code/1990, copyright 1989, Building Officials and Code Administrators International Inc., 4051 West Flossmoor Road, Country Club Hills, Illinois 60478. Published by arrangement with author, all rights reserved.

Southern Building Code Congress International information found within this text has been reproduced from the 1992 edition of the Standard Building Code, with the permission of the copyright holder, Southern Building Code Congress International, Inc., all rights reserved.

All material relating to the American Welding Society has been reproduced with the full authorization of the copyright holder, American Welding Society.

# List of Abbreviations

The following is a legend for the abbreviations of the codes and standards found throughout this publication.

| | |
|---|---|
| AISC | American Institute of Steel Construction-9th Edition |
| AISC, CSP | Code of Standard Practice |
| A2.4 | American Welding Society, A2.4 - 93 |
| A3.0 | American Welding Society, A3.0 - 94 |
| B1.10 | American Welding Society, B1.10 - 86 |
| D1.1 | American Welding Society, D1.1 - 94 |
| D1.1, Comm. | Commentary |
| D1.3 | American Welding Society, D1.3 - 89 |
| D1.4 | American Welding Society, D1.4 - 92 |
| UBC | Uniform Building Code - 1994 |
| SBCCI | Southern Building Code Congress International |
| BOCA | Building Officials & Code Administrators International |

# A General Information

## Abbreviations

| | | | | |
|---|---|---|---|---|
| AB | Anchor Bolt | | COP | Coefficient of Performance |
| AE | Air Eliminator | | CP | Complete Penetration |
| AFF | Above Finished Floor | | DEG | Degree(s) |
| ANC | Anchor | | DN | Down |
| APPROX | Approximately | | EA | Each |
| ARI | Air Conditioning and Refrigeration Institute | | EF | Exhaust Fan |
| BC | Boom Cabinet | | EL or ELEV | Elevation |
| BC1E | Bevel Cut One End | | ENT | Entering |
| BC2E | Bevel Cut Two Ends | | ES | Emergency Shower |
| BEV-1-E | Bevel Cut One End | | EST | Estimated |
| BEV | Bevel | | EXT | Exterior |
| BI | Backward Inclined | | FB | Flat Bar |
| BM | Beam | | FC | Forward Curved / Fueling Crane |
| BNT | Bent | | FD | Floor Drain |
| BOD | Bottom of Duct | | FIN | Finish |
| BOP | Bottom of Pipe | | FLG | Flange |
| BOT | Bottom of Tank | | FL | Flow Line |
| BS | Basket Strainer | | FOB | Freight on Board |
| BTU/H | British Thermal Units per Hour | | FP | Full Penetration |
| CC | Center-to-Center | | FPM | Feet per Minute |
| CFM | Cubic Feet per Minute | | FS | Flow Switch |
| CF/H | Cubic Feet per Hour | | FT | Foot (Feet) |
| COL | Column | | GA | Gage |

| | | | |
|---|---|---|---|
| GPM | Gallons per Minute | P | Pump |
| GR | Grade | PDI | Pressure Differential Indicator |
| HP | Horse Power | PE | Plain End |
| HS | High Strength | PL | Plate |
| HSB | High Strength Bolt | PRV/PCV | Pressure Relief Valve / Pressure Control Valve |
| HT | High Tension | QTY | Quantity |
| HTB | High Tension Bolt | R or RAD | Radius |
| HVAC | Heating, Ventilating, Air Conditioning | REBAR | Reinforcing Steel |
| HX | Heat Exchanger | REF | Reference |
| IN | Inch | REV | Revision |
| IN WC | Inches of Water Column | REQ | Required |
| LVG | Leaving | RPZ | Reduced Pressure Zone |
| M | Meter | R/L | Right/Left |
| M BOLT | Machine Bolt | 3/2 R/L | 3 Right and 2 Left |
| MAT$^L$ | Material | SA | Supply Air or Surge Arrestor |
| MAU | Make-up Air Unit | SC1E | Square Cut One End |
| MAX | Maximum | SECT | Section |
| MD | Motorized Damper | SF | Square Feet |
| MIN | Minimum | SF | Supply Fan |
| M-1-E | Mill One End | SP | Static Pressure or Set Point |
| M1E-SC1E | Mill One End, Square Cut One End | SPEC | Specification |
| M-2-E | Mill Two Ends | SQ | Square |
| NC | Normally Closed | SS | Stainless Steel |
| NIC | Not in Contract | STL | Steel |
| NO | Normally Open | T & B | Top and Bottom |
| NO | Number | T-1-E 6″ | Thread One End 6″ |
| NP | No Paint | THRU | Through |
| OA | Outside Air | TOC | Top of Concrete |
| OC | On Center | TOD | Top of Duct |
| OPP | Opposite | TOS | Top of Steel |
| OPP HD | Opposite Hand | | |

| | | | | |
|---|---|---|---|---|
| TS | Temperature Sensor | | UNO | Unless Noted Otherwise |
| TS | Tube Steel | | VD | Volume Damper |
| TSG | Tapered Steel Girder | | VTR | Vent Through Roof |
| TW | Tack Weld | | WB | Wet Bulb |
| TYP | Typical | | WF | Wide Flange |
| UH | Unit Heater | | WFB | Wide Flange Beam |
| UN | Unless Noted | | W/ | With |

## Prequalified Weld Legend

**Symbols for joint types**
> B — butt joint
> C — corner joint
> T — T-joint
> BC — butt or corner joint
> TC — T- or corner joint
> BTC — butt, T-, or corner joint

**Symbols for base metal thickness and penetration**
> L — limited thickness–complete joint penetration
> U — unlimited thickness–complete joint penetration
> P — partial joint penetration

**Symbols for weld types**
> 1 — square-groove
> 2 — single-V-groove
> 3 — double-V-groove
> 4 — single-bevel-groove
> 5 — double-bevel-groove
> 6 — single-U-groove
> 7 — double-U-groove
> 8 — single-J-groove
> 9 — double-J-groove
> 10 — flare-bevel-groove

**Symbols for welding processes if not shielded metal arc**
> S — submerged arc welding
> G — gas metal arc welding
> F — flux cored arc welding

**Welding processes**
> SMAW — shielded metal arc welding
> GMAW — gas metal arc welding
> FCAW — flux cored arc welding
> SAW — submerged arc welding

**Welding positions**
> F — flat
> H — horizontal
> V — vertical
> OH — overhead

The lower case letters, e.g., a, b, c, etc., are used to differentiate between joints that would otherwise have the same joint designation.

# Prequalified Weld Tables

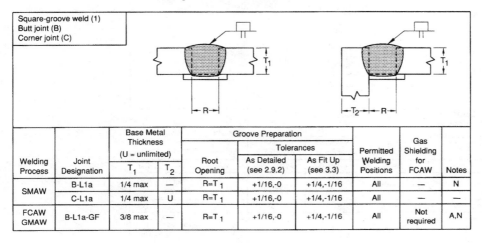

Square-groove weld (1)
Butt joint (B)
Corner joint (C)

| Welding Process | Joint Designation | Base Metal Thickness (U = unlimited) | | Groove Preparation | | | Permitted Welding Positions | Gas Shielding for FCAW | Notes |
| | | $T_1$ | $T_2$ | Root Opening | As Detailed (see 2.9.2) | As Fit Up (see 3.3) | | | |
|---|---|---|---|---|---|---|---|---|---|
| SMAW | B-L1a | 1/4 max | — | $R=T_1$ | +1/16,-0 | +1/4,-1/16 | All | — | N |
| | C-L1a | 1/4 max | U | $R=T_1$ | +1/16,-0 | +1/4,-1/16 | All | — | — |
| FCAW GMAW | B-L1a-GF | 3/8 max | — | $R=T_1$ | +1/16,-0 | +1/4,-1/16 | All | Not required | A,N |

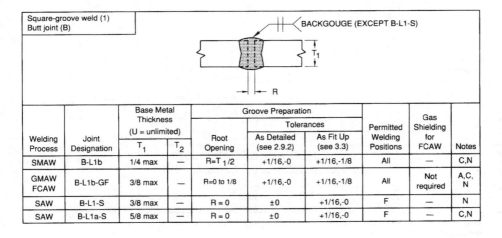

Square-groove weld (1)
Butt joint (B)

BACKGOUGE (EXCEPT B-L1-S)

| Welding Process | Joint Designation | Base Metal Thickness (U = unlimited) | | Groove Preparation | | | Permitted Welding Positions | Gas Shielding for FCAW | Notes |
| | | $T_1$ | $T_2$ | Root Opening | As Detailed (see 2.9.2) | As Fit Up (see 3.3) | | | |
|---|---|---|---|---|---|---|---|---|---|
| SMAW | B-L1b | 1/4 max | — | $R=T_1/2$ | +1/16,-0 | +1/16,-1/8 | All | — | C,N |
| GMAW FCAW | B-L1b-GF | 3/8 max | — | R=0 to 1/8 | +1/16,-0 | +1/16,-1/8 | All | Not required | A,C,N |
| SAW | B-L1-S | 3/8 max | — | R = 0 | ±0 | +1/16,-0 | F | — | N |
| SAW | B-L1a-S | 5/8 max | — | R = 0 | ±0 | +1/16,-0 | F | — | C,N |

Note A:  Not prequalified for gas tungsten arc and gas metal arc welding using short circuiting transfer. Refer to Appendix A.
Note C:  Backgouge root to sound metal before welding second side.
Note N:  The orientation of the two members in the joints may vary from 135° to 180° provided that the basic joint configuration (groove angle, root face, root opening) remain the same and that the design weld size is maintained.

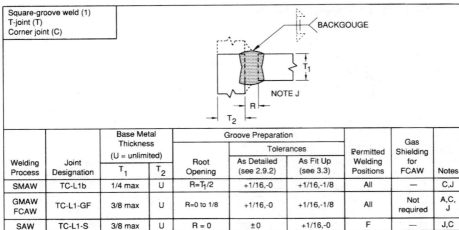

**Square-groove weld (1)**
**T-joint (T)**
**Corner joint (C)**

BACKGOUGE

$T_1$

NOTE J

R

$T_2$

| Welding Process | Joint Designation | Base Metal Thickness (U = unlimited) | | Groove Preparation | | | Permitted Welding Positions | Gas Shielding for FCAW | Notes |
|---|---|---|---|---|---|---|---|---|---|
| | | $T_1$ | $T_2$ | Root Opening | As Detailed (see 2.9.2) | As Fit Up (see 3.3) | | | |
| | | | | | Tolerances | | | | |
| SMAW | TC-L1b | 1/4 max | U | $R=T_1/2$ | +1/16,-0 | +1/16,-1/8 | All | — | C,J |
| GMAW FCAW | TC-L1-GF | 3/8 max | U | R=0 to 1/8 | +1/16,-0 | +1/16,-1/8 | All | Not required | A,C, J |
| SAW | TC-L1-S | 3/8 max | U | R = 0 | ±0 | +1/16,-0 | F | — | J,C |

**Single-V-groove weld (2)**
**Butt joint (B)**

$\alpha$

$T_1$

R

| Tolerances | |
|---|---|
| As Detailed (see 2.9.2) | As Fit Up (see 3.3) |
| R=+1/16,-0 | +1/4,-1/16 |
| $\alpha = +10°, -0°$ | +10°,- 5° |

| Welding Process | Joint Designation | Base Metal Thickness (U = unlimited) | | Groove Preparation | | Permitted Welding Positions | Gas Shielding for FCAW | Notes |
|---|---|---|---|---|---|---|---|---|
| | | $T_1$ | $T_2$ | Root Opening | Groove Angle | | | |
| SMAW | B-U2a | U | — | R=1/4 | $\alpha = 45°$ | All | — | N |
| | | | | R=3/8 | $\alpha = 30°$ | F,V,OH | — | N |
| | | | | R=1/2 | $\alpha = 20°$ | F,V,OH | — | N |
| GMAW FCAW | B-U2a-GF | U | | R=3/16 | $\alpha = 30°$ | F,V,OH | Required | A,N |
| | | | | R=3/8 | $\alpha = 30°$ | F,V,OH | Not req. | A,N |
| | | | | R=1/4 | $\alpha = 45°$ | F,V,OH | Not req. | A,N |
| SAW | B-L2a-S | 2 max | — | R=1/4 | $\alpha = 30°$ | F | — | N |
| SAW | B-U2-s | U | — | R=5/8 | $\alpha = 20°$ | F | — | N |

Note A: Not prequalified for gas tungsten arc and gas metal arc welding using short circuiting transfer. Refer to Appendix A.

Note C: Backgouge root to sound metal before welding second side.

Note J: If fillet welds are used in statically loaded structures to reinforce groove welds in corner and T-joints, they shall be equal to 1/4 $T_1$, but need not exceed 3/8 in. Groove welds in corner and T-joints of dynamically loaded structures shall be reinforced with fillet welds equal to 1/4 $T_1$, but not more than 3/8 in.

Note N: The orientation of the two members in the joints may vary from 135° to 180° provided that the basic joint configuration (groove angle, root face, root opening) remain the same and that the design weld size is maintained.

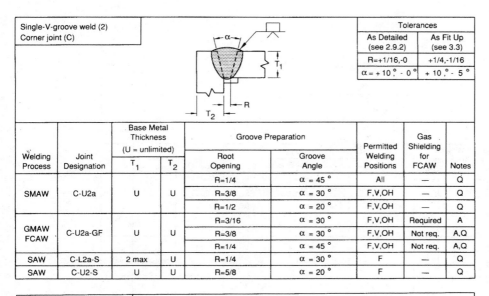

| Single-V-groove weld (2) Corner joint (C) | | | | | | | |
|---|---|---|---|---|---|---|---|

**Tolerances**

| | As Detailed (see 2.9.2) | As Fit Up (see 3.3) |
|---|---|---|
| | R=+1/16,-0 | +1/4,-1/16 |
| | $\alpha = +10^\circ,-0^\circ$ | $+10^\circ,-5^\circ$ |

| Welding Process | Joint Designation | Base Metal Thickness (U = unlimited) T₁ | T₂ | Groove Preparation Root Opening | Groove Angle | Permitted Welding Positions | Gas Shielding for FCAW | Notes |
|---|---|---|---|---|---|---|---|---|
| SMAW | C-U2a | U | U | R=1/4 | $\alpha = 45^\circ$ | All | — | Q |
| | | | | R=3/8 | $\alpha = 30^\circ$ | F,V,OH | — | Q |
| | | | | R=1/2 | $\alpha = 20^\circ$ | F,V,OH | — | Q |
| GMAW FCAW | C-U2a-GF | U | U | R=3/16 | $\alpha = 30^\circ$ | F,V,OH | Required | A |
| | | | | R=3/8 | $\alpha = 30^\circ$ | F,V,OH | Not req. | A,Q |
| | | | | R=1/4 | $\alpha = 45^\circ$ | F,V,OH | Not req. | A,Q |
| SAW | C-L2a-S | 2 max | U | R=1/4 | $\alpha = 30^\circ$ | F | — | Q |
| SAW | C-U2-S | U | U | R=5/8 | $\alpha = 20^\circ$ | F | — | Q |

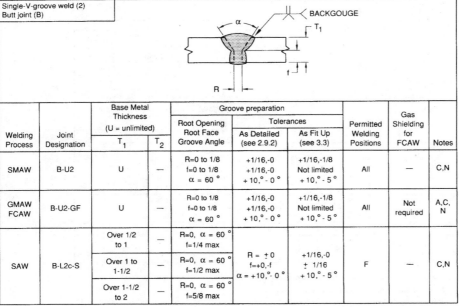

| Single-V-groove weld (2) Butt joint (B) | | | |
|---|---|---|---|

BACKGOUGE

| Welding Process | Joint Designation | Base Metal Thickness (U = unlimited) T₁ | T₂ | Groove preparation Root Opening Root Face Groove Angle | Tolerances As Detailed (see 2.9.2) | As Fit Up (see 3.3) | Permitted Welding Positions | Gas Shielding for FCAW | Notes |
|---|---|---|---|---|---|---|---|---|---|
| SMAW | B-U2 | U | — | R=0 to 1/8 f=0 to 1/8 $\alpha = 60^\circ$ | +1/16,-0 +1/16,-0 $+10^\circ,-0^\circ$ | +1/16,-1/8 Not limited $+10^\circ,-5^\circ$ | All | — | C,N |
| GMAW FCAW | B-U2-GF | U | — | R=0 to 1/8 f=0 to 1/8 $\alpha = 60^\circ$ | +1/16,-0 +1/16,-0 $+10^\circ,-0^\circ$ | +1/16,-1/8 Not limited $+10^\circ,-5^\circ$ | All | Not required | A,C, N |
| SAW | B-L2c-S | Over 1/2 to 1 | — | R=0, $\alpha = 60^\circ$ f=1/4 max | R = ±0 f=+0,-f $\alpha=+10^\circ,-0^\circ$ | +1/16,-0 ± 1/16 $+10^\circ,-5^\circ$ | F | — | C,N |
| | | Over 1 to 1-1/2 | — | R=0, $\alpha = 60^\circ$ f=1/2 max | | | | | |
| | | Over 1-1/2 to 2 | — | R=0, $\alpha = 60^\circ$ f=5/8 max | | | | | |

Note A: Not prequalified for gas tungsten arc and gas metal arc welding using short circuiting transfer. Refer to Appendix A.

Note C: Backgouge root to sound metal before welding second side.

Note N: The orientation of the two members in the joints may vary from 135° to 180° provided that the basic joint configuration (groove angle, root face, root opening) remain the same and that the design weld size is maintained.

Note Q: For corner and T-joints, the member orientation may be changed provided the groove angle is maintained as specified.

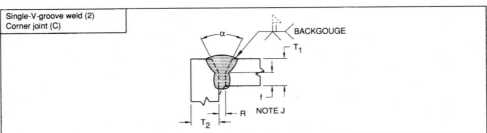

**Single-V-groove weld (2)**
**Corner joint (C)**

| Welding Process | Joint Designation | Base Metal Thickness (U = unlimited) | | Groove Preparation | | | Permitted Welding Positions | Gas Shielding for FCAW | Notes |
| | | $T_1$ | $T_2$ | Root Opening Root Face Groove Angle | Tolerances | | | | |
| | | | | | As Detailed (see 2.9.2) | As Fit Up (see 3.3) | | | |
| SMAW | C-U2 | U | U | R=0 to 1/8<br>f=0 to 1/8<br>$\alpha$ = 60° | +1/16,-0<br>+1/16,-0<br>+10,°- 0° | +1/16,-1/8<br>Not limited<br>+10,°- 5° | All | — | C,J, R |
| GMAW FCAW | C-U2-GF | U | U | R=0 to 1/8<br>f=0 to 1/8<br>$\alpha$ =60° | +1/16,-0<br>+1/16,-0<br>+10,°- 0° | +1/16,-1/8<br>Not limited<br>+10,°- 5° | All | Not required | A,C, J,R |
| SAW | C-U2b-S | U | U | R = 0<br>f=1/4 max<br>$\alpha$ = 60° | ± 0<br>+0,-1/4<br>+10,°- 0° | +1/16,-0<br>± 1/16<br>+10,°- 5° | F | — | C,J, R |

**Double-V-groove weld (3)**
**Butt joint (B)**

| | Tolerances | |
| | As Detailed (see 2.9.2) | As Fit Up (see 3.3) |
| R = ±0 | +1/4,-0 |
| f = ±0 | +1/16,-0 |
| $\alpha$ = +10,°- 0° | +10,°- 5° |
| Spacer SAW | ±0 | +1/16,-0 |
| Spacer SMAW | ±0 | +1/8,-0 |

| Welding Process | Joint Designation | Base Metal Thickness (U = unlimited) | | Groove Preparation | | | Permitted Welding Positions | Gas Shielding for FCAW | Notes |
| | | $T_1$ | $T_2$ | Root Opening | Root Face | Groove Angle | | | |
| SMAW | B-U3a | U<br>Spacer=1/8 X R | — | R=1/4 | f=0 to 1/8 | $\alpha$ = 45° | All | — | C,M, N |
| | | | | R=3/8 | f=0 to 1/8 | $\alpha$ = 30° | F,V,OH | — | |
| | | | | R=1/2 | f=0 to 1/8 | $\alpha$ = 20° | F,V,OH | — | |
| SAW | B-U3a-S | U<br>Spacer=1/4 x R | — | R=5/8 | f=0 to 1/4 | $\alpha$ = 20° | F | — | C,M, N |

Note A: Not prequalified for gas tungsten arc and gas metal arc welding using short circuiting transfer. Refer to Appendix A.

Note C: Backgouge root to sound metal before welding second side.

Note J: If fillet welds are used in statically loaded structures to reinforce groove welds in corner and T-joints, they shall be equal to 1/4 $T_1$, but need not exceed 3/8 in. Groove welds in corner and T-joints of dynamically loaded structures shall be reinforced with fillet welds equal to 1/4 $T_1$, but not more than 3/8 in.

Note M: Double-groove welds may have grooves of unequal depth, but the depth of the shallower groove shall be no less than one-fourth of the thickness of the thinner part joined.

Note N: The orientation of the two members in the joints may vary from 135° to 180° provided that the basic joint configuration (groove angle, root face, root opening) remain the same and that the design weld size is maintained.

Note R: The orientation of two members in the joints may vary from 45° to 135° for corner joints and from 45° to 90° for T-joints, provided that the basic joint configuration (groove angle, root face, root opening) remain the same and that the design weld size is maintained.

| Double-V-groove weld (3)<br>Butt joint (B) | | | | | | | | | For B-U3c-S only | | |

**For B-U3c-S only**

| $T_1$ | | $S_1$ |
|---|---|---|
| Over | to | |
| 2 | 2-1/2 | 1-3/8 |
| 2-1/2 | 3 | 1-3/4 |
| 3 | 3-5/8 | 2-1/8 |
| 3-5/8 | 4 | 2-3/8 |
| 4 | 4-3/4 | 2-3/4 |
| 4-3/4 | 5-1/2 | 3-1/4 |
| 5-1/2 | 6-1/4 | 3-3/4 |

For $T_1 > 6\text{-}1/4$, or $T_1 \leq 2$
$S_1 = 2/3\,(T_1 - 1/4)$

| Welding Process | Joint Designation | Base Metal Thickness (U = unlimited) $T_1$ | $T_2$ | Groove Preparation Root Opening Root Face Groove Angle | Tolerances As Detailed (see 2.9.2) | As Fit Up (see 3.3) | Permitted Welding Positions | Gas Shielding for FCAW | Notes |
|---|---|---|---|---|---|---|---|---|---|
| SMAW | B-U3b | U | — | R=0 to 1/8<br>f=0 to 1/8<br>$\alpha = \beta = 60°$ | +1/16,-0<br>+1/16,-0<br>+10°, -0° | +1/16,-1/8<br>Not limited<br>+10°, -5° | All | — | C,M,N |
| GMAW<br>FCAW | B-U3-GF | | | | | | All | Not required | A,C,M,N |
| SAW | B-U3c-S | U | — | R=0<br>f=1/4 min<br>$\alpha = \beta = 60°$ | +1/16,-0<br>+1/4,-0<br>+10°, -0° | +1/16,-0<br>+1/4,-0<br>+10°, -5° | F | — | C,M,N |
| | | | | To find $S_1$ see table above; $\quad S_2 = T_1 - (S_1 + f)$ | | | | | |

Note A:  Not prequalified for gas tungsten arc and gas metal arc welding using short circuiting transfer. Refer to Appendix A.

Note C:  Backgouge root to sound metal before welding second side.

Note M:  Double-groove welds may have grooves of unequal depth, but the depth of the shallower groove shall be no less than one-fourth of the thickness of the thinner part joined.

Note N:  The orientation of the two members in the joints may vary from 135° to 180° provided that the basic joint configuration (groove angle, root face, root opening) remain the same and that the design weld size is maintained.

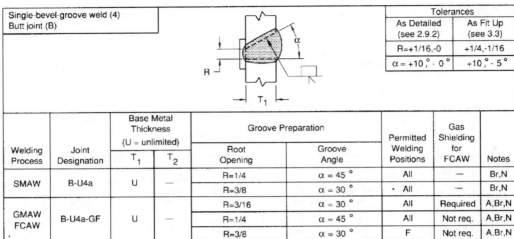

| Single-bevel-groove weld (4)<br>Butt joint (B) | | | | | Tolerances | | |
| --- | --- | --- | --- | --- | --- | --- | --- |
| | | | | | As Detailed<br>(see 2.9.2) | | As Fit Up<br>(see 3.3) |
| | | | | | R=+1/16,-0 | | +1/4,-1/16 |
| | | | | | $\alpha = +10°- 0°$ | | $+10°- 5°$ |

| Welding<br>Process | Joint<br>Designation | Base Metal Thickness (U = unlimited) | | Groove Preparation | | Permitted<br>Welding<br>Positions | Gas<br>Shielding<br>for<br>FCAW | Notes |
| --- | --- | --- | --- | --- | --- | --- | --- | --- |
| | | $T_1$ | $T_2$ | Root<br>Opening | Groove<br>Angle | | | |
| SMAW | B-U4a | U | — | R=1/4 | $\alpha = 45°$ | All | — | Br,N |
| | | | | R=3/8 | $\alpha = 30°$ | All | — | Br,N |
| GMAW<br>FCAW | B-U4a-GF | U | — | R=3/16 | $\alpha = 30°$ | All | Required | A,Br,N |
| | | | | R=1/4 | $\alpha = 45°$ | All | Not req. | A,Br,N |
| | | | | R=3/8 | $\alpha = 30°$ | F | Not req. | A,Br,N |

| Single-bevel-groove-weld (4)<br>T-joint (T)<br>Corner joint (C) | | | | | Tolerances | | |
| --- | --- | --- | --- | --- | --- | --- | --- |
| | | | | | As Detailed<br>(see 2.9.2) | | As Fit Up<br>(see 3.3) |
| | | | | | R=+1/16,-0 | | +1/4,-1/16 |
| | | | | | $\alpha = +10°-0°$ | | $+10°- 5°$ |

NOTE J
NOTE V

| Welding<br>Process | Joint<br>Designation | Base Metal Thickness (U = unlimited) | | Groove Preparation | | Permitted<br>Welding<br>Positions | Gas<br>Shielding<br>for<br>FCAW | Notes |
| --- | --- | --- | --- | --- | --- | --- | --- | --- |
| | | $T_1$ | $T_2$ | Root<br>Opening | Groove<br>Angle | | | |
| SMAW | TC-U4a | U | U | R=1/4 | $\alpha = 45°$ | All | — | J,Q,V |
| | | | | R=3/8 | $\alpha = 30°$ | F,V,OH | — | J,Q,V |
| GMAW<br>FCAW | TC-U4a-GF | U | U | R=3/16 | $\alpha = 30°$ | All | Required | A,J,Q,V |
| | | | | R=3/8 | $\alpha = 30°$ | F | Not req. | A,J,Q,V |
| | | | | R=1/4 | $\alpha = 45°$ | All | Not req. | A,J,Q,V |
| SAW | TC-U4a-S | U | U | R=3/8 | $\alpha = 30°$ | F | — | J,Q,V |
| | | | | R=1/4 | $\alpha = 45°$ | | | |

Note A: Not prequalified for gas tungsten arc and gas metal arc welding using short circuiting transfer. Refer to Appendix A.

Note Br: Dynamic load application limits these joints to the horizontal position (see 9.12.5).

Note C: Backgouge root to sound metal before welding second side.

Note J: If fillet welds are used in statically loaded structures to reinforce groove welds in corner and T-joints, they shall be equal to 1/4 $T_1$, but need not exceed 3/8 in. Groove welds in corner and T-joints of dynamically loaded structures shall be reinforced with fillet welds equal to 1/4 $T_1$, but not more than 3/8 in.

Note N: The orientation of the two members in the joints may vary from 135° to 180° provided that the basic joint configuration (groove angle, root face, root opening) remain the same and that the design weld size is maintained.

Note Q: For corner and T-joints, the member orientation may be changed provided the groove angle is maintained as specified.

Note V: For corner joints, the outside groove preparation may be in either or both members, provided the basic groove configuration is not changed and adequate edge distance is maintained to support the welding operations without excessive edge melting.

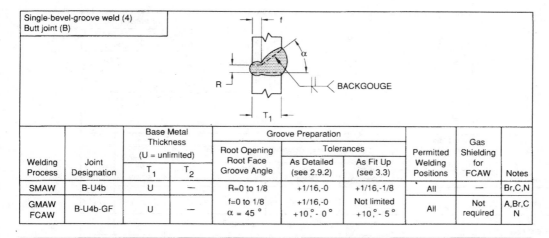

Single-bevel-groove weld (4)
Butt joint (B)

| Welding Process | Joint Designation | Base Metal Thickness (U = unlimited) | | Groove Preparation | | | Permitted Welding Positions | Gas Shielding for FCAW | Notes |
|---|---|---|---|---|---|---|---|---|---|
| | | | | Root Opening Root Face Groove Angle | Tolerances | | | | |
| | | $T_1$ | $T_2$ | | As Detailed (see 2.9.2) | As Fit Up (see 3.3) | | | |
| SMAW | B-U4b | U | — | R=0 to 1/8 | +1/16,-0 | +1/16,-1/8 | All | — | Br,C,N |
| GMAW FCAW | B-U4b-GF | U | — | f=0 to 1/8 $\alpha$ = 45 ° | +1/16,-0 +10,°- 0 ° | Not limited +10,°- 5 ° | All | Not required | A,Br,C N |

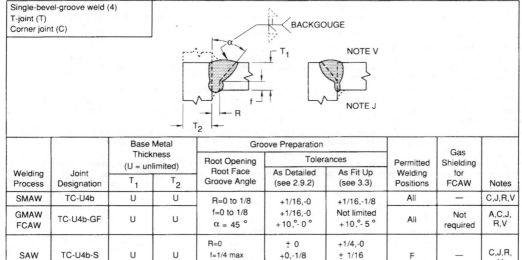

Single-bevel-groove weld (4)
T-joint (T)
Corner joint (C)

| Welding Process | Joint Designation | Base Metal Thickness (U = unlimited) | | Groove Preparation | | | Permitted Welding Positions | Gas Shielding for FCAW | Notes |
|---|---|---|---|---|---|---|---|---|---|
| | | | | Root Opening Root Face Groove Angle | Tolerances | | | | |
| | | $T_1$ | $T_2$ | | As Detailed (see 2.9.2) | As Fit Up (see 3.3) | | | |
| SMAW | TC-U4b | U | U | R=0 to 1/8 | +1/16,-0 | +1/16,-1/8 | All | — | C,J,R,V |
| GMAW FCAW | TC-U4b-GF | U | U | f=0 to 1/8 $\alpha$ = 45 ° | +1/16,-0 +10,°- 0 ° | Not limited +10,°- 5 ° | All | Not required | A,C,J, R,V |
| SAW | TC-U4b-S | U | U | R=0 f=1/4 max $\alpha$ = 60 ° | ± 0 +0,-1/8 +10,°- 0 ° | +1/4,-0 ± 1/16 +10,°- 5 ° | F | — | C,J,R, V |

Note A:  Not prequalified for gas tungsten arc and gas metal arc welding using short circuiting transfer. Refer to Appendix A.

Note Br: Dynamic load application limits these joints to the horizontal position (see 9.12.5).

Note C:  Backgouge root to sound metal before welding second side.

Note J:  If fillet welds are used in statically loaded structures to reinforce groove welds in corner and T-joints, they shall be equal to 1/4 $T_1$, but need not exceed 3/8 in. Groove welds in corner and T-joints of dynamically loaded structures shall be reinforced with fillet welds equal to 1/4 $T_1$, but not more than 3/8 in.

Note N:  The orientation of the two members in the joints may vary from 135° to 180° provided that the basic joint configuration (groove angle, root face, root opening) remain the same and that the design weld size is maintained.

Note R:  The orientation of two members in the joints may vary from 45° to 135° for corner joints and from 45° to 90° for T-joints, provided that the basic joint configuration (groove angle, root face, root opening) remain the same and that the design weld size is maintained.

Note V:  For corner joints, the outside groove preparation may be in either or both members, provided the basic groove configuration is not changed and adequate edge distance is maintained to support the welding operations without excessive edge melting.

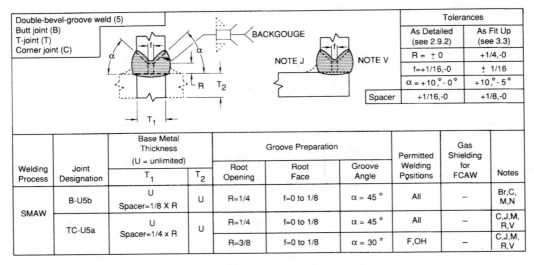

| | | Double-bevel-groove weld (5) Butt joint (B) T-joint (T) Corner joint (C) | | | | | | | Tolerances | |
|---|---|---|---|---|---|---|---|---|---|---|
| | | | | | | | | | As Detailed (see 2.9.2) | As Fit Up (see 3.3) |
| | | | | | | | | | R = ± 0 | +1/4,-0 |
| | | | | | | | | | f=+1/16,-0 | ± 1/16 |
| | | | | | | | | | α = +10,°- 0° | +10,°- 5° |
| | | | | | | | | Spacer | +1/16,-0 | +1/8,-0 |

| Welding Process | Joint Designation | Base Metal Thickness (U = unlimited) | | Groove Preparation | | | Permitted Welding Positions | Gas Shielding for FCAW | Notes |
|---|---|---|---|---|---|---|---|---|---|
| | | $T_1$ | $T_2$ | Root Opening | Root Face | Groove Angle | | | |
| SMAW | B-U5b | U Spacer=1/8 X R | U | R=1/4 | f=0 to 1/8 | α = 45° | All | – | Br,C, M,N |
| | TC-U5a | U Spacer=1/4 x R | U | R=1/4 | f=0 to 1/8 | α = 45° | All | – | C,J,M, R,V |
| | | | | R=3/8 | f=0 to 1/8 | α = 30° | F,OH | – | C,J,M, R,V |

Note Br: Dynamic load application limits these joints to the horizontal position (see 9.12.5).

Note C: Backgouge root to sound metal before welding second side.

Note J:  If fillet welds are used in statically loaded structures to reinforce groove welds in corner and T-joints, they shall be equal to 1/4 $T_1$, but need not exceed 3/8 in. Groove welds in corner and T-joints of dynamically loaded structures shall be reinforced with fillet welds equal to 1/4 $T_1$, but not more than 3/8 in.

Note M: Double-groove welds may have grooves of unequal depth, but the depth of the shallower groove shall be no less than one-fourth of the thickness of the thinner part joined.

Note N: The orientation of the two members in the joints may vary from 135° to 180° provided that the basic joint configuration (groove angle, root face, root opening) remain the same and that the design weld size is maintained.

Note R: The orientation of two members in the joints may vary from 45° to 135° for corner joints and from 45° to 90° for T-joints, provided that the basic joint configuration (groove angle, root face, root opening) remain the same and that the design weld size is maintained.

Note V: For corner joints, the outside groove preparation may be in either or both members, provided the basic groove configuration is not changed and adequate edge distance is maintained to support the welding operations without excessive edge melting.

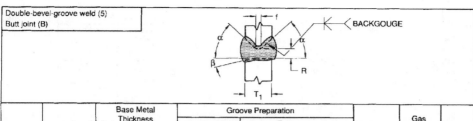

Double-bevel-groove weld (5)
Butt joint (B)

| Welding Process | Joint Designation | Base Metal Thickness (U = unlimited) | | Groove Preparation | | | Permitted Welding Positions | Gas Shielding for FCAW | Notes |
|---|---|---|---|---|---|---|---|---|---|
| | | $T_1$ | $T_2$ | Root Opening Root Face Groove Angle | Tolerances | | | | |
| | | | | | As Detailed (see 2.9.2) | As Fit Up (see 3.3) | | | |
| SMAW | B-U5a | U | — | R=0 to 1/8<br>f=0 to 1/8<br>$\alpha = 45°$<br>$\beta = 0°$ to $15°$ | +1/16,-0<br>+1/16,-0<br>$\alpha + \beta$ $^{+10°}_{-0°}$ | +1/16,-1/8<br>Not limited<br>$\alpha + \beta$ $^{+10°}_{-5°}$ | All | — | Br, C,M,N |
| GMAW FCAW | B-U5-GF | U | — | R=0 to 1/8<br>f=0 to 1/8<br>$\alpha = 45°$<br>$\beta = 0°$ to $15°$ | +1/16,-0<br>+1/16,-0<br>$\alpha + \beta =$<br>$+10°,-0°$ | +1/16,-1/8<br>Not limited<br>$\alpha + \beta =$<br>$+10°,-5°$ | All | Not req. | A,Br,C, M,N |

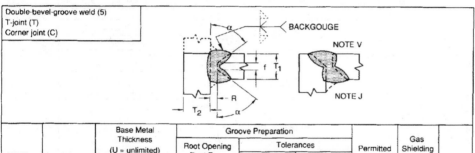

Double-bevel-groove weld (5)
T-joint (T)
Corner joint (C)

| Welding Process | Joint Designation | Base Metal Thickness (U = unlimited) | | Groove Preparation | | | Permitted Welding Positions | Gas Shielding for FCAW | Notes |
|---|---|---|---|---|---|---|---|---|---|
| | | $T_1$ | $T_2$ | Root Opening Root Face Groove Angle | Tolerances | | | | |
| | | | | | As Detailed (see 2.9.2) | As Fit Up (see 3.3) | | | |
| SMAW | TC-U5b | U | U | R=0 to 1/8<br>f=0 to 1/8<br>$\alpha = 45°$ | +1/16,-0<br>+1/16,-0<br>$+10°,-0°$ | +1/16,-1/8<br>Not limited<br>$+10°,-5°$ | All | — | C,J,M, R,V |
| GMAW FCAW | TC-U5-GF | U | U | | | | All | Not required | A,C,J, M,R,V |
| SAW | TC-U5-S | U | U | R=0<br>f=3/16 max<br>$\alpha = 60°$ | ± 0<br>+0,-3/16<br>$+10°,-0°$ | +1/16,-0<br>± 1/16<br>$+10°,-5°$ | F | — | C,J,M, R,V |

Note A:  Not prequalified for gas tungsten arc and gas metal arc welding using short circuiting transfer. Refer to Appendix A.

Note Br: Dynamic load application limits these joints to the horizontal position (see 9.12.5).

Note C:  Backgouge root to sound metal before welding second side.

Note J:  If fillet welds are used in statically loaded structures to reinforce groove welds in corner and T-joints, they shall be equal to 1/4 $T_1$, but need not exceed 3/8 in. Groove welds in corner and T-joints of dynamically loaded structures shall be reinforced with fillet welds equal to 1/4 $T_1$, but not more than 3/8 in.

Note M:  Double-groove welds may have grooves of unequal depth, but the depth of the shallower groove shall be no less than one-fourth of the thickness of the thinner part joined.

Note N:  The orientation of the two members in the joints may vary from 135° to 180° provided that the basic joint configuration (groove angle, root face, root opening) remain the same and that the design weld size is maintained.

Note R:  The orientation of two members in the joints may vary from 45° to 135° for corner joints and from 45° to 90° for T-joints, provided that the basic joint configuration (groove angle, root face, root opening) remain the same and that the design weld size is maintained.

Note V:  For corner joints, the outside groove preparation may be in either or both members, provided the basic groove configuration is not changed and adequate edge distance is maintained to support the welding operations without excessive edge melting.

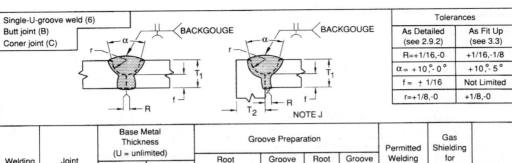

**Single-U-groove weld (6)**
**Butt joint (B)**
**Coner joint (C)**

| Tolerances | | |
|---|---|---|
| | As Detailed (see 2.9.2) | As Fit Up (see 3.3) |
| R=+1/16,-0 | | +1/16,-1/8 |
| $\alpha = +10°,-0°$ | | +10°,-5° |
| f = ±1/16 | | Not Limited |
| r=+1/8,-0 | | +1/8,-0 |

| Welding Process | Joint Designation | Base Metal Thickness (U = unlimited) | | Groove Preparation | | | | Permitted Welding Positions | Gas Shielding for FCAW | Notes |
|---|---|---|---|---|---|---|---|---|---|---|
| | | $T_1$ | $T_2$ | Root Opening | Groove Angle | Root Face | Groove Radius | | | |
| SMAW | B-U6 | U | U | R=0 to 1/8 | $\alpha = 45°$ | f=1/8 | r=1/4 | All | — | C,N |
| | | | | R=0 to 1/8 | $\alpha = 20°$ | f=1/8 | r=1/4 | F,OH | — | C,N |
| | C-U6 | U | U | R=0 to 1/8 | $\alpha = 45°$ | f=1/8 | r=1/4 | All | — | C,J,R |
| | | | | R=0 to 1/8 | $\alpha = 20°$ | f=1/8 | r=1/4 | F,OH | — | C,J,R |
| GMAW FCAW | B-U6-GF | U | U | R=0 to 1/8 | $\alpha = 20°$ | f=1/8 | r=1/4 | All | Not req. | A,C,N |
| | C-U6-GF | U | U | R=0 to 1/8 | $\alpha = 20°$ | f=1/8 | r=1/4 | All | Not req. | A,C,J,R |

**Double-U-groove weld (7)**
**Butt joint (B)**

| Tolerances | | | | |
|---|---|---|---|---|
| For B-U7 and B-U7-GF | | | For B-U7-S | |
| As Detailed (see 2.9.2) | As Fit Up (see 3.3) | | As Detailed (see 2.9.2) | As Fit Up (see 3.3) |
| R=+1/16,-0 | +1/16,-1/8 | | R=±0 | +1/16,-0 |
| $\alpha = +10°,-0°$ | +10°,-5° | | f=+0,-1/4 | ±1/16 |
| f=+1/16,-0 | Not Limited | | | |
| r=+1/4,-0 | ±1/16 | | | |

| Welding Process | Joint Designation | Base Metal Thickness (U = unlimited) | | Groove Preparation | | | | Permitted Welding Positions | Gas Shielding for FCAW | Notes |
|---|---|---|---|---|---|---|---|---|---|---|
| | | $T_1$ | $T_2$ | Root Opening | Groove Angle | Root Face | Groove Radius | | | |
| SMAW | B-U7 | U | — | R=0 to 1/8 | $\alpha = 45°$ | f=1/8 | r=1/4 | All | — | C,M,N |
| | | | | R=0 to 1/8 | $\alpha = 20°$ | f=1/8 | r=1/4 | F,OH | — | C,M,N |
| GMAW FCAW | B-U7-GF | U | — | R=0 to 1/8 | $\alpha = 20°$ | f=1/8 | r=1/4 | All | Not Required | A,C, N,M |
| SAW | B-U7-S | U | — | R=0 | $\alpha = 20°$ | f=1/4 max | r=1/4 | F | — | C,M,N |

Note A: Not prequalified for gas tungsten arc and gas metal arc welding using short circuiting transfer. Refer to Appendix A.

Note C: Backgouge root to sound metal before welding second side.

Note J: If fillet welds are used in statically loaded structures to reinforce groove welds in corner and T-joints, they shall be equal to 1/4 $T_1$, but need not exceed 3/8 in. Groove welds in corner and T-joints of dynamically loaded structures shall be reinforced with fillet welds equal to 1/4 $T_1$, but not more than 3/8 in.

Note M: Double-groove welds may have grooves of unequal depth, but the depth of the shallower groove shall be no less than one-fourth of the thickness of the thinner part joined.

Note N: The orientation of the two members in the joints may vary from 135° to 180° provided that the basic joint configuration (groove angle, root face, root opening) remain the same and that the design weld size is maintained.

Note R: The orientation of two members in the joints may vary from 45° to 135° for corner joints and from 45° to 90° for T-joints, provided that the basic joint configuration (groove angle, root face, root opening) remain the same and that the design weld size is maintained.

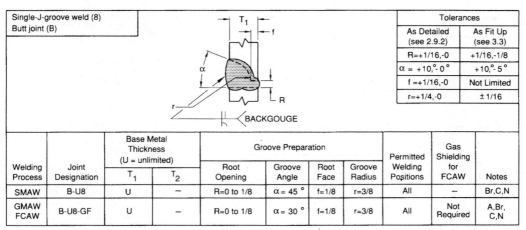

| | | Base Metal Thickness (U = unlimited) | | Groove Preparation | | | | Permitted Welding Positions | Gas Shielding for FCAW | |
|---|---|---|---|---|---|---|---|---|---|---|
| Welding Process | Joint Designation | $T_1$ | $T_2$ | Root Opening | Groove Angle | Root Face | Groove Radius | | | Notes |
| SMAW | B-U8 | U | — | R=0 to 1/8 | $\alpha$ = 45 ° | f=1/8 | r=3/8 | All | — | Br,C,N |
| GMAW FCAW | B-U8-GF | U | — | R=0 to 1/8 | $\alpha$ = 30 ° | f=1/8 | r=3/8 | All | Not Required | A,Br, C,N |

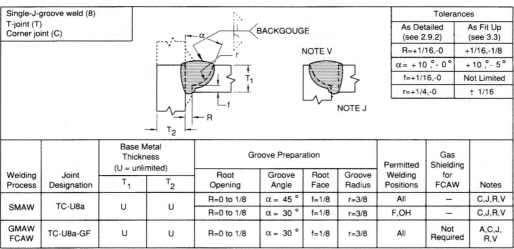

| | | Base Metal Thickness (U = unlimited) | | Groove Preparation | | | | Permitted Welding Positions | Gas Shielding for FCAW | |
|---|---|---|---|---|---|---|---|---|---|---|
| Welding Process | Joint Designation | $T_1$ | $T_2$ | Root Opening | Groove Angle | Root Face | Groove Radius | | | Notes |
| SMAW | TC-U8a | U | U | R=0 to 1/8 | $\alpha$ = 45 ° | f=1/8 | r=3/8 | All | — | C,J,R,V |
| | | | | R=0 to 1/8 | $\alpha$ = 30 ° | f=1/8 | r=3/8 | F,OH | — | C,J,R,V |
| GMAW FCAW | TC-U8a-GF | U | U | R=0 to 1/8 | $\alpha$ = 30 ° | f=1/8 | r=3/8 | All | Not Required | A,C,J, R,V |

Note A: Not prequalified for gas tungsten arc and gas metal arc welding using short circuiting transfer. Refer to Appendix A.

Note Br: Dynamic load application limits these joints to the horizontal position (see 9.12.5).

Note C: Backgouge root to sound metal before welding second side.

Note J: If fillet welds are used in statically loaded structures to reinforce groove welds in corner and T-joints, they shall be equal to 1/4 $T_1$, but need not exceed 3/8 in. Groove welds in corner and T-joints of dynamically loaded structures shall be reinforced with fillet welds equal to 1/4 $T_1$, but not more than 3/8 in.

Note N: The orientation of the two members in the joints may vary from 135° to 180° provided that the basic joint configuration (groove angle, root face, root opening) remain the same and that the design weld size is maintained.

Note R: The orientation of two members in the joints may vary from 45° to 135° for corner joints and from 45° to 90° for T-joints, provided that the basic joint configuration (groove angle, root face, root opening) remain the same and that the design weld size is maintained.

Note V: For corner joints, the outside groove preparation may be in either or both members, provided the basic groove configuration is not changed and adequate edge distance is maintained to support the welding operations without excessive edge melting.

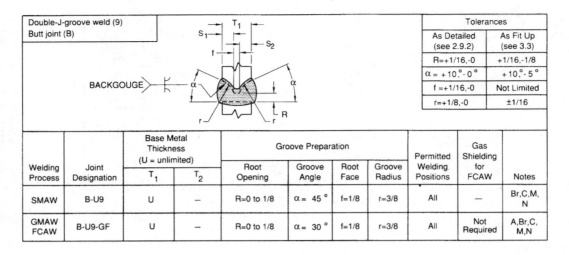

| Double-J-groove weld (9) | | | | | | | | Tolerances | | |
| Butt joint (B) | | | | | | | | As Detailed (see 2.9.2) | As Fit Up (see 3.3) | |

Tolerances:
- R=+1/16,-0 / +1/16,-1/8
- α = +10°-0° / +10°-5°
- f =+1/16,-0 / Not Limited
- r=+1/8,-0 / ±1/16

| Welding Process | Joint Designation | Base Metal Thickness (U = unlimited) | | Groove Preparation | | | | Permitted Welding Positions | Gas Shielding for FCAW | Notes |
|---|---|---|---|---|---|---|---|---|---|---|
| | | $T_1$ | $T_2$ | Root Opening | Groove Angle | Root Face | Groove Radius | | | |
| SMAW | B-U9 | U | — | R=0 to 1/8 | α = 45° | f=1/8 | r=3/8 | All | — | Br,C,M,N |
| GMAW FCAW | B-U9-GF | U | — | R=0 to 1/8 | α = 30° | f=1/8 | r=3/8 | All | Not Required | A,Br,C,M,N |

Note A: Not prequalified for gas tungsten arc and gas metal arc welding using short circuiting transfer. Refer to Appendix A.

Note Br: Dynamic load application limits these joints to the horizontal position (see 9.12.5).

Note C: Backgouge root to sound metal before welding second side.

Note M: Double-groove welds may have grooves of unequal depth, but the depth of the shallower groove shall be no less than one-fourth of the thickness of the thinner part joined.

Note N: The orientation of the two members in the joints may vary from 135° to 180° provided that the basic joint configuration (groove angle, root face, root opening) remain the same and that the design weld size is maintained.

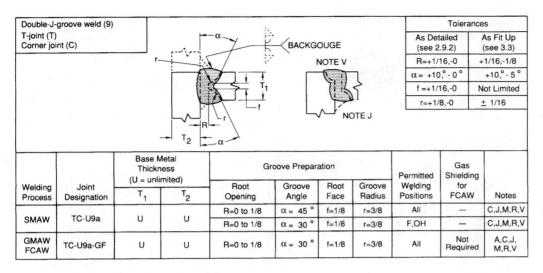

| Double-J-groove weld (9) T-joint (T) Corner joint (C) | | | | Tolerances | |
| --- | --- | --- | --- | --- | --- |
| | | | | As Detailed (see 2.9.2) | As Fit Up (see 3.3) |
| | | | | R=+1/16,-0 | +1/16,-1/8 |
| | | | | $\alpha$ = +10,°- 0 ° | +10,°- 5 ° |
| | | | | f =+1/16,-0 | Not Limited |
| | | | | r=+1/8,-0 | ± 1/16 |

| Welding Process | Joint Designation | Base Metal Thickness (U = unlimited) | | Groove Preparation | | | | Permitted Welding Positions | Gas Shielding for FCAW | Notes |
| --- | --- | --- | --- | --- | --- | --- | --- | --- | --- | --- |
| | | $T_1$ | $T_2$ | Root Opening | Groove Angle | Root Face | Groove Radius | | | |
| SMAW | TC-U9a | U | U | R=0 to 1/8 | $\alpha$ = 45 ° | f=1/8 | r=3/8 | All | — | C,J,M,R,V |
| | | | | R=0 to 1/8 | $\alpha$ = 30 ° | f=1/8 | r=3/8 | F,OH | — | C,J,M,R,V |
| GMAW FCAW | TC-U9a-GF | U | U | R=0 to 1/8 | $\alpha$ = 30 ° | f=1/8 | r=3/8 | All | Not Required | A,C,J, M,R,V |

Note A:  Not prequalified for gas tungsten arc and gas metal arc welding using short circuiting transfer. Refer to Appendix A.

Note C:  Backgouge root to sound metal before welding second side.

Note J:  If fillet welds are used in statically loaded structures to reinforce groove welds in corner and T-joints, they shall be equal to 1/4 $T_1$, but need not exceed 3/8 in. Groove welds in corner and T-joints of dynamically loaded structures shall be reinforced with fillet welds equal to 1/4 $T_1$, but not more than 3/8 in.

Note M:  Double-groove welds may have grooves of unequal depth, but the depth of the shallower groove shall be no less than one-fourth of the thickness of the thinner part joined.

Note R:  The orientation of two members in the joints may vary from 45° to 135° for corner joints and from 45° to 90° for T-joints, provided that the basic joint configuration (groove angle, root face, root opening) remain the same and that the design weld size is maintained.

Note V:  For corner joints, the outside groove preparation may be in either or both members, provided the basic groove configuration is not changed and adequate edge distance is maintained to support the welding operations without excessive edge melting.

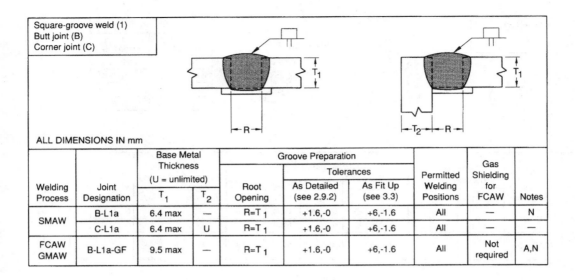

Square-groove weld (1)
Butt joint (B)
Corner joint (C)

ALL DIMENSIONS IN mm

| Welding Process | Joint Designation | Base Metal Thickness (U = unlimited) | | Groove Preparation | | | Permitted Welding Positions | Gas Shielding for FCAW | Notes |
| | | $T_1$ | $T_2$ | Root Opening | Tolerances | | | | |
| | | | | | As Detailed (see 2.9.2) | As Fit Up (see 3.3) | | | |
| SMAW | B-L1a | 6.4 max | — | $R=T_1$ | +1.6,-0 | +6,-1.6 | All | — | N |
| | C-L1a | 6.4 max | U | $R=T_1$ | +1.6,-0 | +6,-1.6 | All | — | — |
| FCAW GMAW | B-L1a-GF | 9.5 max | — | $R=T_1$ | +1.6,-0 | +6,-1.6 | All | Not required | A,N |

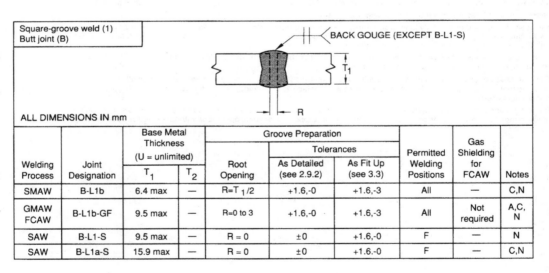

Square-groove weld (1)
Butt joint (B)

BACK GOUGE (EXCEPT B-L1-S)

ALL DIMENSIONS IN mm

| Welding Process | Joint Designation | Base Metal Thickness (U = unlimited) | | Groove Preparation | | | Permitted Welding Positions | Gas Shielding for FCAW | Notes |
| | | $T_1$ | $T_2$ | Root Opening | Tolerances | | | | |
| | | | | | As Detailed (see 2.9.2) | As Fit Up (see 3.3) | | | |
| SMAW | B-L1b | 6.4 max | — | $R=T_1/2$ | +1.6,-0 | +1.6,-3 | All | — | C,N |
| GMAW FCAW | B-L1b-GF | 9.5 max | — | R=0 to 3 | +1.6,-0 | +1.6,-3 | All | Not required | A,C, N |
| SAW | B-L1-S | 9.5 max | — | R = 0 | ±0 | +1.6,-0 | F | — | N |
| SAW | B-L1a-S | 15.9 max | — | R = 0 | ±0 | +1.6,-0 | F | — | C,N |

Note A:  Not prequalified for gas tungsten arc and gas metal arc welding using short circuiting transfer. Refer to Appendix A.

Note C:  Backgouge root to sound metal before welding second side.

Note N:  The orientation of the two members in the joints may vary from 135° to 180° provided that the basic joint configuration (groove angle, root face, root opening) remain the same and that the design weld size is maintained.

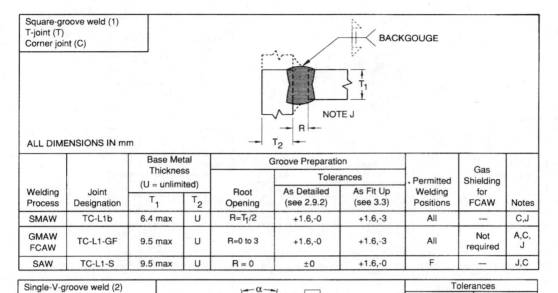

Square-groove weld (1)
T-joint (T)
Corner joint (C)

BACKGOUGE

NOTE J

ALL DIMENSIONS IN mm

| Welding Process | Joint Designation | Base Metal Thickness (U = unlimited) | | Groove Preparation | | | Permitted Welding Positions | Gas Shielding for FCAW | Notes |
| | | $T_1$ | $T_2$ | Root Opening | Tolerances | | | | |
| | | | | | As Detailed (see 2.9.2) | As Fit Up (see 3.3) | | | |
| SMAW | TC-L1b | 6.4 max | U | $R=T_1/2$ | +1.6,-0 | +1.6,-3 | All | — | C,J |
| GMAW FCAW | TC-L1-GF | 9.5 max | U | R=0 to 3 | +1.6,-0 | +1.6,-3 | All | Not required | A,C, J |
| SAW | TC-L1-S | 9.5 max | U | R = 0 | ±0 | +1.6,-0 | F | — | J,C |

Single-V-groove weld (2)
Butt joint (B)

ALL DIMENSIONS IN mm

| | Tolerances | |
| | As Detailed (see 2.9.2) | As Fit Up (see 3.3) |
| | R=+1.6,-0 | +6,-1.6 |
| | $\alpha = +10°,-0°$ | $+10°,-5°$ |

| Welding Process | Joint Designation | Base Metal Thickness (U = unlimited) | | Groove Preparation | | Permitted Welding Positions | Gas Shielding for FCAW | Notes |
| | | $T_1$ | $T_2$ | Root Opening | Groove Angle | | | |
| SMAW | B-U2a | U | — | R=6 | $\alpha = 45°$ | All | — | N |
| | | | | R=10 | $\alpha = 30°$ | F,V,OH | — | N |
| | | | | R=13 | $\alpha = 20°$ | F,V,OH | — | N |
| GMAW FCAW | B-U2a-GF | U | — | R=5 | $\alpha = 30°$ | F,V,OH | Required | A,N |
| | | | | R=10 | $\alpha = 30°$ | F,V,OH | Not req. | A,N |
| | | | | R=6 | $\alpha = 45°$ | F,V,OH | Not req. | A,N |
| SAW | B-L2a-S | 50.8 max | — | R=6 | $\alpha = 30°$ | F | — | N |
| SAW | B-U2-s | U | — | R=16 | $\alpha = 20°$ | F | — | N |

Note A:  Not prequalified for gas tungsten arc and gas metal arc welding using short circuiting transfer. Refer to Appendix A.

Note C:  Backgouge root to sound metal before welding second side.

Note J:  If fillet welds are used in statically loaded structures to reinforce groove welds in corner and T-joints, they shall be equal to 1/4 $T_1$, but need not exceed 10 mm. Groove welds in corner and T-joints of dynamically loaded structures shall be reinforced with fillet welds equal to 1/4 $T_1$, but not more than 10 mm.

Note N:  The orientation of the two members in the joints may vary from 135° to 180° provided that the basic joint configuration (groove angle, root face, root opening) remain the same and that the design weld size is maintained.

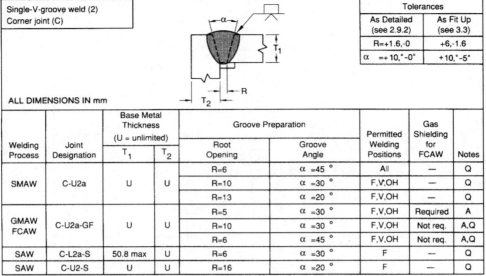

| Single-V-groove weld (2) Corner joint (C) | Tolerances | | |
|---|---|---|---|
| | | As Detailed (see 2.9.2) | As Fit Up (see 3.3) |
| | | R=+1.6,-0 | +6,-1.6 |
| | | $\alpha$ =+10,°-0° | +10,°-5° |

ALL DIMENSIONS IN mm

| Welding Process | Joint Designation | Base Metal Thickness (U = unlimited) | | Groove Preparation | | Permitted Welding Positions | Gas Shielding for FCAW | Notes |
|---|---|---|---|---|---|---|---|---|
| | | $T_1$ | $T_2$ | Root Opening | Groove Angle | | | |
| SMAW | C-U2a | U | U | R=6 | $\alpha$ =45 ° | All | — | Q |
| | | | | R=10 | $\alpha$ =30 ° | F,V,OH | — | Q |
| | | | | R=13 | $\alpha$ =20 ° | F,V,OH | — | Q |
| GMAW FCAW | C-U2a-GF | U | U | R=5 | $\alpha$ =30 ° | F,V,OH | Required | A |
| | | | | R=10 | $\alpha$ =30 ° | F,V,OH | Not req. | A,Q |
| | | | | R=6 | $\alpha$ =45 ° | F,V,OH | Not req. | A,Q |
| SAW | C-L2a-S | 50.8 max | U | R=6 | $\alpha$ =30 ° | F | — | Q |
| SAW | C-U2-S | U | U | R=16 | $\alpha$ =20 ° | F | — | Q |

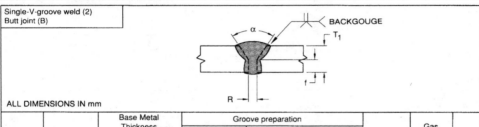

| Single-V-groove weld (2) Butt joint (B) | | | | | | | | |
|---|---|---|---|---|---|---|---|---|

ALL DIMENSIONS IN mm

| Welding Process | Joint Designation | Base Metal Thickness (U = unlimited) | | Groove preparation | | | Permitted Welding Positions | Gas Shielding for FCAW | Notes |
|---|---|---|---|---|---|---|---|---|---|
| | | $T_1$ | $T_2$ | Root Opening Root Face Groove Angle | Tolerances | | | | |
| | | | | | As Detailed (see 2.9.2) | As Fit Up (see 3.3) | | | |
| SMAW | B-U2 | U | — | R=0 to 3 f=0 to 3 $\alpha$ =60 ° | +1.6,-0 +1.6,-0 +10,°0 ° | +1.6,-3 Not limited +10,°5 ° | All | — | C,N |
| GMAW FCAW | B-U2-GF | U | — | R=0 to 3 f=0 to 3 $\alpha$ =60 ° | +1.6,-0 +1.6,-0 +10,°0 ° | +1.6,-3 Not limited +10,°5 ° | All | Not required | A,C, N |
| SAW | B-L2c-S | Over 12.7 to 25.4 | — | R=0, $\alpha$ =60 ° f=6 max | R = ±0 f=+0,-f $\alpha$ =+10,°0 ° | +1.6,-0 ±1.6 +10,°5 ° | F | — | C,N |
| | | Over 25.4 to 38.1 | — | R=0, $\alpha$ =60 ° f=13 max | | | | | |
| | | Over 38.1 to 50.8 | — | R=0, $\alpha$ =60 ° f=16 max | | | | | |

Note A: Not prequalified for gas tungsten arc and gas metal arc welding using short circuiting transfer. Refer to Appendix A.

Note C: Backgouge root to sound metal before welding second side.

Note N: The orientation of the two members in the joints may vary from 135° to 180° provided that the basic joint configuration (groove angle, root face, root opening) remain the same and that the design weld size is maintained.

Note Q: For corner and T-joints, the member orientation may be changed provided the groove angle is maintained as specified.

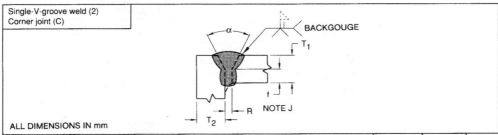

Single-V-groove weld (2)
Corner joint (C)

BACKGOUGE

ALL DIMENSIONS IN mm

| Welding Process | Joint Designation | Base Metal Thickness (U = unlimited) $T_1$ | Base Metal Thickness (U = unlimited) $T_2$ | Groove Preparation Root Opening Root Face Groove Angle | Groove Preparation Tolerances As Detailed (see 2.9.2) | Groove Preparation Tolerances As Fit Up (see 3.3) | Permitted Welding Positions | Gas Shielding for FCAW | Notes |
|---|---|---|---|---|---|---|---|---|---|
| SMAW | C-U2 | U | U | R=0 to 3<br>f=0 to 3<br>$\alpha$ =60 ° | +1.6,-0<br>+1.6,-0<br>+10,0 ° | +1.6,-3<br>Not limited<br>+10,5 ° | All | — | C,J, R |
| GMAW FCAW | C-U2-GF | U | U | R=0 to 3<br>f=0 to 3<br>$\alpha$ =60 ° | +1.6,-0<br>+1.6,-0<br>+10,0 ° | +1.6,-3<br>Not limited<br>+10,5 ° | All | Not required | A,C, J,R |
| SAW | C-U2b-S | U | U | R = 0<br>f=6 max<br>$\alpha$ =60 ° | ±0<br>+0,-6<br>+10,0 ° | +1.6,-0<br>± 1.6<br>+10,5 ° | F | — | C,J, R |

Double-V-groove weld (3)
Butt joint (B)

BACKGOUGE

| | Tolerances As Detailed (see 2.9.2) | Tolerances As Fit Up (see 3.3) |
|---|---|---|
| | R = ±0 | +6,-0 |
| | f = ±0 | +1.6,-0 |
| | $\alpha$ =+10,0 ° | +10,5 ° |
| Spacer   SAW | ±0 | +1.6,-0 |
| Spacer   SMAW | ±0 | +3,-0 |

ALL DIMENSIONS IN mm

| Welding Process | Joint Designation | Base Metal Thickness (U = unlimited) $T_1$ | Base Metal Thickness (U = unlimited) $T_2$ | Groove Preparation Root Opening | Groove Preparation Root Face | Groove Preparation Groove Angle | Permitted Welding Positions | Gas Shielding for FCAW | Notes |
|---|---|---|---|---|---|---|---|---|---|
| SMAW | B-U3a | U<br>Spacer=3 X R | — | R=6<br>R=10<br>R=13 | f=0 to 3<br>f=0 to 3<br>f=0 to 3 | $\alpha$ =45 °<br>$\alpha$ =30 °<br>$\alpha$ =20 ° | All<br>F,V,OH<br>F,V,OH | —<br>—<br>— | C,M, N |
| SAW | B-U3a-S | U<br>Spacer=6 x R | — | R=16 | f=0 to 6 | $\alpha$ =20 ° | F | — | C,M, N |

Note A:  Not prequalified for gas tungsten arc and gas metal arc welding using short circuiting transfer. Refer to Appendix A.

Note C:  Backgouge root to sound metal before welding second side.

Note J:  If fillet welds are used in statically loaded structures to reinforce groove welds in corner and T-joints, they shall be equal to 1/4 $T_1$, but need not exceed 10 mm. Groove welds in corner and T-joints of dynamically loaded structures shall be reinforced with fillet welds equal to 1/4 $T_1$, but not more than 10 mm.

Note M:  Double-groove welds may have grooves of unequal depth, but the depth of the shallower groove shall be no less than one-fourth of the thickness of the thinner part joined.

Note N:  The orientation of the two members in the joints may vary from 135° to 180° provided that the basic joint configuration (groove angle, root face, root opening) remain the same and that the design weld size is maintained.

Note R:  The orientation of two members in the joints may vary from 45° to 135° for corner joints and from 45° to 90° for T-joints, provided that the basic joint configuration (groove angle, root face, root opening) remain the same and that the design weld size is maintained.

| Double-V-groove weld (3) Butt joint (B) | | | | | | | | For B-U3c-S only | | |

| T₁ | | S₁ |
|---|---|---|
| Over | to | |
| 50.8 | 63.5 | 35 |
| 63.5 | 76.2 | 44 |
| 76.2 | 92.1 | 54 |
| 92.1 | 101.6 | 60 |
| 101.6 | 120.7 | 70 |
| 120.7 | 139.7 | 83 |
| 139.7 | 158.8 | 95 |

ALL DIMENSIONS IN mm

For $T_1 > 158.8$, or $T_1 \leq 50.8$
$S_1 = 2/3 (T_1 - 6)$

| Welding Process | Joint Designation | Base Metal Thickness (U = unlimited) | | Groove Preparation | | | | Permitted Welding Positions | Gas Shielding for FCAW | Notes |
|---|---|---|---|---|---|---|---|---|---|---|
| | | T₁ | T₂ | Root Opening Root Face Groove Angle | Tolerances | | | | | |
| | | | | | As Detailed (see 2.9.2) | As Fit Up (see 3.3) | | | | |
| SMAW | B-U3b | | | R=0 to 3 | +1.6,-0 | +1.6,-3 | | All | — | C,M,N |
| GMAW FCAW | B-U3-GF | U | — | f=0 to 3 $\alpha = \beta = 60°$ | +1.6,-0 +10°,-0° | Not limited +10°,-5° | | All | Not required | A,C,M, N |
| SAW | B-U3c-S | U | — | R=0 f=6 min $\alpha = \beta = 60°$ To find $S_1$ see table above; $S_1 = T_1 - (S_1 + f)$ | +1.6,-0 +6,-0 +10°,-0° | +1.6,-0 +6,-0 +10°,-5° | | F | — | C,M, N |

Note A: Not prequalified for gas tungsten arc and gas metal arc welding using short circuiting transfer. Refer to Appendix A.

Note C: Backgouge root to sound metal before welding second side.

Note M: Double-groove welds may have grooves of unequal depth, but the depth of the shallower groove shall be no less than one-fourth of the thickness of the thinner part joined.

Note N: The orientation of the two members in the joints may vary from 135° to 180° provided that the basic joint configuration (groove angle, root face, root opening) remain the same and that the design weld size is maintained.

**Single-bevel-groove weld (4)**
**Butt joint (B)**

| Tolerances | |
|---|---|
| As Detailed (see 2.9.2) | As Fit Up (see 3.3) |
| R=+1.6,-0 | +6,-1.6 |
| $\alpha$ =+10,°0 ° | +10°,-5 ° |

ALL DIMENSIONS IN mm

| Welding Process | Joint Designation | Base Metal Thickness (U = unlimited) $T_1$ | $T_2$ | Groove Preparation — Root Opening | Groove Preparation — Groove Angle | Permitted Welding Positions | Gas Shielding for FCAW | Notes |
|---|---|---|---|---|---|---|---|---|
| SMAW | B-U4a | U | — | R=6 | $\alpha$ =45° | All | — | Br,N |
| | | | | R=10 | $\alpha$ =30° | All | — | Br,N |
| GMAW FCAW | B-U4a-GF | U | — | R=5 | $\alpha$ =30° | All | Required | A,Br,N |
| | | | | R=6 | $\alpha$ =45° | All | Not req. | A,Br,N |
| | | | | R=10 | $\alpha$ =30° | F | Not req. | A,Br,N |

**Single-bevel-groove-weld (4)**
**T-joint (T)**
**Corner joint (C)**

NOTE J
NOTE V

| Tolerances | |
|---|---|
| As Detailed (see 2.9.2) | As Fit Up (see 3.3) |
| R=+1.6,-0 | +6,-1.6 |
| $\alpha$ = +10° - 0 ° | +10° - 5 ° |

ALL DIMENSIONS IN mm

| Welding Process | Joint Designation | Base Metal Thickness (U = unlimited) $T_1$ | $T_2$ | Groove Preparation — Root Opening | Groove Preparation — Groove Angle | Permitted Welding Positions | Gas Shielding for FCAW | Notes |
|---|---|---|---|---|---|---|---|---|
| SMAW | TC-U4a | U | U | R=6 | $\alpha$ = 45° | All | — | J,Q,V |
| | | | | R=10 | $\alpha$ = 30° | F,V,OH | — | J,Q,V |
| GMAW FCAW | TC-U4a-GF | U | U | R=5 | $\alpha$ = 30° | All | Required | A,J,Q,V |
| | | | | R=10 | $\alpha$ = 30° | F | Not req. | A,J,Q,V |
| | | | | R=6 | $\alpha$ = 45° | All | Not req. | A,J,Q,V |
| SAW | TC-U4a-S | U | U | R=10 | $\alpha$ = 30° | F | — | J,Q,V |
| | | | | R=6 | $\alpha$ = 45° | | | |

Note A: Not prequalified for gas tungsten arc and gas metal arc welding using short circuiting transfer. Refer to Appendix A.

Note Br: Dynamic load application limits these joints to the horizontal position (see 9.12.5).

Note C: Backgouge root to sound metal before welding second side.

Note J: If fillet welds are used in statically loaded structures to reinforce groove welds in corner and T-joints, they shall be equal to 1/4 $T_1$, but need not exceed 10 mm. Groove welds in corner and T-joints of dynamically loaded structures shall be reinforced with fillet welds equal to 1/4 $T_1$, but not more than 10 mm.

Note N: The orientation of the two members in the joints may vary from 135° to 180° provided that the basic joint configuration (groove angle, root face, root opening) remain the same and that the design weld size is maintained.

Note Q: For corner and T-joints, the member orientation may be changed provided the groove angle is maintained as specified.

Note V: For corner joints, the outside groove preparation may be in either or both members, provided the basic groove configuration is not changed and adequate edge distance is maintained to support the welding operations without excessive edge melting.

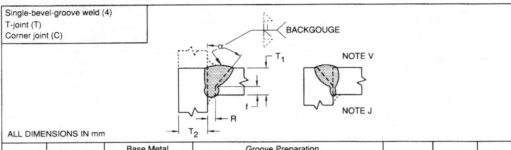

| Single-bevel-groove weld (4) Butt joint (B) | | | | | | | | | |

ALL DIMENSIONS IN mm

| Welding Process | Joint Designation | Base Metal Thickness (U = unlimited) | | Groove Preparation | | | Permitted Welding Positions | Gas Shielding for FCAW | Notes |
| | | $T_1$ | $T_2$ | Root Opening Root Face Groove Angle | Tolerances | | | | |
| | | | | | As Detailed (see 2.9.2) | As Fit Up (see 3.3) | | | |
| SMAW | B-U4b | U | — | R=0 to 3 | +1.6,-0 | +1.6,-3 | All | — | Br,C,N |
| GMAW FCAW | B-U4b-GF | U | — | f=0 to 3<br>$\alpha = 45°$ | +1.6,-0<br>+10,°-0° | Not limited<br>+10,°-5° | All | Not required | A,Br,C N |

| Single-bevel-groove weld (4) T-joint (T) Corner joint (C) | | | | | | | | | |

BACKGOUGE

NOTE V

NOTE J

ALL DIMENSIONS IN mm

| Welding Process | Joint Designation | Base Metal Thickness (U = unlimited) | | Groove Preparation | | | Permitted Welding Positions | Gas Shielding for FCAW | Notes |
| | | $T_1$ | $T_2$ | Root Opening Root Face Groove Angle | Tolerances | | | | |
| | | | | | As Detailed (see 2.9.2) | As Fit Up (see 3.3) | | | |
| SMAW | TC-U4b | U | U | R=0 to 3 | +1.6,-0 | +1.6,-3 | All | — | C,J,R,V |
| GMAW FCAW | TC-U4b-GF | U | U | f=0 to 3<br>$\alpha = 45°$ | +1.6,-0<br>+10,°-0° | Not limited<br>+10,°-5° | All | Not required | A,C,J, R,V |
| SAW | TC-U4b-S | U | U | R=0<br>f=6 max<br>$\alpha = 60°$ | ± 0<br>+0,-3<br>+10,°-0° | +6,-0<br>± 1.6<br>+10,°-5° | F | — | C,J,R, V |

Note A:  Not prequalified for gas tungsten arc and gas metal arc welding using short circuiting transfer. Refer to Appendix A.

Note Br: Dynamic load application limits these joints to the horizontal position (see 9.12.5).

Note C:  Backgouge root to sound metal before welding second side.

Note J:  If fillet welds are used in statically loaded structures to reinforce groove welds in corner and T-joints, they shall be equal to $1/4\,T_1$, but need not exceed 10 mm. Groove welds in corner and T-joints of dynamically loaded structures shall be reinforced with fillet welds equal to $1/4\,T_1$, but not more than 10 mm.

Note N:  The orientation of the two members in the joints may vary from 135° to 180° provided that the basic joint configuration (groove angle, root face, root opening) remain the same and that the design weld size is maintained.

Note R:  The orientation of two members in the joints may vary from 45° to 135° for corner joints and from 45° to 90° for T-joints, provided that the basic joint configuration (groove angle, root face, root opening) remain the same and that the design weld size is maintained.

Note V:  For corner joints, the outside groove preparation may be in either or both members, provided the basic groove configuration is not changed and adequate edge distance is maintained to support the welding operations without excessive edge melting.

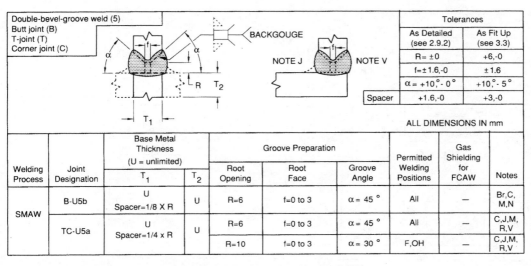

| Double-bevel-groove weld (5) Butt joint (B) T-joint (T) Corner joint (C) | | Tolerances | | |
|---|---|---|---|---|
| | | | As Detailed (see 2.9.2) | As Fit Up (see 3.3) |
| | | R = ±0 | +6,-0 |
| | | f=±1.6,-0 | ±1.6 |
| | | α = +10,°- 0° | +10,°- 5° |
| | Spacer | +1.6,-0 | +3,-0 |

ALL DIMENSIONS IN mm

| Welding Process | Joint Designation | Base Metal Thickness (U = unlimited) | | Groove Preparation | | | Permitted Welding Positions | Gas Shielding for FCAW | Notes |
|---|---|---|---|---|---|---|---|---|---|
| | | $T_1$ | $T_2$ | Root Opening | Root Face | Groove Angle | | | |
| SMAW | B-U5b | U Spacer=1/8 X R | U | R=6 | f=0 to 3 | α = 45° | All | — | Br,C, M,N |
| | TC-U5a | U Spacer=1/4 x R | U | R=6 | f=0 to 3 | α = 45° | All | — | C,J,M, R,V |
| | | | | R=10 | f=0 to 3 | α = 30° | F,OH | — | C,J,M, R,V |

Note Br: Dynamic load application limits these joints to the horizontal position (see 9.12.5).

Note C: Backgouge root to sound metal before welding second side.

Note J: If fillet welds are used in statically loaded structures to reinforce groove welds in corner and T-joints, they shall be equal to 1/4 $T_1$, but need not exceed 10 mm. Groove welds in corner and T-joints of dynamically loaded structures shall be reinforced with fillet welds equal to 1/4 $T_1$, but not more than 10 mm.

Note M: Double-groove welds may have grooves of unequal depth, but the depth of the shallower groove shall be no less than one-fourth of the thickness of the thinner part joined.

Note N: The orientation of the two members in the joints may vary from 135° to 180° provided that the basic joint configuration (groove angle, root face, root opening) remain the same and that the design weld size is maintained.

Note R: The orientation of two members in the joints may vary from 45° to 135° for corner joints and from 45° to 90° for T-joints, provided that the basic joint configuration (groove angle, root face, root opening) remain the same and that the design weld size is maintained.

Note V: For corner joints, the outside groove preparation may be in either or both members, provided the basic groove configuration is not changed and adequate edge distance is maintained to support the welding operations without excessive edge melting.

**Double-bevel-groove weld (5)**
**Butt joint (B)**

ALL DIMENSIONS IN mm

| Welding Process | Joint Designation | Base Metal Thickness (U = unlimited) | | Groove Preparation | | | Permitted Welding Positions | Gas Shielding for FCAW | Notes |
|---|---|---|---|---|---|---|---|---|---|
| | | $T_1$ | $T_2$ | Root Opening Root Face Groove Angle | Tolerances As Detailed (see 2.9.2) | As Fit Up (see 3.3) | | | |
| SMAW | B-U5a | U | — | R=0 to 3<br>f=0 to 3<br>$\alpha$ = 45°<br>$\beta$ = 0° to 15° | +1.6,-0<br>+1.6,-0<br>$\alpha + \beta \begin{array}{l}+10°\\-0°\end{array}$ | +1.6,-3<br>Not limited<br>$\alpha + \beta \begin{array}{l}+10°\\-5°\end{array}$ | All | — | Br, C,M,N |
| GMAW FCAW | B-U5-GF | U | — | R=0 to 3<br>f=0 to 3<br>$\alpha$ = 45°<br>$\beta$ = 0° to 15° | +1.6,-0<br>+1.6,-0<br>$\alpha + \beta =$<br>+10°- 0° | +1.6,-3<br>Not limited<br>$\alpha + \beta =$<br>+10°- 5° | All | Not req. | A,Br,C, M,N |

**Double-bevel-groove weld (5)**
**T-joint (T)**
**Corner joint (C)**

ALL DIMENSIONS IN mm

| Welding Process | Joint Designation | Base Metal Thickness (U = unlimited) | | Groove Preparation | | | Permitted Welding Positions | Gas Shielding for FCAW | Notes |
|---|---|---|---|---|---|---|---|---|---|
| | | $T_1$ | $T_2$ | Root Opening Root Face Groove Angle | Tolerances As Detailed (see 2.9.2) | As Fit Up (see 3.3) | | | |
| SMAW | TC-U5b | U | U | R=0 to 3<br>f=0 to 3<br>$\alpha$ = 45° | +1.6,-0<br>+1.6,-0<br>+10°- 0° | +1.6,-3<br>Not limited<br>+10°- 5° | All | — | C,J,M, R,V |
| GMAW FCAW | TC-U5-GF | U | U | | | | All | Not required | A,C,J, M,R,V |
| SAW | TC-U5-S | U | U | R=0<br>f=5 max<br>$\alpha$ = 60° | ± 0<br>+0,-5<br>+10°- 0° | +1.6,-0<br>± 1.6<br>+10°- 5° | F | — | C,J,M, R,V |

Note A: Not prequalified for gas tungsten arc and gas metal arc welding using short circuiting transfer. Refer to Appendix A.

Note Br: Dynamic load application limits these joints to the horizontal position (see 9.12.5).

Note C: Backgouge root to sound metal before welding second side.

Note J: If fillet welds are used in statically loaded structures to reinforce groove welds in corner and T-joints, they shall be equal to 1/4 $T_1$, but need not exceed 10 mm. Groove welds in corner and T-joints of dynamically loaded structures shall be reinforced with fillet welds equal to 1/4 $T_1$, but not more than 10 mm.

Note M: Double-groove welds may have grooves of unequal depth, but the depth of the shallower groove shall be no less than one-fourth of the thickness of the thinner part joined.

Note N: The orientation of the two members in the joints may vary from 135° to 180° provided that the basic joint configuration (groove angle, root face, root opening) remain the same and that the design weld size is maintained.

Note R: The orientation of two members in the joints may vary from 45° to 135° for corner joints and from 45° to 90° for T-joints, provided that the basic joint configuration (groove angle, root face, root opening) remain the same and that the design weld size is maintained.

Note V: For corner joints, the outside groove preparation may be in either or both members, provided the basic groove configuration is not changed and adequate edge distance is maintained to support the welding operations without excessive edge melting.

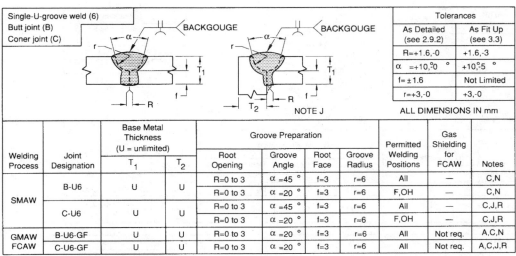

| Welding Process | Joint Designation | Base Metal Thickness (U = unlimited) | | Groove Preparation | | | | Permitted Welding Positions | Gas Shielding for FCAW | Notes |
|---|---|---|---|---|---|---|---|---|---|---|
| | | $T_1$ | $T_2$ | Root Opening | Groove Angle | Root Face | Groove Radius | | | |
| SMAW | B-U6 | U | U | R=0 to 3 | $\alpha$ =45 ° | f=3 | r=6 | All | — | C,N |
| | | | | R=0 to 3 | $\alpha$ =20 ° | f=3 | r=6 | F,OH | — | C,N |
| | C-U6 | U | U | R=0 to 3 | $\alpha$ =45 ° | f=3 | r=6 | All | — | C,J,R |
| | | | | R=0 to 3 | $\alpha$ =20 ° | f=3 | r=6 | F,OH | — | C,J,R |
| GMAW FCAW | B-U6-GF | U | U | R=0 to 3 | $\alpha$ =20 ° | f=3 | r=6 | All | Not req. | A,C,N |
| | C-U6-GF | U | U | R=0 to 3 | $\alpha$ =20 ° | f=3 | r=6 | All | Not req. | A,C,J,R |

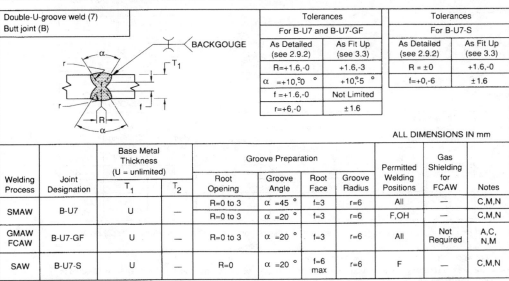

| Welding Process | Joint Designation | Base Metal Thickness (U = unlimited) | | Groove Preparation | | | | Permitted Welding Positions | Gas Shielding for FCAW | Notes |
|---|---|---|---|---|---|---|---|---|---|---|
| | | $T_1$ | $T_2$ | Root Opening | Groove Angle | Root Face | Groove Radius | | | |
| SMAW | B-U7 | U | — | R=0 to 3 | $\alpha$ =45 ° | f=3 | r=6 | All | — | C,M,N |
| | | | | R=0 to 3 | $\alpha$ =20 ° | f=3 | r=6 | F,OH | — | C,M,N |
| GMAW FCAW | B-U7-GF | U | — | R=0 to 3 | $\alpha$ =20 ° | f=3 | r=6 | All | Not Required | A,C, N,M |
| SAW | B-U7-S | U | — | R=0 | $\alpha$ =20 ° | f=6 max | r=6 | F | — | C,M,N |

Note A: Not prequalified for gas tungsten arc and gas metal arc welding using short circuiting transfer. Refer to Appendix A.

Note C: Backgouge root to sound metal before welding second side.

Note J: If fillet welds are used in statically loaded structures to reinforce groove welds in corner and T-joints, they shall be equal to 1/4 $T_1$, but need not exceed 10 mm. Groove welds in corner and T-joints of dynamically loaded structures shall be reinforced with fillet welds equal to 1/4 $T_1$, but not more than 10 mm.

Note M: Double-groove welds may have grooves of unequal depth, but the depth of the shallower groove shall be no less than one-fourth of the thickness of the thinner part joined.

Note N: The orientation of the two members in the joints may vary from 135° to 180° provided that the basic joint configuration (groove angle, root face, root opening) remain the same and that the design weld size is maintained.

Note R: The orientation of two members in the joints may vary from 45° to 135° for corner joints and from 45° to 90° for T-joints, provided that the basic joint configuration (groove angle, root face, root opening) remain the same and that the design weld size is maintained.

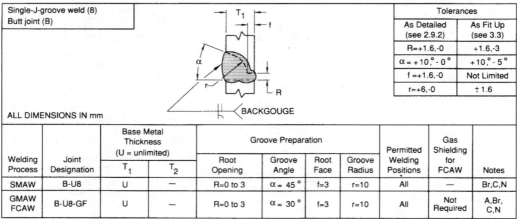

Single-J-groove weld (8)
Butt joint (B)

ALL DIMENSIONS IN mm

| | | Tolerances | |
|---|---|---|---|
| | | As Detailed (see 2.9.2) | As Fit Up (see 3.3) |
| | | R=+1.6,-0 | +1.6,-3 |
| | | $\alpha = +10°,-0°$ | $+10°,-5°$ |
| | | f =+1.6,-0 | Not Limited |
| | | r=+6,-0 | ±1.6 |

| Welding Process | Joint Designation | Base Metal Thickness (U = unlimited) | | Groove Preparation | | | | Permitted Welding Positions | Gas Shielding for FCAW | Notes |
|---|---|---|---|---|---|---|---|---|---|---|
| | | $T_1$ | $T_2$ | Root Opening | Groove Angle | Root Face | Groove Radius | | | |
| SMAW | B-U8 | U | — | R=0 to 3 | $\alpha = 45°$ | f=3 | r=10 | All | — | Br,C,N |
| GMAW FCAW | B-U8-GF | U | — | R=0 to 3 | $\alpha = 30°$ | f=3 | r=10 | All | Not Required | A,Br, C,N |

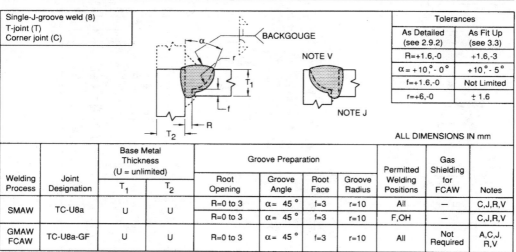

Single-J-groove weld (8)
T-joint (T)
Corner joint (C)

NOTE V

NOTE J

ALL DIMENSIONS IN mm

| | | Tolerances | |
|---|---|---|---|
| | | As Detailed (see 2.9.2) | As Fit Up (see 3.3) |
| | | R=+1.6,-0 | +1.6,-3 |
| | | $\alpha = +10°,-0°$ | $+10°,-5°$ |
| | | f=+1.6,-0 | Not Limited |
| | | r=+6,-0 | ±1.6 |

| Welding Process | Joint Designation | Base Metal Thickness (U = unlimited) | | Groove Preparation | | | | Permitted Welding Positions | Gas Shielding for FCAW | Notes |
|---|---|---|---|---|---|---|---|---|---|---|
| | | $T_1$ | $T_2$ | Root Opening | Groove Angle | Root Face | Groove Radius | | | |
| SMAW | TC-U8a | U | U | R=0 to 3 | $\alpha = 45°$ | f=3 | r=10 | All | — | C,J,R,V |
| | | | | R=0 to 3 | $\alpha = 45°$ | f=3 | r=10 | F,OH | — | C,J,R,V |
| GMAW FCAW | TC-U8a-GF | U | U | R=0 to 3 | $\alpha = 45°$ | f=3 | r=10 | All | Not Required | A,C,J, R,V |

Note A: Not prequalified for gas tungsten arc and gas metal arc welding using short circuiting transfer. Refer to Appendix A.

Note Br: Dynamic load application limits these joints to the horizontal position (see 9.12.5).

Note C: Backgouge root to sound metal before welding second side.

Note J: If fillet welds are used in statically loaded structures to reinforce groove welds in corner and T-joints, they shall be equal to 1/4 $T_1$, but need not exceed 10 mm. Groove welds in corner and T-joints of dynamically loaded structures shall be reinforced with fillet welds equal to 1/4 $T_1$, but not more than 10 mm.

Note N: The orientation of the two members in the joints may vary from 135° to 180° provided that the basic joint configuration (groove angle, root face, root opening) remain the same and that the design weld size is maintained.

Note R: The orientation of two members in the joints may vary from 45° to 135° for corner joints and from 45° to 90° for T-joints, provided that the basic joint configuration (groove angle, root face, root opening) remain the same and that the design weld size is maintained.

Note V: For corner joints, the outside groove preparation may be in either or both members, provided the basic groove configuration is not changed and adequate edge distance is maintained to support the welding operations without excessive edge melting.

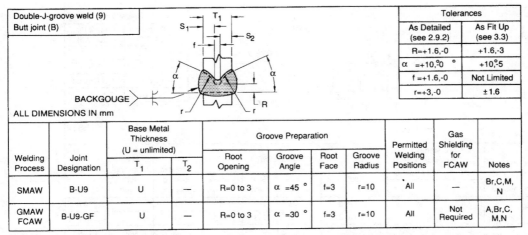

| | | Double-J-groove weld (9) Butt joint (B) | | | | | | Tolerances | | |
|---|---|---|---|---|---|---|---|---|---|---|

| | | | | | | | | Tolerances | |
|---|---|---|---|---|---|---|---|---|---|
| | | | | | | | | As Detailed (see 2.9.2) | As Fit Up (see 3.3) |
| | | | | | | | | R=+1.6,-0 | +1.6,-3 |
| | | | | | | | | α =+10,°0  ° | +10,°5 |
| | | | | | | | | f =+1.6,-0 | Not Limited |
| | | | | | | | | r=+3,-0 | ±1.6 |

BACKGOUGE

ALL DIMENSIONS IN mm

| Welding Process | Joint Designation | Base Metal Thickness (U = unlimited) | | Groove Preparation | | | | Permitted Welding Positions | Gas Shielding for FCAW | Notes |
|---|---|---|---|---|---|---|---|---|---|---|
| | | T₁ | T₂ | Root Opening | Groove Angle | Root Face | Groove Radius | | | |
| SMAW | B-U9 | U | — | R=0 to 3 | α =45 ° | f=3 | r=10 | All | — | Br,C,M, N |
| GMAW FCAW | B-U9-GF | U | — | R=0 to 3 | α =30 ° | f=3 | r=10 | All | Not Required | A,Br,C, M,N |

Note A: Not prequalified for gas tungsten arc and gas metal arc welding using short circuiting transfer. Refer to Appendix A.

Note Br: Dynamic load application limits these joints to the horizontal position (see 9.12.5).

Note C: Backgouge root to sound metal before welding second side.

Note M: Double-groove welds may have grooves of unequal depth, but the depth of the shallower groove shall be no less than one-fourth of the thickness of the thinner part joined.

Note N: The orientation of the two members in the joints may vary from 135° to 180° provided that the basic joint configuration (groove angle, root face, root opening) remain the same and that the design weld size is maintained.

| Double-J-groove weld (9) T-joint (T) Corner joint (C) | | | | | | | Tolerances | | |
|---|---|---|---|---|---|---|---|---|---|
| | | | | | | | **As Detailed** (see 2.9.2) | **As Fit Up** (see 3.3) | |
| | | | | | | | R=+1.6,-0 | +1.6,-3 | |
| | | | | | | | α = +10,°- 0° | +10,°- 5° | |
| | | | | | | | f =+1.6,-0 | Not Limited | |
| | | | | | | | r=+3,-0 | ± 1.6 | |

ALL DIMENSIONS IN mm

| Welding Process | Joint Designation | Base Metal Thickness (U = unlimited) | | Groove Preparation | | | | Permitted Welding Positions | Gas Shielding for FCAW | Notes |
|---|---|---|---|---|---|---|---|---|---|---|
| | | $T_1$ | $T_2$ | Root Opening | Groove Angle | Root Face | Groove Radius | | | |
| SMAW | TC-U9a | U | U | R=0 to 3 | α = 45 ° | f=3 | r=10 | All | — | C,J,M,R,V |
| | | | | R=0 to 3 | α = 30 ° | f=3 | r=10 | F,OH | — | C,J,M,R,V |
| GMAW FCAW | TC-U9a-GF | U | U | R=0 to 3 | α = 30 ° | f=3 | r=10 | All | Not Required | A,C,J, M,R,V |

Note A:  Not prequalified for gas tungsten arc and gas metal arc welding using short circuiting transfer. Refer to Appendix A.

Note C:  Backgouge root to sound metal before welding second side.

Note J:  If fillet welds are used in statically loaded structures to reinforce groove welds in corner and T-joints, they shall be equal to 1/4 $T_1$, but need not exceed 10 mm. Groove welds in corner and T-joints of dynamically loaded structures shall be reinforced with fillet welds equal to 1/4 $T_1$, but not more than 10 mm.

Note M:  Double-groove welds may have grooves of unequal depth, but the depth of the shallower groove shall be no less than one-fourth of the thickness of the thinner part joined.

Note R:  The orientation of two members in the joints may vary from 45° to 135° for corner joints and from 45° to 90° for T-joints, provided that the basic joint configuration (groove angle, root face, root opening) remain the same and that the design weld size is maintained.

Note V:  For corner joints, the outside groove preparation may be in either or both members, provided the basic groove configuration is not changed and adequate edge distance is maintained to support the welding operations without excessive edge melting.

| Square-groove weld (1) Butt joint (B) | | | | | | | | | | |
|---|---|---|---|---|---|---|---|---|---|---|
| 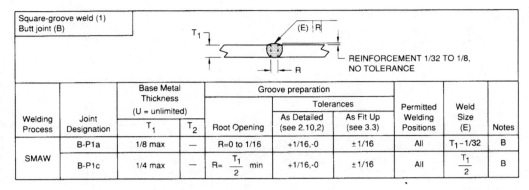 | | | | | | | | | | |

| Welding Process | Joint Designation | Base Metal Thickness (U = unlimited) | | Groove preparation | | | Permitted Welding Positions | Weld Size (E) | Notes |
|---|---|---|---|---|---|---|---|---|---|
| | | $T_1$ | $T_2$ | Root Opening | Tolerances | | | | |
| | | | | | As Detailed (see 2.10,2) | As Fit Up (see 3.3) | | | |
| SMAW | B-P1a | 1/8 max | — | R=0 to 1/16 | +1/16,-0 | ±1/16 | All | $T_1 - 1/32$ | B |
| | B-P1c | 1/4 max | — | $R= \dfrac{T_1}{2}$ min | +1/16,-0 | ±1/16 | All | $\dfrac{T_1}{2}$ | B |

| Square-groove weld (1) Butt joint (B) | | | | | | | | | | |
|---|---|---|---|---|---|---|---|---|---|---|
| $E_1 + E_2$ MUST NOT EXCEED $\dfrac{3T_1}{4}$ | | 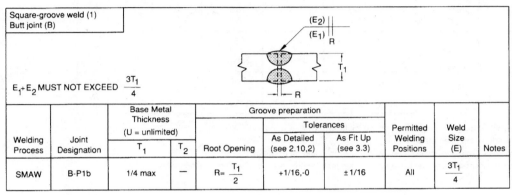 | | | | | | | | |

| Welding Process | Joint Designation | Base Metal Thickness (U = unlimited) | | Groove preparation | | | Permitted Welding Positions | Weld Size (E) | Notes |
|---|---|---|---|---|---|---|---|---|---|
| | | $T_1$ | $T_2$ | Root Opening | Tolerances | | | | |
| | | | | | As Detailed (see 2.10,2) | As Fit Up (see 3.3) | | | |
| SMAW | B-P1b | 1/4 max | — | $R= \dfrac{T_1}{2}$ | +1/16,-0 | ±1/16 | All | $\dfrac{3T_1}{4}$ | |

Note B:  Joint is welded from one side only.

Single-V-groove weld (2)
Butt joint (B)
Corner joint (C)

| Welding Process | Joint Designation | Base Metal Thickness (U = unlimited) | | Root Opening Root Face Groove Angle | Tolerances | | Permitted Welding Positions | Weld Size (E) | Notes |
|---|---|---|---|---|---|---|---|---|---|
| | | $T_1$ | $T_2$ | | As Detailed (see 2.10.2) | As Fit Up (see 3.3) | | | |
| SMAW | BC-P2 | 1/4 min | U | R=0 f=1/32 min α =60 ° | 0,+1/16 +U,-0 +10°,0 ° | +1/8,-1/16 ±1/16 +10°,5 ° | All | S | B,E, Q2 |
| GMAW FCAW | BC-P2-GF | 1/4 min | U | R=0 f=1/8 min α =60 ° | 0,+1/16 +U,-0 +10°,0 ° | +1/8,-1/16 ±1/16 +10°,5 ° | All | S | A,B, E,Q2 |
| SAW | BC-P2-S | 7/16 min | U | R=0 f=1/4 min α =60 ° | ±0 +U,-0 +10°,0 ° | +1/16,-0 ±1/16 +10°,5 ° | F | S | B,E, Q2 |

Note A:   Not prequalified for gas tungsten arc and gas metal arc welding using short circuiting transfer. Refer to Appendix A.
Note B:   Joint is welded from one side only.
Note E:   Minimum weld size (E) as shown in Table 2.3; S as specified on drawings.
Note Q2: The member orientation may be changed provided that the groove dimensions are maintained as specified.

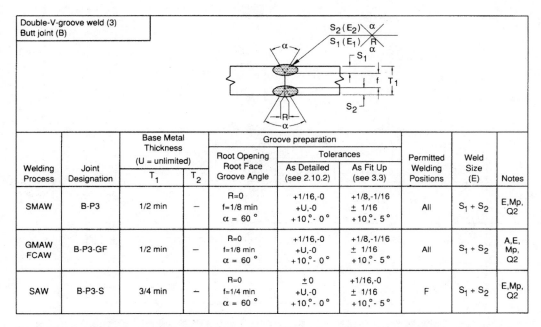

| Welding Process | Joint Designation | Base Metal Thickness (U = unlimited) | | Groove preparation | | | Permitted Welding Positions | Weld Size (E) | Notes |
| | | $T_1$ | $T_2$ | Root Opening Root Face Groove Angle | As Detailed (see 2.10.2) | As Fit Up (see 3.3) | | | |
|---|---|---|---|---|---|---|---|---|---|
| SMAW | B-P3 | 1/2 min | — | R=0<br>f=1/8 min<br>$\alpha = 60°$ | +1/16,-0<br>+U,-0<br>+10°,-0° | +1/8,-1/16<br>± 1/16<br>+10°,- 5° | All | $S_1 + S_2$ | E,Mp, Q2 |
| GMAW FCAW | B-P3-GF | 1/2 min | — | R=0<br>f=1/8 min<br>$\alpha = 60°$ | +1/16,-0<br>+U,-0<br>+10°,-0° | +1/8,-1/16<br>± 1/16<br>+10°,- 5° | All | $S_1 + S_2$ | A,E, Mp, Q2 |
| SAW | B-P3-S | 3/4 min | — | R=0<br>f=1/4 min<br>$\alpha = 60°$ | ± 0<br>+U,-0<br>+10°,-0° | +1/16,-0<br>± 1/16<br>+10°,- 5° | F | $S_1 + S_2$ | E,Mp, Q2 |

Note A:  Not prequalified for gas tungsten arc and gas metal arc welding using short circuiting transfer. Refer to Appendix A.

Note E:  Minimum weld size (E) as shown in Table 2.3; S as specified on drawings.

Note Mp: Double-groove welds may have grooves of unequal depth, provided they conform to the limitations of Note E. Also the weld size (E), less any reduction applies individually to each groove.

Note Q2: The member orientation may be changed provided that the groove dimensions are maintained as specified.

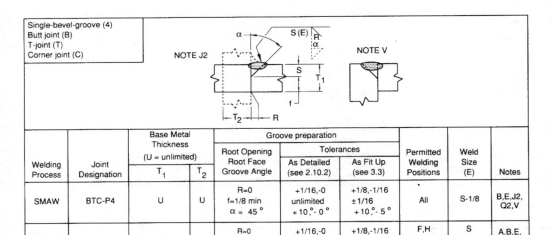

| Welding Process | Joint Designation | Base Metal Thickness (U = unlimited) | | Groove preparation | | | Permitted Welding Positions | Weld Size (E) | Notes |
|---|---|---|---|---|---|---|---|---|---|
| | | | | Root Opening Root Face Groove Angle | Tolerances | | | | |
| | | $T_1$ | $T_2$ | | As Detailed (see 2.10.2) | As Fit Up (see 3.3) | | | |
| SMAW | BTC-P4 | U | U | R=0<br>f=1/8 min<br>$\alpha = 45°$ | +1/16,-0<br>unlimited<br>+10°,-0° | +1/8,-1/16<br>±1/16<br>+10°,-5° | All | S-1/8 | B,E,J2,<br>Q2,V |
| GMAW FCAW | BTC-P4-GF | 1/4 min | U | R=0<br>f=1/8 min<br>$\alpha = 45°$ | +1/16,-0<br>unlimited*<br>+10°,-0° | +1/8,-1/16<br>±1/16<br>+10°,-5° | F,H | S | A,B,E,<br>J2,Q2,<br>V |
| | | | | | | | V,OH | S-1/8 | |
| SAW | TC-P4-S | 7/16 min | U | R=0<br>f=1/4 min<br>$\alpha = 60°$ | ±0<br>+U,-0<br>+10°,-0° | +1/16,-0<br>±1/16<br>+10°,-5° | F | S | B,E,J2,<br>Q2,V |

Note A:    Not prequalified for gas tungsten arc and gas metal arc welding using short circuiting transfer. Refer to Appendix A.

Note B:    Joint is welded from one side only.

Note E:    Minimum weld size (E) as shown in Table 2.3; S as specified on drawings.

Note J2:   If fillet welds are used in statically loaded structures to reinforce groove welds in corner and T-joints, they shall be equal to 1/4 $T_1$, but need not exceed 3/8 in.

Note Q2:   The member orientation may be changed provided that the groove dimensions are maintained as specified.

Note V:    For corner joints, the outside groove preparation may be in either or both members, provided the basic groove configuration is not changed and adequate edge distance is maintained to support the welding operations without excessive edge melting.

*For flat and horizontal positions, f=+U,-0

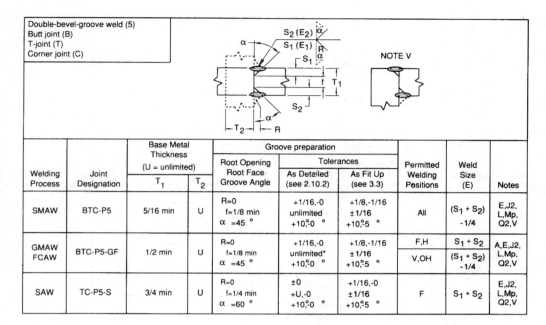

Double-bevel-groove weld (5)
Butt joint (B)
T-joint (T)
Corner joint (C)

| Welding Process | Joint Designation | Base Metal Thickness (U = unlimited) | | Groove preparation | | | | Permitted Welding Positions | Weld Size (E) | Notes |
|---|---|---|---|---|---|---|---|---|---|---|
| | | $T_1$ | $T_2$ | Root Opening Root Face Groove Angle | Tolerances | | | | | |
| | | | | | As Detailed (see 2.10.2) | As Fit Up (see 3.3) | | | | |
| SMAW | BTC-P5 | 5/16 min | U | R=0 f=1/8 min α =45 ° | +1/16,-0 unlimited +10°,-0 ° | +1/8,-1/16 ±1/16 +10°,-5 ° | | All | $(S_1 + S_2)$ -1/4 | E,J2, L,Mp, Q2,V |
| GMAW FCAW | BTC-P5-GF | 1/2 min | U | R=0 f=1/8 min α =45 ° | +1/16,-0 unlimited* +10°,-0 ° | +1/8,-1/16 ±1/16 +10°,-5 ° | | F,H | $S_1 + S_2$ | A,E,J2, L,Mp, Q2,V |
| | | | | | | | | V,OH | $(S_1 + S_2)$ -1/4 | |
| SAW | TC-P5-S | 3/4 min | U | R=0 f=1/4 min α =60 ° | ±0 +U,-0 +10°,-0 ° | +1/16,-0 ±1/16 +10°,-5 ° | | F | $S_1 + S_2$ | E,J2, L,Mp, Q2,V |

Note A:   Not prequalified for gas tungsten arc and gas metal arc welding using short circuiting transfer. Refer to Appendix A.

Note E:   Minimum weld size (E) as shown in Table 2.3; S as specified on drawings.

Note J2:   If fillet welds are used in statically loaded structures to reinforce groove welds in corner and T-joints, they shall be equal to 1/4 $T_1$, but need not exceed 3/8 in.

Note L:   Butt and T-Joints are not prequalified for dynamically loaded structures.

Note Mp: Double-groove welds may have grooves of unequal depth, provided they conform to the limitations of Note E. Also the weld size (E), less any reduction applies individually to each groove.

Note Q2: The member orientation may be changed provided that the groove dimensions are maintained as specified.

Note V:   For corner joints, the outside groove preparation may be in either or both members, provided the basic groove configuration is not changed and adequate edge distance is maintained to support the welding operations without excessive edge melting.

*For flat and horizontal positions, f=+U,-0

| Single-U-groove weld (6) Butt joint (B) Corner joint (C) | | | | | | | | | |
|---|---|---|---|---|---|---|---|---|---|

| | | Base Metal Thickness (U = unlimited) | | Groove preparation | | | | | | |
| | | | | Root Opening Root Face Groove Radius Groove Angle | Tolerances | | Permitted Welding Positions | Weld Size (E) | Notes |
| Welding Process | Joint Designation | $T_1$ | $T_2$ | | As Detailed (see 2.10.2) | As Fit Up (see 3.3) | | | |
|---|---|---|---|---|---|---|---|---|---|
| SMAW | BC-P6 | 1/4 min | U | R=0<br>f=1/32 min<br>r=1/4<br>$\alpha$ =45° | +1/16,-0<br>+U,-0<br>+1/4,-0<br>+10°,0° | +1/8,-1/16<br>±1/16<br>±1/16<br>+10°,5° | All | S | B,E,Q2 |
| GMAW FCAW | BC-P6-GF | 1/4 min | U | R=0<br>f=1/8 min<br>r=1/4<br>$\alpha$ =20° | +1/16,-0<br>+U,-0<br>+1/4,-0<br>+10°,0° | +1/8,-1/16<br>±1/16<br>±1/16<br>+10°,5° | All | S | A,B, E,Q2 |
| SAW | BC-P6-S | 7/16 min | U | R=0<br>f=1/4 min<br>r=1/4<br>$\alpha$ =20° | ±0<br>+U,-0<br>+1/4,-0<br>+10°,0° | +1/16,-0<br>±1/16<br>±1/16<br>+10°,5° | F | S | B,E,Q2 |

Note A:   Not prequalified for gas tungsten arc and gas metal arc welding using short circuiting transfer. Refer to Appendix A.

Note B:   Joint is welded from one side only.

Note E:   Minimum weld size (E) as shown in Table 2.3; S as specified on drawings.

Note Q2: The member orientation may be changed provided that the groove dimensions are maintained as specified.

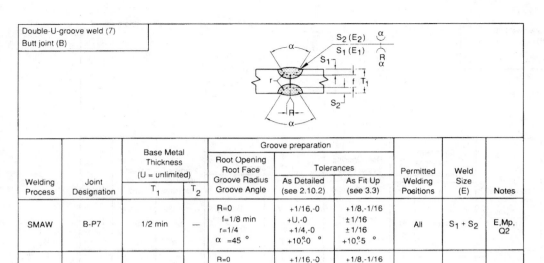

| Welding Process | Joint Designation | Base Metal Thickness (U = unlimited) | | Groove preparation | | | Permitted Welding Positions | Weld Size (E) | Notes |
|---|---|---|---|---|---|---|---|---|---|
| | | $T_1$ | $T_2$ | Root Opening Root Face Groove Radius Groove Angle | As Detailed (see 2.10.2) | As Fit Up (see 3.3) | | | |
| SMAW | B-P7 | 1/2 min | — | R=0<br>f=1/8 min<br>r=1/4<br>$\alpha$ =45 ° | +1/16,-0<br>+U,-0<br>+1/4,-0<br>+10°,-0 ° | +1/8,-1/16<br>± 1/16<br>± 1/16<br>+10°,-5 ° | All | $S_1 + S_2$ | E,Mp,<br>Q2 |
| GMAW FCAW | B-P7-GF | 1/2 min | — | R=0<br>f=1/8 min<br>r=1/4<br>$\alpha$ =20 ° | +1/16,-0<br>+U,-0<br>+1/4,-0<br>+10°,-0 ° | +1/8,-1/16<br>± 1/16<br>± 1/16<br>+10°,-5 ° | All | $S_1 + S_2$ | A,E,<br>Mp,Q2 |
| SAW | B-P7-S | 3/4 min | — | R=0<br>f=1/4 min<br>r=1/4<br>$\alpha$ =20 ° | ±0<br>+U,-0<br>+1/4,-0<br>+10°,-0 ° | +1/16,-0<br>± 1/16<br>± 1/16<br>+10°,-5 ° | F | $S_1 + S_2$ | E,Mp,<br>Q2 |

Note A:  Not prequalified for gas tungsten arc and gas metal arc welding using short circuiting transfer. Refer to Appendix A.

Note E:  Minimum weld size (E) as shown in Table 2.3; S as specified on drawings.

Note Mp: Double-groove welds may have grooves of unequal depth, provided they conform to the limitations of Note E. Also the weld size (E), less any reduction applies individually to each groove.

Note Q2: The member orientation may be changed provided that the groove dimensions are maintained as specified.

Single-J-groove weld (8)
Butt joint (B)
T-joint (T)
Corner joint (C)

α   S (E)   R
r   α
S
T₁
NOTE V
f
T₂   R

| Welding Process | Joint Designation | Base Metal Thickness (U = unlimited) | | Groove preparation | | | Permitted Welding Positions | Weld Size (E) | Notes |
|---|---|---|---|---|---|---|---|---|---|
| | | | | Root Opening Root Face Groove Radius Groove Angle | Tolerances | | | | |
| | | $T_1$ | $T_2$ | | As Detailed (see 2.10.2) | As Fit Up (see 3.3) | | | |
| SMAW | TC-P8* | 1/4 min | U | R=0<br>f=1/8 min<br>r=3/8<br>α =45 ° | +1/16,-0<br>+U,-0<br>+1/4,-0<br>+10°,-0 ° | +1/8,-1/16<br>±1/16<br>±1/16<br>+10°,-5 ° | All | S | E,J2,<br>Q2,V |
| SMAW | BC-P8** | 1/4 min | U | R=0<br>f=1/8 min<br>r=3/8<br>α =30 ° | +1/16,-0<br>+U,-0<br>+1/4,-0<br>+10°,-0 ° | +1/8,-1/16<br>±1/16<br>±1/16<br>+10°,-5 ° | All | S | E,J2,<br>Q2,V |
| GMAW FCAW | TC-P8-GF* | 1/4 min | U | R=0<br>f=1/8 min<br>r=3/8<br>α =45 ° | +1/16,-0<br>+U,-0<br>+1/4,-0<br>+10°,-0 ° | +1/8,-1/16<br>±1/16<br>±1/16<br>+10°,-5 ° | All | S | A,E,<br>J2,Q2,<br>V |
| GMAW FCAW | BC-P8-GF** | 1/4 min | U | R=0<br>f=1/8 min<br>r=3/8<br>α =30 ° | +1/16,-0<br>+U,-0<br>+1/4,-0<br>+10°,-0 ° | +1/8,-1/16<br>±1/16<br>±1/16<br>+10°,-5 ° | All | S | A,E,<br>J2,Q2,<br>V |
| SAW | TC-P8-S* | 7/16 min | U | R=0<br>f=1/4 min<br>r=1/2<br>α =45 ° | ±0<br>+U,-0<br>+1/4,-0<br>+10°,-0 ° | +1/16,-0<br>±1/16<br>±1/16<br>+10°,-5 ° | F | S | E,J2,<br>Q2,V |
| SAW | C-P8-S** | 7/16 min | U | R=0<br>f=1/4 min<br>r=1/2<br>α =20 ° | ±0<br>+U,-0<br>+1/4,-0<br>+10°,-0 ° | +1/16,-0<br>±1/16<br>±1/16<br>+10°,-5 ° | F | S | E,J2,<br>Q2,V |

Note A:   Not prequalified for gas tungsten arc and gas metal arc welding using short circuiting transfer. Refer to Appendix A.

Note E:   Minimum weld size (E) as shown in Table 2.3; S as specified on drawings.

Note J2:  If fillet welds are used in statically loaded structures to reinforce groove welds in corner and T-joints, they shall be equal to 1/4 $T_1$, but need not exceed 3/8 in.

Note Q2:  The member orientation may be changed provided that the groove dimensions are maintained as specified.

Note V:   For corner joints, the outside groove preparation may be in either or both members, provided the basic groove configuration is not changed and adequate edge distance is maintained to support the welding operations without excessive edge melting.

* Applies to inside corner joints.
** Applies to outside corner joints.

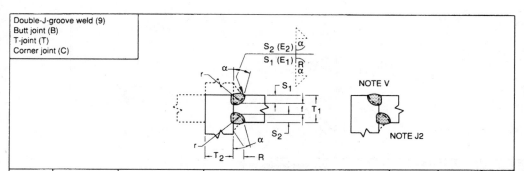

Double-J-groove weld (9)
Butt joint (B)
T-joint (T)
Corner joint (C)

| Welding Process | Joint Designation | Base Metal Thickness (U = unlimited) | | Groove preparation | | | Permitted Welding Positions | Weld Size (E) | Notes |
| | | $T_1$ | $T_2$ | Root Opening Root Face Groove Radius Groove Angle | Tolerances | | | | |
| | | | | | As Detailed (see 2.10.2) | As Fit Up (see 3.3) | | | |
| SMAW | BTC-P9* | 1/2 min | U | R=0<br>f=1/8 min<br>r=3/8<br>$\alpha$ =45 ° | +1/16,-0<br>+U,-0<br>+1/4,-0<br>+10°,0 ° | +1/8,-1/16<br>±1/16<br>±1/16<br>+10°,5 ° | All | $S_1 + S_2$ | E,J2,<br>Mp,Q2,<br>V |
| GMAW FCAW | BTC-P9-GF** | 1/2 min | U | R=0<br>f=1/8 min<br>r=3/8<br>$\alpha$ =30 ° | +1/16,-0<br>+U,-0<br>+1/4,-0<br>+10°,0 ° | +1/8,-1/16<br>±1/16<br>±1/16<br>+10°,5 ° | All | $S_1 + S_2$ | A,J2,<br>Mp,Q2,<br>V |
| SAW | C-P9-S* | 3/4 min | U | R=0<br>f=1/4 min<br>r=1/2<br>$\alpha$ =45 ° | ±0<br>+U,-0<br>+1/4,-0<br>+10°,0 ° | +1/16,-0<br>±1/16<br>±1/16<br>+10°,5 ° | F | $S_1 + S_2$ | E,J2,<br>Mp,Q2,<br>V |
| SAW | C-P9-S** | 3/4 min | U | R=0<br>f=1/4 min<br>r=1/2<br>$\alpha$ =20 ° | ±0<br>+U,-0<br>+1/4,-0<br>+10°,0 ° | +1/16,-0<br>±1/16<br>±1/16<br>+10°,5 ° | F | $S_1 + S_2$ | E,J2,<br>Mp,Q2,<br>V |
| SAW | T-P9-S | 3/4 min | U | R=0<br>f=1/4 min<br>r=1/2<br>$\alpha$ =45 ° | ±0<br>+U,-0<br>+1/4,-0<br>+10°,0 ° | +1/16,-0<br>±1/16<br>±1/16<br>+10°,5 ° | F | $S_1 + S_2$ | E,J2,<br>Mp,Q2 |

Note A:   Not prequalified for gas tungsten arc and gas metal arc welding using short circuiting transfer. Refer to Appendix A.

Note E:   Minimum weld size (E) as shown in Table 2.3; S as specified on drawings.

Note J2:  If fillet welds are used in statically loaded structures to reinforce groove welds in corner and T-joints, they shall be equal to 1/4 $T_1$, but need not exceed 3/8 in.

Note Mp: Double-groove welds may have grooves of unequal depth, provided they conform to the limitations of Note E. Also the weld size (E), less any reduction applies individually to each groove.

Note Q2:  The member orientation may be changed provided that the groove dimensions are maintained as specified.

Note V:   For corner joints, the outside groove preparation may be in either or both members, provided the basic groove configuration is not changed and adequate edge distance is maintained to support the welding operations without excessive edge melting.

 * Applies to inside corner joints.
** Applies to outside corner joints.

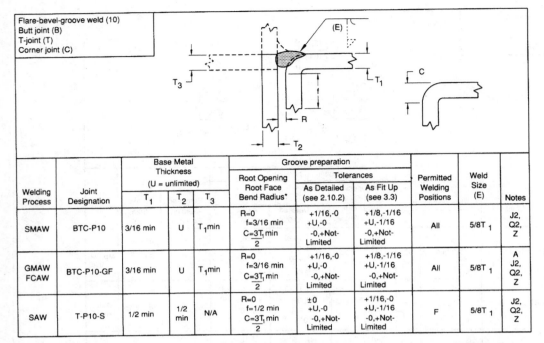

Flare-bevel-groove weld (10)
Butt joint (B)
T-joint (T)
Corner joint (C)

| Welding Process | Joint Designation | Base Metal Thickness (U = unlimited) | | | Groove preparation | | | Permitted Welding Positions | Weld Size (E) | Notes |
|---|---|---|---|---|---|---|---|---|---|---|
| | | $T_1$ | $T_2$ | $T_3$ | Root Opening Root Face Bend Radius* | As Detailed (see 2.10.2) | As Fit Up (see 3.3) | | | |
| SMAW | BTC-P10 | 3/16 min | U | $T_1$min | R=0 f=3/16 min C=$\frac{3T_1\ min}{2}$ | +1/16,-0 +U,-0 -0,+Not-Limited | +1/8,-1/16 +U,-1/16 -0,+Not-Limited | All | 5/8$T_1$ | J2, Q2, Z |
| GMAW FCAW | BTC-P10-GF | 3/16 min | U | $T_1$min | R=0 f=3/16 min C=$\frac{3T_1\ min}{2}$ | +1/16,-0 +U,-0 -0,+Not-Limited | +1/8,-1/16 +U,-1/16 -0,+Not-Limited | All | 5/8$T_1$ | A J2, Q2, Z |
| SAW | T-P10-S | 1/2 min | 1/2 min | N/A | R=0 f=1/2 min C=$\frac{3T_1\ min}{2}$ | ±0 +U,-0 -0,+Not-Limited | +1/16,-0 +U,-1/16 -0,+Not-Limited | F | 5/8$T_1$ | J2, Q2, Z |

Note A:   Not prequalified for gas tungsten arc and gas metal arc welding using short circuiting transfer. Refer to Appendix A.
Note J2:  If fillet welds are used in statically loaded structures to reinforce groove welds in corner and T-joints, they shall be equal to 1/4 $T_1$, but need not exceed 3/8 in.
Note Q2: The member orientation may be changed provided that the groove dimensions are maintained as specified.
Note Z:   Weld size (E) is based on joints welded flush.
*For cold formed (A500) rectangular tubes, C dimension is not limited (see commentary).

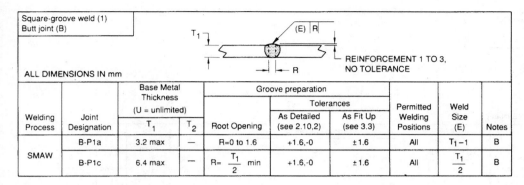

| Welding Process | Joint Designation | Base Metal Thickness (U = unlimited) $T_1$ | $T_2$ | Root Opening | As Detailed (see 2.10,2) | As Fit Up (see 3.3) | Permitted Welding Positions | Weld Size (E) | Notes |
|---|---|---|---|---|---|---|---|---|---|
| SMAW | B-P1a | 3.2 max | — | R=0 to 1.6 | +1.6,-0 | ±1.6 | All | $T_1-1$ | B |
| | B-P1c | 6.4 max | — | $R=\dfrac{T_1}{2}$ min | +1.6,-0 | ±1.6 | All | $\dfrac{T_1}{2}$ | B |

Square-groove weld (1) — Butt joint (B)

$E_1 + E_2$ MUST NOT EXCEED $\dfrac{3T_1}{4}$     ALL DIMENSIONS IN mm

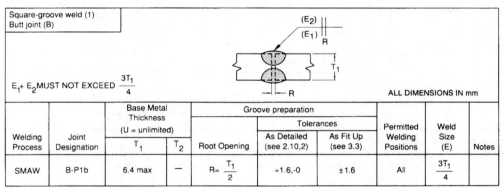

| Welding Process | Joint Designation | Base Metal Thickness (U = unlimited) $T_1$ | $T_2$ | Root Opening | As Detailed (see 2.10,2) | As Fit Up (see 3.3) | Permitted Welding Positions | Weld Size (E) | Notes |
|---|---|---|---|---|---|---|---|---|---|
| SMAW | B-P1b | 6.4 max | — | $R=\dfrac{T_1}{2}$ | +1.6,-0 | ±1.6 | All | $\dfrac{3T_1}{4}$ | |

Note B:  Joint is welded from one side only.

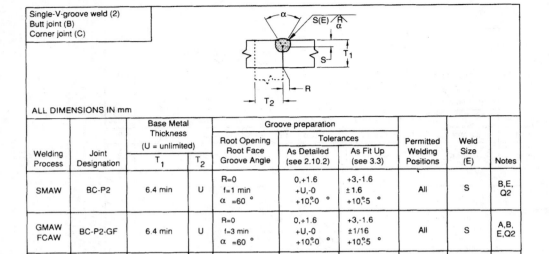

| | | Base Metal Thickness (U = unlimited) | | Groove preparation | | | Permitted Welding Positions | Weld Size (E) | Notes |
|---|---|---|---|---|---|---|---|---|---|
| Welding Process | Joint Designation | $T_1$ | $T_2$ | Root Opening Root Face Groove Angle | Tolerances | | | | |
| | | | | | As Detailed (see 2.10.2) | As Fit Up (see 3.3) | | | |
| SMAW | BC-P2 | 6.4 min | U | R=0<br>f=1 min<br>$\alpha$ =60 ° | 0,+1.6<br>+U,-0<br>+10°,0 ° | +3,-1.6<br>±1.6<br>+10°,5 ° | All | S | B,E,<br>Q2 |
| GMAW FCAW | BC-P2-GF | 6.4 min | U | R=0<br>f=3 min<br>$\alpha$ =60 ° | 0,+1.6<br>+U,-0<br>+10°,0 ° | +3,-1.6<br>±1/16<br>+10°,5 ° | All | S | A,B,<br>E,Q2 |
| SAW | BC-P2-S | 11.1 min | U | R=0<br>f=6 min<br>$\alpha$ =60 ° | ±0<br>+U,-0<br>+10°,0 ° | +1.6,-0<br>±1.6<br>+10°,5 ° | F | S | B,E,<br>Q2 |

Note A:   Not prequalified for gas tungsten arc and gas metal arc welding using short circuiting transfer. Refer to Appendix A.

Note B:   Joint is welded from one side only.

Note E:   Minimum weld size (E) as shown in Table 2.3; S as specified on drawings.

Note Q2: The member orientation may be changed provided that the groove dimensions are maintained as specified.

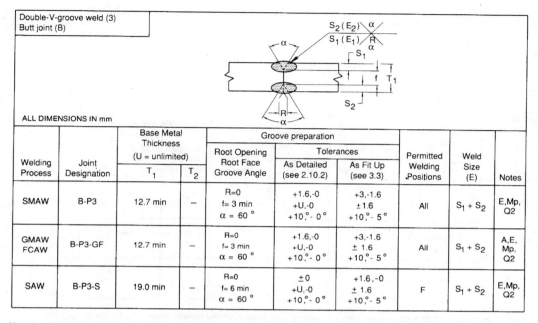

Double-V-groove weld (3)
Butt joint (B)

ALL DIMENSIONS IN mm

| Welding Process | Joint Designation | Base Metal Thickness (U = unlimited) | | Groove preparation | | | Permitted Welding Positions | Weld Size (E) | Notes |
|---|---|---|---|---|---|---|---|---|---|
| | | $T_1$ | $T_2$ | Root Opening Root Face Groove Angle | Tolerances | | | | |
| | | | | | As Detailed (see 2.10.2) | As Fit Up (see 3.3) | | | |
| SMAW | B-P3 | 12.7 min | — | R=0 f= 3 min $\alpha$ = 60 ° | +1.6,-0 +U,-0 +10,°- 0 ° | +3,-1.6 ±1.6 +10,°- 5 ° | All | $S_1 + S_2$ | E,Mp, Q2 |
| GMAW FCAW | B-P3-GF | 12.7 min | — | R=0 f= 3 min $\alpha$ = 60 ° | +1.6,-0 +U,-0 +10,°- 0 ° | +3,-1.6 ± 1.6 +10,°- 5 ° | All | $S_1 + S_2$ | A,E, Mp, Q2 |
| SAW | B-P3-S | 19.0 min | — | R=0 f= 6 min $\alpha$ = 60 ° | ±0 +U,-0 +10,°- 0 ° | +1.6,-0 ± 1.6 +10,°- 5 ° | F | $S_1 + S_2$ | E,Mp, Q2 |

Note A:   Not prequalified for gas tungsten arc and gas metal arc welding using short circuiting transfer. Refer to Appendix A.

Note E:   Minimum weld size (E) as shown in Table 2.3; S as specified on drawings.

Note Mp: Double-groove welds may have grooves of unequal depth, provided they conform to the limitations of Note E. Also the weld size (E), less any reduction applies individually to each groove.

Note Q2: The member orientation may be changed provided that the groove dimensions are maintained as specified.

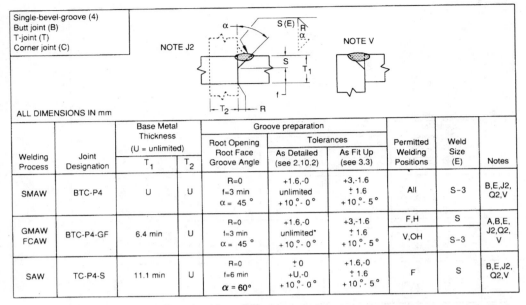

Single-bevel-groove (4)
Butt joint (B)
T-joint (T)
Corner joint (C)

NOTE J2

NOTE V

ALL DIMENSIONS IN mm

| Welding Process | Joint Designation | Base Metal Thickness (U = unlimited) | | Groove preparation | | | Permitted Welding Positions | Weld Size (E) | Notes |
|---|---|---|---|---|---|---|---|---|---|
| | | $T_1$ | $T_2$ | Root Opening Root Face Groove Angle | Tolerances As Detailed (see 2.10.2) | As Fit Up (see 3.3) | | | |
| SMAW | BTC-P4 | U | U | R=0 <br> f=3 min <br> α = 45° | +1.6,-0 <br> unlimited <br> +10°,-0° | +3,-1.6 <br> ±1.6 <br> +10°,-5° | All | S-3 | B,E,J2, Q2,V |
| GMAW FCAW | BTC-P4-GF | 6.4 min | U | R=0 <br> f=3 min <br> α = 45° | +1.6,-0 <br> unlimited* <br> +10°,-0° | +3,-1.6 <br> ±1.6 <br> +10°,-5° | F,H <br> V,OH | S <br> S-3 | A,B,E, J2,Q2, V |
| SAW | TC-P4-S | 11.1 min | U | R=0 <br> f=6 min <br> α = 60° | ±0 <br> +U,-0 <br> +10°,-0° | +1.6,-0 <br> ±1.6 <br> +10°,-5° | F | S | B,E,J2, Q2,V |

Note A:   Not prequalified for gas tungsten arc and gas metal arc welding using short circuiting transfer. Refer to Appendix A.

Note B:   Joint is welded from one side only.

Note E:   Minimum weld size (E) as shown in Table 2.3; S as specified on drawings.

Note J2:  If fillet welds are used in statically loaded structures to reinforce groove welds in corner and T-joints, they shall be equal to 1/4 $T_1$, but need not exceed 10 mm.

Note Q2:  The member orientation may be changed provided that the groove dimensions are maintained as specified.

Note V:   For corner joints, the outside groove preparation may be in either or both members, provided the basic groove configuration is not changed and adequate edge distance is maintained to support the welding operations without excessive edge melting.

*For flat and horizontal positions, f = +U,-0

| Double-bevel-groove weld (5) |
| Butt joint (B) |
| T-joint (T) |
| Corner joint (C) |

ALL DIMENSIONS IN mm

| Welding Process | Joint Designation | Base Metal Thickness (U = unlimited) | | Groove preparation | | | Permitted Welding Positions | Weld Size (E) | Notes |
| | | $T_1$ | $T_2$ | Root Opening Root Face Groove Angle | Tolerances | | | | |
| | | | | | As Detailed (see 2.10.2) | As Fit Up (see 3.3) | | | |
| SMAW | BTC-P5 | 8.0 min | U | R=0 <br> f=3 min <br> $\alpha$ =45 ° | +1.6,-0 <br> unlimited <br> +10°,0 ° | +3,-1.6 <br> ±1.6 <br> +10°,5 ° | All | $(S_1 + S_2)$ <br> -6 | E,J2, L,Mp, Q2,V |
| GMAW FCAW | BTC-P5-GF | 12.7 min | U | R=0 <br> f=3 min <br> $\alpha$ =45 ° | +1.6,-0 <br> unlimited* <br> +10°,0 ° | +3,-1.6 <br> ±1.6 <br> +10°,5 ° | F,H <br><br> V,OH | $(S_1 + S_2)$ <br><br> $(S_1 + S_2)$ <br> -6 | A,E,J2, L,Mp, Q2,V |
| SAW | TC-P5-S | 19.0 min | U | R=0 <br> f=6 min <br> $\alpha$ =60 ° | ±0 <br> +U,-0 <br> +10°,0 ° | +1.6,-0 <br> ±1.6 <br> +10°,5 ° | F | $(S_1 + S_2)$ | E,J2, L,Mp, Q2,V |

Note A:  Not prequalified for gas tungsten arc and gas metal arc welding using short circuiting transfer. Refer to Appendix A.

Note E:  Minimum weld size (E) as shown in Table 2.3; S as specified on drawings.

Note J2:  If fillet welds are used in statically loaded structures to reinforce groove welds in corner and T-joints, they shall be equal to 1/4 $T_1$, but need not exceed 10 mm.

Note L:  Butt and T-Joints are not prequalified for dynamically loaded structures.

Note Mp: Double–groove welds may have grooves of unequal depth, provided they conform to the limitations of Note E. Also the weld size (E), less any reduction applies individually to each groove.

Note Q2:  The member orientation may be changed provided that the groove dimensions are maintained as specified.

Note V:  For corner joints, the outside groove preparation may be in either or both members, provided the basic groove configuration is not changed and adequate edge distance is maintained to support the welding operations without excessive edge melting.

*For flat and horizontal positions, f = +U,-0

| Single-U-groove weld (6) |
|---|
| Butt joint (B) |
| Corner joint (C) |

ALL DIMENSIONS IN mm

| Welding Process | Joint Designation | Base Metal Thickness (U = unlimited) | | Groove preparation | | | Permitted Welding Positions | Weld Size (E) | Notes |
|---|---|---|---|---|---|---|---|---|---|
| | | $T_1$ | $T_2$ | Root Opening Root Face Groove Radius Groove Angle | Tolerances As Detailed (see 2.10.2) | As Fit Up (see 3.3) | | | |
| SMAW | BC-P6 | 6.4 min | U | R=0<br>f=1 min<br>r=6<br>$\alpha$ =45 ° | +1.6,-0<br>+U,-0<br>+6,-0<br>$+10^\circ, -0^\circ$ ° | +3,-1.6<br>±1.6<br>±1.6<br>$+10^\circ, -5^\circ$ ° | All | S | B,E,Q2 |
| GMAW FCAW | BC-P6-GF | 6.4 min | U | R=0<br>f=3 min<br>r=6<br>$\alpha$ =20 ° | +1.6,-0<br>+U,-0<br>+6,-0<br>$+10^\circ, -0^\circ$ ° | +3,-1.6<br>±1.6<br>±1.6<br>$+10^\circ, -5^\circ$ ° | All | S | A,B,<br>E,Q2 |
| SAW | BC-P6-S | 11.1 min | U | R=0<br>f = 6 min<br>r=6<br>$\alpha$ =20 ° | ±0<br>+U,-0<br>+6,-0<br>$+10^\circ, -0^\circ$ ° | +1.6,-0<br>±1.6<br>±1.6<br>$+10^\circ, -5^\circ$ ° | F | S | B,E,Q2 |

Note A:   Not prequalified for gas tungsten arc and gas metal arc welding using short circuiting transfer. Refer to Appendix A.

Note B:   Joint is welded from one side only.

Note E:   Minimum weld size (E) as shown in Table 2.3; S as specified on drawings.

Note Q2: The member orientation may be changed provided that the groove dimensions are maintained as specified.

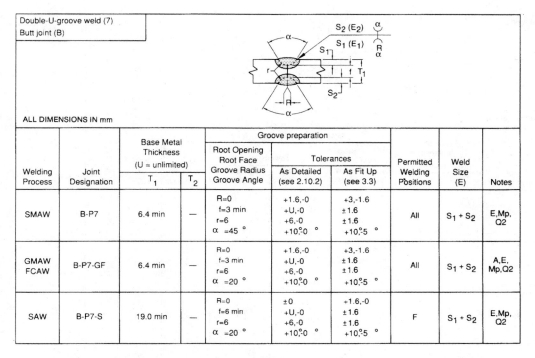

| Double-U-groove weld (7) | | | | | | | | | |
| Butt joint (B) | | | | | | | | | |

ALL DIMENSIONS IN mm

| Welding Process | Joint Designation | Base Metal Thickness (U = unlimited) | | Groove preparation | | | Permitted Welding Positions | Weld Size (E) | Notes |
| | | | | Root Opening Root Face Groove Radius Groove Angle | Tolerances | | | | |
| | | $T_1$ | $T_2$ | | As Detailed (see 2.10.2) | As Fit Up (see 3.3) | | | |
| SMAW | B-P7 | 6.4 min | — | R=0<br>f=3 min<br>r=6<br>$\alpha$ =45 ° | +1.6,-0<br>+U,-0<br>+6,-0<br>+10°,-0 ° | +3,-1.6<br>±1.6<br>±1.6<br>+10°,-5 ° | All | $S_1 + S_2$ | E,Mp, Q2 |
| GMAW FCAW | B-P7-GF | 6.4 min | — | R=0<br>f=3 min<br>r=6<br>$\alpha$ =20 ° | +1.6,-0<br>+U,-0<br>+6,-0<br>+10°,-0 ° | +3,-1.6<br>±1.6<br>±1.6<br>+10°,-5 ° | All | $S_1 + S_2$ | A,E, Mp,Q2 |
| SAW | B-P7-S | 19.0 min | — | R=0<br>f=6 min<br>r=6<br>$\alpha$ =20 ° | ±0<br>+U,-0<br>+6,-0<br>+10°,-0 ° | +1.6,-0<br>±1.6<br>±1.6<br>+10°,-5 ° | F | $S_1 + S_2$ | E,Mp, Q2 |

Note A:   Not prequalified for gas tungsten arc and gas metal arc welding using short circuiting transfer. Refer to Appendix A.

Note E:   Minimum weld size (E) as shown in Table 2.3; S as specified on drawings.

Note Mp: Double-groove welds may have grooves of unequal depth, provided they conform to the limitations of Note E. Also the weld size (E), less any reduction applies individually to each groove.

Note Q2: The member orientation may be changed provided that the groove dimensions are maintained as specified.

Single-J-groove weld (8)
Butt joint (B)
T-joint (T)
Corner joint (C)

NOTE V

ALL DIMENSIONS IN mm

| Welding Process | Joint Designation | Base Metal Thickness (U = unlimited) | | Groove preparation | | | Permitted Welding Positions | Weld Size (E) | Notes |
|---|---|---|---|---|---|---|---|---|---|
| | | $T_1$ | $T_2$ | Root Opening Root Face Groove Radius Groove Angle | Tolerances As Detailed (see 2.10.2) | As Fit Up (see 3.3) | | | |
| SMAW | TC-P8* | 6.4 min | U | R=0<br>f=3 min<br>r=10<br>$\alpha$ =45 ° | +1.6,-0<br>+U,-0<br>+6,-0<br>+10°,0 ° | +3,-1.6<br>±1.6<br>±1.6<br>+10°,5 ° | All | S | E,J2, Q2,V |
| SMAW | BC-P8** | 6.4 min | U | R=0<br>f=3 min<br>r=10<br>$\alpha$ =30 ° | +1.6,-0<br>+U,-0<br>+6,-0<br>+10°,0 ° | +3,-1.6<br>±1.6<br>±1.6<br>+10°,5 ° | All | S | E,J2, Q2,V |
| GMAW FCAW | TC-P8-GF* | 6.4 min | U | R=0<br>f=3 min<br>r=10<br>$\alpha$ =45 ° | +1.6,-0<br>+U,-0<br>+6,-0<br>+10°,0 ° | +3,-1.6<br>±1.6<br>±1.6<br>+10°,5 ° | All | S | A,E, J2,Q2, V |
| GMAW FCAW | BC-P8-GF** | 6.4 min | U | R=0<br>f=3 min<br>r=10<br>$\alpha$ =30 ° | +1.6,-0<br>+U,-0<br>+6,-0<br>+10°,0 ° | +3,-1.6<br>±1.6<br>±1.6<br>+10°,5 ° | All | S | A,E, J2,Q2, V |
| SAW | TC-P8-S* | 11.1 min | U | R=0<br>f=6 min<br>r=13<br>$\alpha$ =45 ° | ±0<br>+U,-0<br>+6,-0<br>+10°,0 ° | +1.6,-0<br>±1.6<br>±1.6<br>+10°,5 ° | F | S | E,J2, Q2,V |
| SAW | C-P8-S** | 11.1 min | U | R=0<br>f=6 min<br>r=13<br>$\alpha$ =20 ° | ±0<br>+U,-0<br>+6,-0<br>+10°,0 ° | +1.6,-0<br>±1.6<br>±1.6<br>+10°,5 ° | F | S | E,J2, Q2,V |

Note A:   Not prequalified for gas tungsten arc and gas metal arc welding using short circuiting transfer. Refer to Appendix A.

Note E:   Minimum weld size (E) as shown in Table 2.3; S as specified on drawings.

Note J2:  If fillet welds are used in statically loaded structures to reinforce groove welds in corner and T-joints, they shall be equal to 1/4 $T_1$, but need not exceed 10 mm.

Note Q2: The member orientation may be changed provided that the groove dimensions are maintained as specified.

Note V:   For corner joints, the outside groove preparation may be in either or both members, provided the basic groove configuration is not changed and adequate edge distance is maintained to support the welding operations without excessive edge melting.

 * Applies to inside corner joints.
** Applies to outside corner joints.

Double-J-groove weld (9)
Butt joint (B)
T-joint (T)
Corner joint (C)

ALL DIMENSIONS IN mm

| Welding Process | Joint Designation | Base Metal Thickness (U = unlimited) | | Groove preparation | | | Permitted Welding Positions | Weld Size (E) | Notes |
|---|---|---|---|---|---|---|---|---|---|
| | | $T_1$ | $T_2$ | Root Opening Root Face Groove Radius Groove Angle | Tolerances | | | | |
| | | | | | As Detailed (see 2.10.2) | As Fit Up (see 3.3) | | | |
| SMAW | BTC-P9* | 12.7 min | U | R=0<br>f=3 min<br>r=10<br>$\alpha$ =45 ° | +1.6,-0<br>+U,-0<br>+6,-0<br>+10°,-0 ° | +3,-1.6<br>±1.6<br>±1.6<br>+10°,-5 ° | All | $S_1 + S_2$ | E,J2,<br>Mp,Q2,<br>V |
| GMAW<br>FCAW | BTC-P9-GF** | 6.4 min | U | R=0<br>f=3 min<br>r=10<br>$\alpha$ =30 ° | +1.6,-0<br>+U,-0<br>+6,-0<br>+10°,-0 ° | +3,-1.6<br>±1.6<br>±1.6<br>+10°,-5 ° | All | $S_1 + S_2$ | A,J2,<br>Mp,Q2,<br>V |
| SAW | C-P9-S* | 19.0 min | U | R=0<br>f=6 min<br>r=13<br>$\alpha$ =45 ° | ±0<br>+U,-0<br>+6,-0<br>+10°,-0 ° | +1.6,-0<br>±1.6<br>±1.6<br>+10°,-5 ° | F | $S_1 + S_2$ | E,J2,<br>Mp,Q2,<br>V |
| SAW | C-P9-S** | 19.0 min | U | R=0<br>f=6 min<br>r=13<br>$\alpha$ =20 ° | ±0<br>+U,-0<br>+6,-0<br>+10°,-0 ° | +1.6,-0<br>±1.6<br>±1.6<br>+10°,-5 ° | F | $S_1 + S_2$ | E,J2,<br>Mp,Q2,<br>V |
| SAW | T-P9-S | 19.0 min | U | R=0<br>f=6 min<br>r=13<br>$\alpha$ =45 ° | ±0<br>+U,-0<br>+6,-0<br>+10°,-0 ° | +1.6,-0<br>±1.6<br>±1.6<br>+10°,-5 ° | F | $S_1 + S_2$ | E,J2,<br>Mp,Q2 |

Note A:   Not prequalified for gas tungsten arc and gas metal arc welding using short circuiting transfer. Refer to Appendix A.

Note E:   Minimum weld size (E) as shown in Table 2.3; S as specified on drawings.

Note J2:   If fillet welds are used in statically loaded structures to reinforce groove welds in corner and T-joints, they shall be equal to 1/4 $T_1$, but need not exceed 10 mm.

Note Mp: Double-groove welds may have grooves of unequal depth, provided they conform to the limitations of Note E. Also the weld size (E), less any reduction applies individually to each groove.

Note Q2: The member orientation may be changed provided that the groove dimensions are maintained as specified.

Note V:   For corner joints, the outside groove preparation may be in either or both members, provided the basic groove configuration is not changed and adequate edge distance is maintained to support the welding operations without excessive edge melting.

*Applies to inside corner joints.
**Applies to outside corner joints.

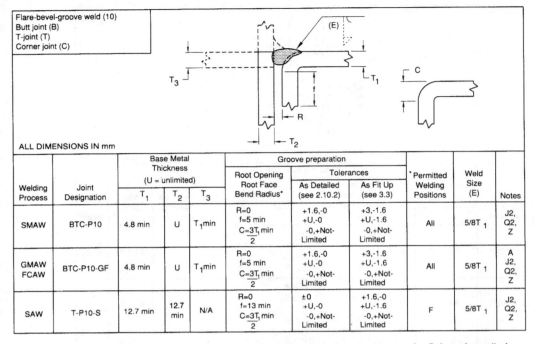

Flare-bevel-groove weld (10)
Butt joint (B)
T-joint (T)
Corner joint (C)

ALL DIMENSIONS IN mm

| Welding Process | Joint Designation | Base Metal Thickness (U = unlimited) | | | Groove preparation | | | *Permitted Welding Positions | Weld Size (E) | Notes |
|---|---|---|---|---|---|---|---|---|---|---|
| | | $T_1$ | $T_2$ | $T_3$ | Root Opening Root Face Bend Radius* | Tolerances | | | | |
| | | | | | | As Detailed (see 2.10.2) | As Fit Up (see 3.3) | | | |
| SMAW | BTC-P10 | 4.8 min | U | $T_1$min | R=0<br>f=5 min<br>$\frac{C=3T_1 \text{ min}}{2}$ | +1.6,-0<br>+U,-0<br>-0,+Not-Limited | +3,-1.6<br>+U,-1.6<br>-0,+Not-Limited | All | 5/8T $_1$ | J2,<br>Q2,<br>Z |
| GMAW FCAW | BTC-P10-GF | 4.8 min | U | $T_1$min | R=0<br>f=5 min<br>$\frac{C=3T_1 \text{ min}}{2}$ | +1.6,-0<br>+U,-0<br>-0,+Not-Limited | +3,-1.6<br>+U,-1.6<br>-0,+Not-Limited | All | 5/8T $_1$ | A<br>J2,<br>Q2,<br>Z |
| SAW | T-P10-S | 12.7 min | 12.7 min | N/A | R=0<br>f=13 min<br>$\frac{C=3T_1 \text{ min}}{2}$ | ±0<br>+U,-0<br>-0,+Not-Limited | +1.6,-0<br>+U,-1.6<br>-0,+Not-Limited | F | 5/8T $_1$ | J2,<br>Q2,<br>Z |

Note A:   Not prequalified for gas tungsten arc and gas metal arc welding using short circuiting transfer. Refer to Appendix A.

Note J2:   If fillet welds are used in statically loaded structures to reinforce groove welds in corner and T-joints, they shall be equal to 1/4 $T_1$, but need not exceed 10 mm.

Note Q2:   The member orientation may be changed provided that the groove dimensions are maintained as specified.

Note Z:   Weld size (E) is based on joints welded flush.

*For cold formed (A500) rectangular tubes, C dimension is not limited (see commentary).

# Structural Shape Size Groupings

## Structural Shape Sizes Per Tensile Group Classifications

| W Shapes | | | | |
|---|---|---|---|---|
| Group 1 | Group 2 | Group 3 | Group 4 | Group 5 |
| W24 x 55 | W44 x 198 | W44 x 248 | W40 x 362 to W40 x 655 | W36 x 848 |
| W24 x 62 | W44 x 224 | W44 x 285 | W36 x 328 to W36 x 798 | W14 x 605 to W14 x 730 |
| W21 x 44 to W21 x 57 | W40 x 149 to W40 x 268 | W40 x 277 to W40 x 328 | W33 x 318 to W33 x 619 | |
| W18 x 35 to W 18 x 71 | W36 x 135 to W36 x 210 | W36 x 230 to W36 x 300 | W30 x 292 to W30 x 581 | |
| W16 x 26 to W16 x 57 | W33 x 118 to W33 x 152 | W33 x 201 to W33 x 291 | W27 x 281 to W27 x 539 | |
| W14 x 22 to W14 x 53 | W30 x 90 to W30 x 211 | W30 x 235 to W30 x 261 | W24 x 250 to W24 x 492 | |
| W12 x 14 to W12 x 58 | W27 x 84 to W27 x 178 | W27 x 194 to W27 x 258 | W21 x 248 to W21 x 402 | |
| W10 x 12 to W10 x 45 | W24 x 68 to W24 x 162 | W24 x 176 to W24 x 229 | W18 x 211 to W18 x 311 | |
| W8 x 10 to W8 x 48 | W21 x 62 to W21 x 147 | W21 x 166 to W21 x 223 | W14 x 233 to W14 x 550 | |
| W6 x 9 to W6 x 25 | W18 x 76 to W18 x 143 | W18 x 158 to W18 x 192 | W12 x 21- to W12 x 336 | |
| W5 x 16 W5 x 19 | W16 x 67 to W16 x 100 | W14 x 145 to W14 x 211 | | |
| W4 x 13 | W14 x 61 to W14 x 132 | W12 x 120 to W12 x 190 | | |
| | W12 x 65 to W12 x 106 | | | |
| | W10 x 49 to W10 x 112 | | | |
| | W8 x 58 | | | |
| | W8 x 67 | | | |

### M Shapes

| to 37.7 lb./ft. | | | |
|---|---|---|---|

### S Shapes

| to 35 lb./ft. | | | |
|---|---|---|---|

### HP Shapes

| | to 102 lb./ft | > 102 lb./ft. | |
|---|---|---|---|

### Standard Channel

| to 20.7 lb./ft. | > 20.7 lb./ft. | | |
|---|---|---|---|

### Miscellaneous Channel

| to 28.5 lb./ft. | >28.5 lb./ft. | | |
|---|---|---|---|

### Angle Iron

| to 1/2 in. | >1/2 to 3/4 in. | >3/4 in. | |
|---|---|---|---|

# Wide Flange Dimensions

## W Shape Dimensions

| Designation | Depth | Web Thickness | Flange | |
| --- | --- | --- | --- | --- |
| | | | Width | Thickness |
| W44 x 285 | 44 | 1 | 11-3/4 | 1-3/4 |
| x248 | 43-5/8 | 7/8 | 11-3/4 | 1-9/16 |
| x224 | 43-1/4 | 13/16 | 11-3/4 | 1-7/16 |
| x198 | 42-7/8 | 11/16 | 11-3/4 | 1-1/4 |
| W40 x 328 | 40 | 15/16 | 17-7/8 | 1-3/4 |
| x298 | 39-3/4 | 13/16 | 17-7/8 | 1-9/16 |
| x268 | 39-3/8 | 3/4 | 17-3/4 | 1-7/16 |
| x244 | 39 | 11/16 | 17-3/4 | 1-1/4 |
| x221 | 38-5/8 | 11/16 | 17-3/4 | 1-1/16 |
| x192 | 38-1/4 | 11/16 | 17-3/4 | 13/16 |
| W40 x 655 | 43-5/8 | 2 | 16-7/8 | 3-9/16 |
| x593 | 43 | 1-13/16 | 16-3/4 | 3-1/5 |
| x531 | 42-3/8 | 1-5/8 | 16-1/2 | 2-15/16 |
| x480 | 41-3/4 | 1-7/16 | 16-3/8 | 2-5/8 |
| x436 | 41-3/8 | 1-5/16 | 16-1/4 | 2-3/8 |
| x397 | 41 | 1-1/4 | 16-1/8 | 2-3/16 |
| x362 | 40-1/2 | 1-1/8 | 16 | 2 |
| x324 | 40-1/8 | 1 | 15-7/8 | 1-13/16 |
| x297 | 39-7/8 | 15/16 | 15-7/8 | 1-5/8 |
| x277 | 39-3/4 | 13/16 | 15-7/8 | 1-9/16 |
| x249 | 39-3/8 | 3/4 | 15-3/4 | 1-7/16 |
| x215 | 39 | 5/8 | 15-3/4 | 1-1/4 |
| x199 | 38-5/8 | 5/8 | 15-3/4 | 1-1/16 |
| W40 x 183 | 39 | 5/8 | 11-3/4 | 1-1/4 |
| x167 | 38-5/8 | 5/8 | 11-3/4 | 1 |
| x149 | 38-1/4 | 5/8 | 11-3/4 | 13/16 |
| | | | | |
| | | | | |
| | | | | |
| | | | | |
| | | | | |

All dimensions in inches.

## W Shape Dimensions

| Designation | Depth | Web Thickness | Flange Width | Flange Thickness |
|---|---|---|---|---|
| W36 x 848 | 42-1/2 | 2-1/2 | 18-1/8 | 4-1/2 |
| x798 | 42 | 2-3/8 | 18 | 4-5/16 |
| x720 | 41-1/4 | 2-3/16 | 17-3/5 | 3-7/8 |
| x650 | 40-1/2 | 2 | 17-5/8 | 3-9/16 |
| x588 | 39-7/8 | 1-13/16 | 17-3/8 | 3-1/4 |
| x527 | 39-1/4 | 1-5/8 | 17-1/4 | 2-15/16 |
| x485 | 38-3/4 | 1-1/2 | 17-1/8 | 2-11/16 |
| x439 | 38-1/4 | 1-3/8 | 17 | 2-7/16 |
| x393 | 37-3/4 | 1-1/4 | 16-7/8 | 2-3/16 |
| x359 | 37-3/8 | 1-1/8 | 16-3/4 | 2 |
| x328 | 37-1/8 | 1 | 16-5/8 | 1-7/8 |
| x300 | 36-3/4 | 15/16 | 16-5/8 | 1-11/16 |
| x280 | 36-1/2 | 7/8 | 16-5/8 | 1-9/16 |
| x260 | 36-1/4 | 13/16 | 16-1/2 | 1-7/16 |
| x245 | 36-1/8 | 13/16 | 16-1/2 | 1-3/8 |
| x230 | 35-7/8 | 3/4 | 16-1/2 | 1-1/4 |
| W36 x 256 | 37-3/8 | 1 | 1-3/4 | 1-3/4 |
| x232 | 37-1/8 | 7/8 | 1-9/16 | 1-9/16 |
| x210 | 36-3/4 | 13/16 | 1-3/8 | 1-3/8 |
| x194 | 36-1/2 | 3/4 | 1-1/4 | 1-1/4 |
| x182 | 36-3/8 | 3/4 | 1-3/16 | 1-3/16 |
| x170 | 36-1/8 | 11/16 | 1-1/8 | 1-1/8 |
| x160 | 36 | 5/8 | 1 | 1 |
| x150 | 35-7/8 | 5/8 | 15/16 | 15/16 |
| x135 | 35-1/2 | 5/8 | 13/16 | 13/16 |

All dimensions in inches.

## W Shape Dimensions

| Designation | Depth | Web Thickness | Flange | |
|---|---|---|---|---|
| | | | Width | Thickness |
| W33 x 619 | 38-1/2 | 2 | 16-7/8 | 3-9/16 |
| x567 | 37-7/8 | 1-13/16 | 16-3/4 | 3-1/4 |
| x515 | 37-3/8 | 1-5/8 | 16-5/8 | 3 |
| x468 | 36-3/4 | 1-1/2 | 16-1/2 | 2-3/4 |
| x424 | 36-3/8 | 1-3/8 | 16-3/8 | 2-1/2 |
| x387 | 36 | 1-1/4 | 16-1/4 | 2-1/4 |
| x354 | 35-1/2 | 1-3/16 | 16-1/8 | 2-1/16 |
| x318 | 35-1/8 | 1-1/16 | 16 | 1-7/8 |
| x291 | 34-7/8 | 1 | 15-7/8 | 1-3/4 |
| x263 | 34-1/2 | 7/8 | 15-3/4 | 1-9/16 |
| x241 | 34-1/8 | 13/16 | 15-7/8 | 1-3/8 |
| x221 | 33-7/8 | 3/4 | 15-3/4 | 1-1/4 |
| x201 | 33-5/8 | 11/16 | 15-3/4 | 1-1/8 |
| W33 x 169 | 33-7/8 | 11/16 | 11-1/2 | 1-1/4 |
| x152 | 33-1/2 | 5/8 | 11-5/8 | 1-1/16 |
| x141 | 33-1/4 | 5/8 | 11-1/2 | 15/16 |
| x130 | 33-1/8 | 9/16 | 11-1/2 | 7/8 |
| x118 | 32-7/8 | 9/16 | 11-1/2 | 3/4 |
| W30 x 581 | 35-3/8 | 2 | 16-1/4 | 3-9/16 |
| x526 | 34-3/4 | 1-13/16 | 16 | 3-1/4 |
| x477 | 34-1/4 | 1-5/8 | 15-7/8 | 3 |
| x433 | 33-5/8 | 1-1/2 | 15-3/4 | 2-11/16 |
| x391 | 33-1/4 | 1-3/8 | 5-5/8 | 2-7/8 |
| x357 | 32-3/4 | 1-1/4 | 15-1/2 | 2-1/4 |
| x326 | 32-3/8 | 1-1/8 | 15-3/8 | 2-1/16 |
| x292 | 32 | 1 | 15-1/4 | 1-7/8 |
| x261 | 31-5/8 | 15/16 | 15-1/8 | 1-5/8 |
| x235 | 31-1/4 | 13/16 | 15 | 1-1/2 |
| x211 | 31 | 3/4 | 15-1/8 | 1-5/16 |
| x191 | 30-5/8 | 11/16 | 15 | 1-3/16 |
| x173 | 30-1/2 | 5/8 | 15 | 1-1/16 |

All dimensions in inches.

## W Shape Dimensions

| Designation | Depth | Web Thickness | Flange Width | Thickness |
|---|---|---|---|---|
| W30 x 148 | 30-5/8 | 5/8 | 10-1/2 | 1-3/16 |
| x132 | 30-1/4 | 5/8 | 10-1/2 | 1 |
| x124 | 30-1/8 | 9/16 | 10-1/2 | 15/16 |
| x116 | 30 | 9/16 | 10-1/2 | 7/8 |
| x108 | 29-7/8 | 9/16 | 10-1/2 | 3/4 |
| x99 | 29-5/8 | 1/2 | 10-1/2 | 11/16 |
| x90 | 29-1/2 | 1/2 | 10-3/8 | 9/16 |
| W27 x 539 | 32-1/2 | 2 | 15-1/4 | 3-9/16 |
| x494 | 32 | 1-13/16 | 15-1/8 | 3-1/4 |
| x448 | 31-3/8 | 1-5/8 | 15 | 3 |
| x407 | 30-7/8 | 1-1/2 | 14-3/4 | 2-3/4 |
| x368 | 30-3/8 | 1-3/8 | 14-5/8 | 2-1/2 |
| x336 | 30 | 1-1/4 | 14-1/2 | 2-1/4 |
| x307 | 29-5/8 | 1-3/16 | 14-1/2 | 2-1/16 |
| x281 | 29-1/4 | 1-1/16 | 14-3/8 | 1-15/16 |
| x258 | 29 | 1 | 14-1/4 | 1-3/4 |
| x235 | 28-5/8 | 15/16 | 14-1/4 | 1-5/8 |
| x217 | 28-3/8 | 13/16 | 14-1/8 | 1-1/2 |
| x194 | 28-1/8 | 3/4 | 14 | 1-5/16 |
| x178 | 27-3/4 | 3/4 | 14-1/8 | 1-3/16 |
| x161 | 27-5/8 | 11/16 | 14 | 1-1/16 |
| x146 | 27-3/8 | 5/8 | 14 | 1 |
| W27 x 129 | 27-5/8 | 5/8 | 10 | 1-1/8 |
| x114 | 27-1/4 | 9/16 | 10-1/8 | 15/16 |
| x102 | 27-1/8 | 1/2 | 10 | 13/16 |
| x94 | 26-7/8 | 1/2 | 10 | 3/4 |
| x84 | 26-3/4 | 7/16 | 10 | 5/8 |

All dimensions in inches.

## W Shape Dimensions

| Designation | Depth | Web Thickness | Flange | |
|---|---|---|---|---|
| | | | Width | Thickness |
| W24 x 492 | 29-5/8 | 2 | 14-1/8 | 3-9/16 |
| x450 | 29-1/8 | 1-13/16 | 14 | 3-1/4 |
| x408 | 28-1/2 | 1-5/8 | 13-3/4 | 3 |
| x370 | 28 | 1-1/2 | 13-5/8 | 2-3/4 |
| x335 | 27-1/2 | 1-3/8 | 13-1/2 | 2-1/2 |
| x306 | 27-1/8 | 1-1/4 | 13-3/8 | 2-1/4 |
| x279 | 26-3/4 | 1-3/16 | 13-1/4 | 2-1/16 |
| x250 | 26-3/8 | 1-1/16 | 13-1/8 | 1-7/8 |
| x229 | 26 | 1 | 13-1/8 | 1-3/4 |
| x207 | 25-3/4 | 7/8 | 13 | 1-9/16 |
| x192 | 25-1/2 | 13/16 | 13 | 1-7/16 |
| x176 | 25-1/4 | 3/4 | 12-7/8 | 1-5/16 |
| x162 | 25 | 11/16 | 13 | 1-1/4 |
| x146 | 24-3/4 | 5/8 | 12-7/8 | 1-1/16 |
| x131 | 24-1/2 | 5/8 | 12-7/8 | 15/16 |
| x117 | 24-1/4 | 9/16 | 12-3/4 | 7/8 |
| x104 | 24 | 1/2 | 12-3/4 | 3/4 |
| W24 x 103 | 24-1/2 | 9/16 | 9 | 1 |
| x94 | 24-1/4 | 1/2 | 9-1/8 | 7/8 |
| x84 | 24-1/8 | 1/2 | 9 | 3/4 |
| x76 | 23-7/8 | 7/16 | 9 | 11/16 |
| x68 | 23-3/4 | 7/16 | 9 | 9/16 |
| W24 x 62 | 23-3/4 | 7/16 | 7 | 9/16 |
| x55 | 23-5/8 | 3/8 | 7 | 1/2 |

All dimensions in inches.

## W Shape Dimensions

| Designation | Depth | Web Thickness | Flange | |
|---|---|---|---|---|
| | | | Width | Thickness |
| W21 x 402 | 26 | 1-3/4 | 13-3/8 | 3-1/8 |
| x364 | 251/2 | 1-9/16 | 13-3/4 | 2-7/8 |
| x333 | 25 | 1-7/16 | 13-1/8 | 2-5/8 |
| x300 | 24-1/2 | 1-5/16 | 13 | 2-3/8 |
| x275 | 24-1/8 | 1-1/4 | 12-7/8 | 2-3/16 |
| x248 | 23-3/4 | 1-1/8 | 12-3/4 | 2 |
| x223 | 23-3/8 | 1 | 12-5/8 | 1-13/16 |
| x201 | 23 | 15/16 | 12-5/8 | 1-5/8 |
| x182 | 22-3/4 | 13/16 | 12-1/2 | 1-1/2 |
| x166 | 22-1/2 | 3/4 | 12-3/8 | 1-3/8 |
| x147 | 22 | 3/4 | 12-1/2 | 1-1/8 |
| x132 | 21-7/8 | 5/8 | 12-1/2 | 1-1/16 |
| x122 | 21-5/8 | 5/8 | 12-3/8 | 15/16 |
| x111 | 21-1/2 | 9/16 | 12-3/8 | 7/8 |
| x101 | 21-3/8 | 1/2 | 12-1/4 | 13/16 |
| W21 x 93 | 21-5/8 | 9/16 | 8-3/8 | 15/16 |
| x83 | 21-3/8 | 1/2 | 8-3/8 | 13/16 |
| x73 | 21-1/4 | 7/16 | 8-1/4 | 3/4 |
| x68 | 21-1/8 | 7/16 | 8-1/4 | 11/16 |
| x62 | 21 | 3/8 | 8-1/4 | 5/8 |
| W21 x 57 | 21 | 3/8 | 6-1/2 | 5/8 |
| x50 | 20-7/8 | 3/8 | 6-1/2 | 9/16 |
| x44 | 20-5/8 | 3/8 | 6-1/2 | 7/16 |

All dimensions in inches.

## W Shape Dimensions

| Designation | Depth | Web Thickness | Flange | |
|---|---|---|---|---|
| | | | Width | Thickness |
| W18 x 311 | 22-3/8 | 1-1/2 | 12 | 2-3/4 |
| x283 | 21-7/8 | 1-3/8 | 11-7/8 | 2-1/2 |
| x258 | 21-1/2 | 1-1/4 | 11-3/4 | 2-5/16 |
| x234 | 21 | 1-3/16 | 11-5/8 | 2-1/8 |
| x211 | 20-5/8 | 1-1/16 | 11-1/2 | 1-15/16 |
| x192 | 20-3/8 | 1 | 11-1/2 | 1-3/4 |
| x175 | 20 | 7/8 | 11-3/8 | 1-9/16 |
| x158 | 10-3/4 | 13/16 | 11-1/4 | 1-7/16 |
| x143 | 19-1/2 | 3/4 | 11-1/4 | 1-5/16 |
| x130 | 19-1/4 | 11/16 | 11-1/8 | 1-3/16 |
| W18 x 119 | 19 | 5/8 | 11-1/4 | 1-1/16 |
| x106 | 18-3/4 | 9/16 | 11-1/4 | 15/16 |
| x97 | 18-5/8 | 9/16 | 11-1/8 | 7/8 |
| x86 | 18-3/8 | 1/2 | 11-1/8 | 3/4 |
| x76 | 18-1/4 | 7/16 | 11 | 11/16 |
| W18 x 71 | 18-1/2 | 1/2 | 7-5/8 | 13/16 |
| x65 | 18-3/8 | 7/16 | 7-5/8 | 3/4 |
| x60 | 18-1/4 | 7/16 | 7-1/2 | 11/16 |
| x55 | 18-1/8 | 3/8 | 7-1/2 | 5/8 |
| x50 | 18 | 3/8 | 7-1/2 | 9/16 |
| W18 x 46 | 18 | 3/8 | 6 | 5/8 |
| x40 | 17-7/8 | 5/16 | 6 | 1/2 |
| x35 | 17-3/4 | 5/16 | 6 | 7/16 |
| W16 x 100 | 17 | 9/16 | 10-3/8 | 1 |
| x89 | 16-3/4 | 1/2 | 10-3/ | 7/8 |
| x77 | 16-1/2 | 7/16 | 10-1/4 | 3/4 |
| x67 | 16-3/8 | 3/8 | 10-1/4 | 11/16 |
| W16 x 57 | 16-3/ | 7/16 | 7-1/8 | 11/16 |
| x50 | 16-1/4 | 3/8 | 7-1/8 | 5/8 |
| x45 | 16-1/8 | 3/8 | 7 | 9/16 |
| x40 | 16 | 5/16 | 7 | 1/2 |
| x36 | 15-7/8 | 5/16 | 7 | 7/16 |

All dimensions in inches.

## W Shape Dimensions

| Designation | Depth | Web Thickness | Flange | |
|---|---|---|---|---|
| | | | Width | Thickness |
| W16 x 31 | 15-7/8 | 1/4 | 5-1/2 | 7/16 |
| x26 | 15-3/4 | 1/4 | 5-1/2 | 3/8 |
| W14 x 730 | 22-3/8 | 3-1/16 | 17-7/8 | 4-15/16 |
| x665 | 21-5/8 | 2-13/16 | 17-5/8 | 4-1/2 |
| x605 | 20-7/8 | 2-5/8 | 17-3/8 | 4-3/16 |
| x550 | 20-1/4 | 2-3/8 | 17-1/4 | 3-13/16 |
| x500 | 19-6/8 | 2-3/16 | 17 | 3-1/2 |
| x455 | 19 | 2 | 16-7/8 | 3-3/16 |
| W14 x 426 | 18-5/8 | 1-7/8 | 16-3/4 | 3-1/16 |
| x398 | 18-1/4 | 1-3/4 | 16-5/8 | 2-7/8 |
| x370 | 17-7/8 | 1-5/8 | 16-1/2 | 2-11/16 |
| x342 | 17-1/2 | 1-9/16 | 16-3/8 | 2-1/2 |
| x311 | 17-1/8 | 1-7/6 | 16-1/4 | 2-1/4 |
| x283 | 16-3/4 | 1-5/16 | 16-1/8 | 2-1/16 |
| x257 | 16-3/8 | 1-3/16 | 16 | 1-7/8 |
| x233 | 16 | 1-1/16 | 15-7/8 | 1-3/4 |
| x211 | 15-3/4 | 1 | 15-3/4 | 1-9/16 |
| x193 | 15-1/2 | 7/8 | 15-3/4 | 1-7/16 |
| x176 | 15-1/4 | 13/16 | 15-5/8 | 1-5/16 |
| x159 | 15 | 3/4 | 15-5/8 | 1-3/16 |
| x145 | 14-3/4 | 11/16 | 15-1/2 | 1-1/16 |
| | | | | |
| | | | | |
| | | | | |
| | | | | |
| | | | | |
| | | | | |
| | | | | |
| | | | | |

All dimensions in inches.

## W Shape Dimensions

| Designation | Depth | Web Thickness | Flange Width | Flange Thickness |
|---|---|---|---|---|
| | | | Width | Thickness |
| W14 x 132 | 14-5/8 | 5/8 | 14-3/4 | 1 |
| x120 | 14-1/2 | 9/16 | 14-5/8 | 15/16 |
| x109 | 14-3/8 | 1/2 | 14-5/8 | 7/8 |
| x99 | 14-1/8 | 1/2 | 14-5/8 | 3/4 |
| x90 | 14 | 7/16 | 14-1/2 | 11/16 |
| W14 x 82 | 14-1/4 | 1/2 | 10-1/8 | 7/8 |
| x74 | 14-1/8 | 7/16 | 10-1/8 | 13/16 |
| x68 | 14 | 7 | 10 | 3/4 |
| x61 | 13-7/8 | 16 | 10 | 5/8 |
| W14 x 53 | 13-7/8 | 3/8 | 8 | 11/16 |
| x48 | 13-3/4 | 3/8 | 8 | 5/8 |
| x43 | 13-5/8 | 5/16 | 8 | 1/2 |
| W14 x 38 | 14-1/8 | 5/16 | 6-3/4 | 1/2 |
| x34 | 14 | 5/16 | 6-3/4 | 7/16 |
| x30 | 13-7/8 | 1/4 | 6-3/4 | 3/8 |
| W14 x 26 | 13-7/8 | 1/4 | 5 | 7/16 |
| x22 | 13-3/4 | 1/4 | 5 | 5/16 |
| | | | | |
| | | | | |
| | | | | |
| | | | | |
| | | | | |
| | | | | |
| | | | | |
| | | | | |
| | | | | |
| | | | | |
| | | | | |
| | | | | |

All dimensions in inches.

## W Shape Dimensions

| Designation | Depth | Web Thickness | Flange | |
|---|---|---|---|---|
| | | | Width | Thickness |
| W12 x 336 | 16-7/8 | 1-3/4 | 13-3/8 | 2-15/16 |
| x305 | 16-3/8 | 1-5/8 | 13-1/4 | 2-11/16 |
| x279 | 15-7/8 | 1-1/2 | 13-1/8 | 2-1/2 |
| x252 | 15-3/8 | 1-3/8 | 13 | 2-1/4 |
| x230 | 15 | 1-5/16 | 12-7/8 | 2-1/16 |
| x210 | 14-3/4 | 1-3/16 | 12-3/4 | 1-7/8 |
| x190 | 14-3/8 | 1-1/16 | 12-5/8 | 1-3/4 |
| x170 | 14 | 15/16 | 12-5/8 | 1-9/16 |
| x152 | 13-3/4 | 7/8 | 12-1/2 | 1-3/8 |
| x136 | 13-3/8 | 13/16 | 12-3/8 | 1-1/4 |
| x120 | 13-1/8 | 11/16 | 12-3/8 | 1-1/8 |
| x106 | 12-7/8 | 5/8 | 12-1/4 | 1 |
| x96 | 12-3/4 | 9/16 | 12-1/8 | 7/8 |
| x87 | 12-1/2 | 1/2 | 12-1/8 | 13/16 |
| x79 | 12-3/8 | 1/2 | 12-1/8 | 3/4 |
| x72 | 12-1/4 | 7/16 | 12 | 11/16 |
| x65 | 12-1/8 | 3/8 | 12 | 5/8 |
| W12 x 58 | 12-1/4 | 3/8 | 10 | 5/8 |
| x53 | 12 | 3/8 | 10 | 9/16 |
| W12 x 50 | 12-1/4 | 3/8 | 8-1/8 | 5/8 |
| x45 | 12 | 5/16 | 8 | 9/16 |
| x40 | 12 | 5/16 | 8 | 1/2 |
| W12 x 35 | 12-1/2 | 5/16 | 6-1/2 | 1/2 |
| x30 | 12-3/8 | 1/4 | 6-1/2 | 7/16 |
| x26 | 12-1/4 | 1/4 | 6-1/2 | 3/8 |
| W12 x 22 | 12-1/4 | 1/4 | 4 | 7/16 |
| x19 | 12-1/8 | 1/4 | 4 | 3/8 |
| x16 | 12 | 1/4 | 4 | 1/4 |
| x14 | 11-7/8 | 3/16 | 4 | 1/4 |

All dimensions in inches.

## W Shape Dimensions

| Designation | Depth | Web Thickness | Flange | |
|---|---|---|---|---|
| | | | Width | Thickness |
| W10 x 112 | 11-3/8 | 3/4 | 10-3/8 | 1-1/4 |
| x100 | 11-1/8 | 11/16 | 10-3/8 | 1-1/8 |
| x88 | 10-7/8 | 5/8 | 10-1/4 | 1 |
| x77 | 10-5/8 | 1/2 | 10-1/4 | 7/8 |
| x68 | 10-3/8 | 1/2 | 10-1/8 | 3/4 |
| x60 | 10-1/4 | 7/16 | 10-1/8 | 11/16 |
| x54 | 10-1/8 | 3/8 | 10 | 5/8 |
| x49 | 10 | 5/16 | 10 | 9/16 |
| W10 x 45 | 10-1/8 | 3/8 | 8 | 5/8 |
| x39 | 9-7/8 | 5/16 | 8 | 1/2 |
| x33 | 9-3/4 | 5/16 | 8 | 7/16 |
| W10 x 30 | 10-1/2 | 5/16 | 5-3/4 | 1/2 |
| x 26 | 10-3/8 | 1/4 | 5-3/4 | 7/16 |
| x22 | 10-1/8 | 1/4 | 5-3/4 | 3/8 |
| W10 x 19 | 10-1/4 | 1/4 | 4 | 3/8 |
| x17 | 10-1/8 | 1/4 | 4 | 5/16 |
| x15 | 10 | 1/4 | 4 | 1/4 |
| x12 | 9-7/8 | 3/16 | 4 | 3/16 |

All dimensions in inches.

## W Shape Dimensions

| Designation | Depth | Web Thickness | Flange | |
| --- | --- | --- | --- | --- |
| | | | Width | Thickness |
| W8 x 67 | 9 | 9/16 | 8-1/4 | 15/16 |
| x 58 | 8-3/4 | 1/2 | 8-1/4 | 13/16 |
| x48 | 8-1/2 | 3/8 | 8-1/8 | 11/16 |
| x40 | 8-1/4 | 3/8 | 8-1/8 | 9/16 |
| x35 | 8-1/8 | 5/16 | 8 | 1/2 |
| x31 | 8 | 5/16 | 8 | 7/16 |
| x28 | 8 | 5/16 | 6-1/2 | 7/16 |
| x24 | 7-7/8 | 1/4 | 6-1/2 | 3/8 |
| x21 | 8-1/4 | 1/4 | 5-1/4 | 3/8 |
| x18 | 8-1/8 | 1/4 | 5-1/4 | 5/16 |
| x15 | 8-1/8 | 1/4 | 4 | 5/16 |
| x13 | 8 | 1/4 | 4 | 1/4 |
| x10 | 7-7/8 | 3/16 | 4 | 3/16 |
| W6 x 25 | 6-3/8 | 5/16 | 6-1/8 | 7/16 |
| x20 | 6-1/4 | 1/4 | 6 | 3/8 |
| x15 | 6 | 1/4 | 6 | 1/4 |
| x16 | 6-14 | 1/4 | 4 | 3/8 |
| x12 | 6 | 1/4 | 4 | 1/4 |
| x9 | 5-7/8 | 3/16 | 4 | 3/˜6 |
| W5 x 19 | 5-1/8 | 1/4 | 5 | 7/16 |
| x16 | 5 | 1/4 | 5 | 3/8 |
| W4 x 13 | 4-1/8 | 1/4 | 4 | 3/8 |

All dimensions in inches.

## M Shape Dimensions

| Designation | Depth | Web Thickness | Flange | |
| --- | --- | --- | --- | --- |
| | | | Width | Thickness |
| M14 x 18 | 14 | 3/16 | 4 | 1/4 |
| M12 x11.8 | 12 | 3/16 | 3-1/8 | 1/4 |
| M12 x 10.8 | 12 | 3/16 | 3-1/8 | 1/4 |
| M12 x 10 | 12 | 3/16 | 3-1/4 | 3/16 |
| M10 x 9 | 10 | 3/16 | 2-3/4 | 3/16 |
| M10 x 8 | 10 | 3/16 | 2-3/4 | 3/16 |
| M10 x 7.5 | 10 | 1/8 | 2-3/4 | 3/16 |
| M8 x 6.5 | 8 | 1/8 | 2-1/4 | 3/16 |
| M6 x 4.4 | 6 | 1/8 | 1-7/8 | 3/16 |
| M5 x 18.9 | 5 | 5/16 | 5 | 7/16 |

All dimensions in inches.

## S Shape Dimensions

| Designation | Depth | Web Thickness | Flange | |
| --- | --- | --- | --- | --- |
| | | | Width | Thickness |
| | | | | |
| S24 x 121 | 24-1/2 | 13/16 | 8 | 1-1/16 |
| x106 | 24-1/2 | 5/8 | 7-7/8 | 1-1/16 |
| S24 x 100 | 24 | 3/4 | 7-1/4 | 7/8 |
| x90 | 24 | 5/8 | 7-1/8 | 7/8 |
| x80 | 24 | 1/2 | 7 | 7/8 |
| S20 x 96 | 20-1/4 | 13/16 | 7-1/4 | 15/16 |
| x86 | 20-1/4 | 11/16 | 7 | 15/16 |
| S20 x 75 | 20 | 5/8 | 6-3/8 | 13/16 |
| x66 | 20 | 1/2 | 6-1/4 | 13/16 |
| S18 x 70 | 18 | 11/16 | 6-1/4 | 11/16 |
| x54.7 | 18 | 7/16 | 6 | 11/16 |
| S15 x 50 | 15 | 9/16 | 5-5/8 | 5/8 |
| x42.9 | 15 | 7/16 | 5-1/2 | 5/8 |
| S12 x 50 | 12 | 11/16 | 5-1/2 | 11/16 |
| x40.8 | 12 | 7/16 | 5-1/4 | 11/16 |
| S12 x 35 | 12 | 7/16 | 5-1/8 | 9/16 |
| x25.4 | 12 | 3/8 | 5 | 9/16 |
| S10 x 35 | 10 | 5/8 | 5 | 1/2 |
| x25.4 | 10 | 5/16 | 4-5/8 | 1/2 |
| S8 x 23 | 8 | 7/16 | 4-1/8 | 7/16 |
| x18.4 | 8 | 1/4 | 4 | 7/16 |
| S7 x 20 | 7 | 7/16 | 3-7/8 | 3/8 |
| x15.3 | 7 | 1/4 | 3-5/8 | 3/8 |
| S6 x 17.25 | 6 | 7/16 | 3-5/8 | 3/8 |
| x12.5 | 6 | 1/4 | 3-3/8 | 3/8 |
| S5 x 14.75 | 5 | 1/2 | 3-1/4 | 5/16 |
| x10 | 5 | 3/16 | 3 | 5/16 |
| S4 x 9.5 | 4 | 5/16 | 2-3/4 | 5/16 |
| x7.7 | 4 | 3/16 | 2-5/8 | 5/16 |
| S3 x 7.5 | 3 | 3/8 | 2-1/2 | 1/4 |
| x5.7 | 3 | 3/16 | 2-3/8 | 1/4 |

All dimensions in inches.

## HP Shape Dimensions

| Designation | Depth | Web Thickness | Flange | |
|---|---|---|---|---|
| | | | Width | Thickness |
| | | | | |
| HP14 x 117 | 14-1/2 | 13/16 | 14-7/8 | 13/16 |
| x102 | 14 | 11/16 | 14-3/4 | 11/16 |
| x89 | 13-7/8 | 5/8 | 14-3/4 | 5/8 |
| x73 | 13-5/8 | 1/2 | 14-5/8 | 1/2 |
| HP13 x 100 | 13-1/8 | 3/4 | 13-1/4 | 3/4 |
| x87 | 13 | 11/16 | 13-1/8 | 11/16 |
| x73 | 12-3/4 | 9/16 | 13 | 9/16 |
| x60 | 12-1/2 | 7/16 | 12-7/8 | 7/16 |
| HP12 x 84 | 12-1/4 | 11/16 | 12-1/4 | 11/16 |
| x74 | 12-1/8 | 5/8 | 12-1/4 | 5/8 |
| x63 | 12 | 1/2 | 12-1/8 | 1/2 |
| x53 | 11-3/4 | 7/16 | 12 | 7/16 |
| HP10 x 57 | 10 | 9/16 | 10-1/4 | 9/16 |
| x42 | 9-3/4 | 7/16 | 10-1/8 | 7/16 |
| HP8 x 36 | 8 | 7/16 | 8-1/8 | 7/16 |

All dimensions in inches.

## Channel Iron
## American Standard Dimensions

| Designation | Depth | Web Thickness | Flange | |
|---|---|---|---|---|
| | | | Width | Thickness |
| C15 x 50 | 15 | 11/16 | 3-3/4 | 5/8 |
| x40 | 15 | 1/2 | 3-1/2 | 5/8 |
| x33.9 | 15 | 3/8 | 3-3/8 | 5/8 |
| C12 x 30 | 12 | 1/2 | 3-1/8 | 1/2 |
| x25 | 12 | 3/8 | 3 | 1/2 |
| x20.7 | 12 | 5/16 | 3 | 1/2 |
| C 10 x 30 | 10 | 11/16 | 3 | 7/16 |
| x25 | 10 | 1/2 | 2-7/8 | 7/16 |
| x20 | 10 | 3/8 | 2-3/4 | 7/16 |
| x15.3 | 10 | 1/4 | 2-5/8 | 7/16 |
| C9 x 20 | 9 | 7/16 | 2-5/8 | 7/16 |
| x15 | 9 | 5/16 | 2-1/2 | 7/16 |
| x13.4 | 9 | 1/4 | 2-3/8 | 7/16 |
| C8 x 18.75 | 8 | 1/2 | 2-1/2 | 3/8 |
| x13.75 | 8 | 5/16 | 2-3/8 | 3/8 |
| x11.5 | 8 | 1/4 | 2-1/4 | 3/8 |
| C7 x 14.75 | 7 | 7/16 | 2-1/4 | 3/8 |
| x12.25 | 7 | 5/16 | 1-1/4 | 3/8 |
| x9.8 | 7 | 3/16 | 1-1/8 | 3/8 |
| C6 x 13 | 6 | 7/16 | 2-1/8 | 5/16 |
| x10.5 | 6 | 5/16 | 2 | 5/16 |
| x8.2 | 6 | 3/16 | 1-7/8 | 5/16 |
| C5 x 9 | 5 | 5/16 | 1-7/8 | 5/16 |
| x6.7 | 5 | 3/16 | 1-3/4 | 5/16 |
| C4 x 7.25 | 4 | 5/16 | 1-3/4 | 5/16 |
| x5.4 | 4 | 3/16 | 1-5/8 | 5/16 |
| C3 x 6 | 3 | 3/8 | 1-5/8 | 1/4 |
| x5 | 3 | 1/4 | 1-1/2 | 1/4 |
| x4.1 | 3 | 3/16 | 1-3/8 | 1/4 |

All dimensions in inches.

## Channel Iron
## Miscellaneous Dimensions

| Designation | Depth | Web Thickness | Flange | |
|---|---|---|---|---|
| | | | Width | Thickness |
| MC18 x 58 | 18 | 11/16 | 4-1/4 | 5/8 |
| x51.9 | 18 | 5/8 | 4-1/8 | 5/8 |
| x45.8 | 18 | 1/2 | 4 | 5/8 |
| x42.7 | 18 | 7/16 | 4 | 5/8 |
| MC13 x 50 | 13 | 13/16 | 4-3/8 | 5/8 |
| x40 | 13 | 9/16 | 4-1/8 | 5/8 |
| x35 | 13 | 7/16 | 4-1/8 | 5/8 |
| x31.8 | 13 | 3/8 | 4 | 5/8 |
| MC12 x 50 | 12 | 13/16 | 4-1/8 | 11/16 |
| x45 | 12 | 11/16 | 4 | 11/16 |
| x40 | 12 | 9/16 | 3-7/8 | 11/16 |
| x35 | 12 | 7/16 | 3-3/4 | 11/16 |
| x31 | 12 | 3/8 | 3-5/8 | 11/16 |
| MC12 x 10.6 | 12 | 3/16 | 1-1/2 | 5/16 |
| MC10 x 41.1 | 10 | 13/16 | 4-3/8 | 9/16 |
| x33.6 | 10 | 9/16 | 4-1/8 | 9/16 |
| x28.5 | 10 | 7/16 | 4 | 9/16 |
| MC10 x 25 | 10 | 3/8 | 3-3/8 | 9/16 |
| x22 | 10 | 5/16 | 3-3/8 | 9/16 |
| MC10 x 8.4 | 10 | 3/16 | 1-1/2 | 1/4 |
| MC10 x 6.5 | 10 | 1/8 | 1-1/8 | 3/16 |
| MC9 x 25.4 | 9 | 7/16 | 3-1/2 | 9/16 |
| x23.9 | 9 | 3/8 | 3-1/2 | 9/16 |
| MC8 x 22.8 | 8 | 7/16 | 3-1/2 | 1/2 |
| x21.4 | 8 | 3/8 | 3-1/2 | 1/2 |
| MC8 x 20 | 8 | 3/8 | 3 | 1/2 |
| x18.7 | 8 | 3/8 | 3 | 1/2 |
| MC8 x 8.5 | 8 | 3/16 | 1-7/8 | 5/16 |
| MC7 x 22.7 | 7 | 1/2 | 3-5/8 | 1/2 |
| x19.1 | 7 | 3/8 | 3-1/2 | 1/2 |
| MC6 x 18 | 6 | 3/8 | 3-1/2 | 1/2 |
| x15.3 | 6 | 5/16 | 3-1/2 | 3/8 |
| MC6 x 16.3 | 6 | 3/8 | 3 | 1/2 |
| x15.1 | 6 | 5/16 | 3 | 1/2 |
| MC6 x 12 | 6 | 5/16 | 2-1/2 | 3/8 |

All dimensions in inches.

## Pipe Dimensions
## Standard Weight

**( All Dimensions Are In Inches)**

| Nominal Diameter | Outside Diameter | Inside Diameter | Wall Thickness | Weight per Foot |
|---|---|---|---|---|
| 1/2 | .840 | .622 | .109 | .85 |
| 3/4 | 1.050 | .824 | .113 | 1.13 |
| 1 | 1.315 | 1.049 | .133 | 1.68 |
| 1-1/4 | 1.660 | 1.380 | .140 | 2.27 |
| 1-1/2 | 1.900 | 1.610 | .145 | 2.72 |
| 2 | 2.375 | 2.067 | .154 | 3.65 |
| 2-1/2 | 2.875 | 2.469 | .203 | 5.79 |
| 3 | 3.500 | 3.068 | .216 | 7.58 |
| 3-1/2 | 4.000 | 3.548 | .226 | 9.11 |
| 4 | 4.500 | 4.026 | .237 | 10.79 |
| 5 | 5.563 | 5.047 | .258 | 14.62 |
| 6 | 6.625 | 6.065 | .280 | 18.97 |
| 8 | 8.625 | 7.981 | .322 | 28.55 |
| 10 | 10.750 | 10.020 | .365 | 40.48 |
| 12 | 12.750 | 12.000 | .375 | 49.56 |

## Pipe Dimensions
## Extra Strong

**( All Dimensions Are In Inches)**

| Nominal Diameter | Outside Diameter | Inside Diameter | Wall Thickness | Weight per Foot |
|---|---|---|---|---|
| 1/2 | .840 | .546 | .147 | 1.09 |
| 3/4 | 1.050 | .742 | .154 | 1.47 |
| 1 | 1.315 | .957 | .179 | 2.17 |
| 1-1/4 | 1.660 | 1.278 | .191 | 3.00 |
| 1-1/2 | 1.900 | 1.500 | .200 | 3.63 |
| 2 | 2.375 | 1.939 | .218 | 5.02 |
| 2-1/2 | 2.875 | 2.323 | .276 | 7.66 |
| 3 | 3.500 | 2.900 | .300 | 10.25 |
| 3-1/2 | 4.000 | 3.364 | .318 | 12.50 |
| 4 | 4.500 | 3.826 | .337 | 14.98 |
| 5 | 5.563 | 4.813 | .375 | 20.78 |
| 6 | 6.625 | 5.761 | .432 | 28.57 |
| 8 | 8.625 | 7.625 | .500 | 43.39 |
| 10 | 10.750 | 9.750 | .500 | 54.74 |
| 12 | 12.750 | 11.750 | .500 | 65.42 |

## Pipe Dimensions
## Double Extra Strong

**( All Dimensions Are In Inches)**

| Nominal Diameter | Outside Diameter | Inside Diameter | Wall Thickness | Weight per Foot |
|---|---|---|---|---|
| 2 | 2.375 | 1.503 | .436 | 9.03 |
| 2-1/2 | 2.875 | 1.771 | .552 | 13.69 |
| 3 | 3.500 | 2.300 | .600 | 18.58 |
| 4 | 4.500 | 3.152 | .674 | 27.54 |
| 5 | 5.563 | 4.063 | .750 | 38.55 |
| 6 | 6.625 | 4.897 | .864 | 53.16 |
| 8 | 8.625 | 6.875 | .875 | 72.42 |

# Structural Tubing Dimensions (Square and Rectangular)
## Square Tubing Dimensions

| Dimensions | | |
|---|---|---|
| Nominal Size in Inches | Wall Thickness in Inches | Weight Per Foot |
| 16 x 16 | 5/8 | 127.37 |
| | 1/2 | 103.30 |
| | 3/8 | 78.52 |
| | 5/16 | 65.87 |
| 14 x 14 | 5/8 | 110.36 |
| | 1/2 | 89.68 |
| | 3/8 | 68.31 |
| | 5/16 | 57.36 |
| 12 x 12 | 5/8 | 93.34 |
| | 1/2 | 76.07 |
| | 3/8 | 58.10 |
| | 5/16 | 48.86 |
| | 1/4 | 39.43 |
| | 3/16 | 29.84 |
| 10 x 10 | 5/8 | 76.33 |
| | 9/16 | 69.48 |
| | 1/2 | 62.46 |
| | 3/8 | 47.90 |
| | 5/16 | 40.35 |
| | 1/4 | 32.63 |
| | 3/16 | 24.73 |
| 9 x 9 | 5/8 | 67.82 |
| | 9/16 | 61.83 |
| | 1/2 | 55.66 |
| | 3/8 | 42.79 |
| | 5/16 | 36.10 |
| | 1/4 | 29.23 |
| | 3/16 | 22.18 |

## Square Tubing
## Dimensions

| Dimensions | | |
|---|---|---|
| **Nominal Size in Inches** | **Wall Thickness in Inches** | **Weight Per Foot** |
| 8 x 8 | 5/8 | 59.32 |
|  | 9/16 | 54.17 |
|  | 1/2 | 48.85 |
|  | 3/8 | 37.69 |
|  | 5/16 | 31.84 |
|  | 1/4 | 25.82 |
|  | 3/16 | 19.63 |
| 7 x 7 | 9/16 | 46.51 |
|  | 1/2 | 42.05 |
|  | 3/8 | 32.58 |
|  | 5/16 | 27.59 |
|  | 1/4 | 22.42 |
|  | 3/16 | 17.08 |
| 6 x 6 | 9/16 | 38.86 |
|  | 1/2 | 35.24 |
|  | 3/8 | 27.48 |
|  | 5/16 | 23.34 |
|  | 1/4 | 19.02 |
|  | 3/16 | 14.53 |
| 5 x 5 | 1/2 | 28.43 |
|  | 3/8 | 22.37 |
|  | 5/16 | 19.08 |
|  | 1/4 | 15.62 |
|  | 3/16 | 11.97 |

## Square Tubing
## Dimensions

| Dimensions | | |
|---|---|---|
| **Nominal Size in Inches** | **Wall Thickness in Inches** | **Weight Per Foot** |
| 4.5 x 4.5 | 1/4 | 13.91 |
| | 3/16 | 10.70 |
| 4 x 4 | 1/2 | 21.63 |
| | 3/8 | 17.27 |
| | 5/16 | 14.83 |
| | 1/4 | 12.21 |
| | 3/16 | 9.42 |
| 3.5 x 3.5 | 5/16 | 12.70 |
| | 1/4 | 10.51 |
| | 3/16 | 8.15 |
| 3 x 3 | 5/16 | 10.58 |
| | 1/4 | 8.81 |
| | 3/16 | 6.87 |
| 2.5 x 2.5 | 5/16 | 8.45 |
| | 1/4 | 7.11 |
| | 3/16 | 5.59 |
| 2 x 2 | 5/16 | 6.32 |
| | 1/4 | 5.41 |
| | 3/16 | 4.32 |

## Rectangular Tubing
## Dimensions

| Dimensions | | |
|---|---|---|
| **Nominal Size in Inches** | **Wall Thickness in Inches** | **Weight Per Foot** |
| 20 x 12 | 1/2 | 103.30 |
| | 3/8 | 78.52 |
| | 5/16 | 65.87 |
| 20 x 8 | 1/2 | 89.68 |
| | 3/8 | 68.31 |
| | 5/16 | 57.36 |
| 20 x 4 | 1/2 | 76.07 |
| | 3/8 | 58.10 |
| | 5/16 | 48.86 |
| 18 x 6 | 1/2 | 76.07 |
| | 3/8 | 58.10 |
| | 5/16 | 48.86 |
| 16 x 12 | 5/8 | 110.36 |
| | 1/2 | 89.68 |
| | 3/8 | 68.31 |
| | 5/16 | 57.36 |
| 16 x 8 | 1/2 | 76.07 |
| | 3/8 | 58.10 |
| | 5/16 | 48.86 |
| 16 x 4 | 1/2 | 62.46 |
| | 3/8 | 47.90 |
| | 5/16 | 40.35 |
| 14 x 10 | 5/8 | 93.34 |
| | 1/2 | 76.07 |
| | 3/8 | 58.10 |
| | 5/16 | 48.86 |

**Rectangular Tubing
Dimensions**

| Dimensions | | |
|---|---|---|
| **Nominal Size in Inches** | **Wall Thickness in Inches** | **Weight Per Foot** |
| 14 x 16 | 1/2 | 62.46 |
| | 3/8 | 47.90 |
| | 5/16 | 40.35 |
| | 1/4 | 32.63 |
| | | |
| 14 x 4 | 1/2 | 55.66 |
| | 3/8 | 42.79 |
| | 5/16 | 36.10 |
| | 1/4 | 29.23 |
| | | |
| 12 x 8 | 5/8 | 76.33 |
| | 9/16 | 69.48 |
| | 1/2 | 62.46 |
| | 3/8 | 47.90 |
| | 5/16 | 40.35 |
| | 1/4 | 32.63 |
| | 3/16 | 24.73 |
| | | |
| 12 x 6 | 5/8 | 67.82 |
| | 9/16 | 61.83 |
| | 1/2 | 55.66 |
| | 3/8 | 42.79 |
| | 5/16 | 36.10 |
| | 1/4 | 29.23 |
| | 3/16 | 22.18 |
| | | |
| 12 x 4 | 5/8 | 59.32 |
| | 9/16 | 54.17 |
| | 1/2 | 48.85 |
| | 3/8 | 37.69 |
| | 5/16 | 31.84 |
| | 1/4 | 25.82 |
| | 3/16 | 19.63 |

## Rectangular Tubing
## Dimensions

| Dimensions | | |
|---|---|---|
| **Nominal Size in Inches** | **Wall Thickness in Inches** | **Weight Per Foot** |
| 12 x 2 | 1/4 | 22.42 |
| | 3/16 | 17.08 |
| | | |
| 10 x 8 | 5/8 | 67.82 |
| | 9/16 | 61.83 |
| | 1/2 | 55.66 |
| | 3/8 | 42.79 |
| | 5/16 | 36.10 |
| | 1/4 | 29.23 |
| | 3/16 | 22.18 |
| | | |
| 10 x 6 | 5/8 | 59.32 |
| | 9/16 | 54.17 |
| | 1/2 | 48.85 |
| | 3/8 | 37.69 |
| | 5/16 | 31.84 |
| | 1/4 | 25.82 |
| | 3/16 | 19.63 |
| | | |
| 10 x 5 | 5/8 | 55.06 |
| | 9/16 | 50.34 |
| | 1/2 | 45.45 |
| | 3/8 | 35.13 |
| | 5/16 | 29.72 |
| | 1/4 | 24.12 |
| | 3/16 | 18.35 |
| | | |
| 10 x 4 | 9/16 | 46.51 |
| | 1/2 | 42.05 |
| | 3/8 | 32.58 |
| | 5/16 | 27.59 |
| | 1/4 | 22.42 |
| | 3/16 | 17.08 |

## Rectangular Tubing
## Dimensions

| Dimensions | | |
|---|---|---|
| Nominal Size in Inches | Wall Thickness in Inches | Weight Per Foot |
| 10 x 2 | 3/8 | 27.48 |
| | 5/16 | 23.34 |
| | 1/4 | 19.02 |
| | 3/16 | 14.53 |
| 9 x 7 | 5/8 | 59.32 |
| | 9/16 | 54.17 |
| | 1/2 | 48.85 |
| | 3/8 | 37.69 |
| | 5/16 | 31.84 |
| | 1/4 | 25.82 |
| | 3/16 | 19.63 |
| 9 x 6 | 5/8 | 55.06 |
| | 9/16 | 50.34 |
| | 1/2 | 45.45 |
| | 3/8 | 35.13 |
| | 5/16 | 29.72 |
| | 1/4 | 24.12 |
| | 3/16 | 18.35 |
| 9 x 5 | 9/16 | 46.51 |
| | 1/2 | 42.05 |
| | 3/8 | 32.58 |
| | 5/16 | 27.59 |
| | 1/4 | 22.42 |
| | 3/16 | 17.08 |
| 9 x 3 | 1/2 | 35.24 |
| | 3/8 | 27.48 |
| | 5/16 | 23.34 |
| | 1/4 | 19.02 |
| | 3/16 | 14.53 |

## Rectangular Tubing
## Dimensions

| Dimensions | | |
|---|---|---|
| **Nominal Size in Inches** | **Wall Thickness in Inches** | **Weight Per Foot** |
| 8 x 6 | 9/16 | 46.51 |
|  | 1/2 | 42.05 |
|  | 3/8 | 32.58 |
|  | 5/16 | 27.59 |
|  | 1/4 | 22.42 |
|  | 3/16 | 17.08 |
| 8 x 4 | 9/16 | 38.86 |
|  | 1/2 | 35.24 |
|  | 3/8 | 27.48 |
|  | 5/16 | 23.34 |
|  | 1/4 | 19.02 |
|  | 3/16 | 14.53 |
| 8 x 3 | 1/2 | 31.84 |
|  | 3/8 | 24.93 |
|  | 5/16 | 21.21 |
|  | 1/4 | 17.32 |
|  | 3/16 | 13.25 |
| 8 x 2 | 3/8 | 22.37 |
|  | 5/16 | 19.08 |
|  | 1/4 | 15.62 |
|  | 3/16 | 11.97 |
| 7 x 5 | 1/2 | 35.24 |
|  | 3/8 | 27.48 |
|  | 5/16 | 23.34 |
|  | 1/4 | 19.02 |
|  | 3/16 | 14.53 |

## Rectangular Tubing
## Dimensions

| Dimensions | | |
|---|---|---|
| Nominal Size in Inches | Wall Thickness in Inches | Weight Per Foot |
| 7 x 4 | 1/2 | 31.84 |
|  | 3/8 | 24.93 |
|  | 5/16 | 21.21 |
|  | 1/4 | 17.32 |
|  | 3/16 | 13.25 |
| 7 x 3 | 1/2 | 28.43 |
|  | 3/8 | 22.37 |
|  | 5/16 | 19.08 |
|  | 1/4 | 15.62 |
|  | 3/16 | 11.97 |
| 7 x 2 | 1/4 | 13.91 |
|  | 3/16 | 10.70 |
| 6 x 5 | 1/2 | 31.84 |
|  | 3/8 | 24.93 |
|  | 5/16 | 21.21 |
|  | 1/4 | 17.32 |
|  | 3/16 | 13.25 |
| 6 x 4 | 1/2 | 28.43 |
|  | 3/8 | 22.37 |
|  | 5/16 | 19.08 |
|  | 1/4 | 15.62 |
|  | 3/16 | 11.97 |
| 6 x 3 | 3/8 | 19.82 |
|  | 5/16 | 16.96 |
|  | 1/4 | 13.91 |
|  | 3/16 | 10.70 |

## Rectangular Tubing
## Dimensions

| Dimensions | | |
|---|---|---|
| Nominal Size in Inches | Wall Thickness in Inches | Weight Per Foot |
| | | |
| 6 x 2 | 3/8 | 17.27 |
| | 5/16 | 14.83 |
| | 1/4 | 12.21 |
| | 3/16 | 9.42 |
| | | |
| 5 x 4 | 3/8 | 19.82 |
| | 5/16 | 16.96 |
| | 1/4 | 13.91 |
| | 3/16 | 10.70 |
| | | |
| 5 x 3 | 1/2 | 21.63 |
| | 3/8 | 17.27 |
| | 5/16 | 14.83 |
| | 1/4 | 12.21 |
| | 3/16 | 9.42 |
| | | |
| 5 x 2 | 5/16 | 12.70 |
| | 1/4 | 10.51 |
| | 3/16 | 8.15 |
| | | |
| 4 x 3 | 5/16 | 12.70 |
| | 1/4 | 10.51 |
| | 3/16 | 8.15 |
| | | |
| 4 x 2 | 5/16 | 10.58 |
| | 1/4 | 8.81 |
| | 3/16 | 6.87 |
| | | |
| 3.5 x 3.5 | 1/4 | 8.81 |
| | 3/16 | 6.87 |
| | | |
| 3 x 2 | 1/4 | 7.11 |
| | 3/16 | 5.59 |

# Sheet Steel Thickness

| Table D1 Gage Numbers and Equivalent Thicknesses Hot-Rolled and Cold-Rolled Sheet | | |
|---|---|---|
| Manufacturers' Standard Gage Number | Thickness Equivalent in. | mm |
| 3 | 0.2391 | 6.073 |
| 4 | 0.2242 | 5.695 |
| 5 | 0.2092 | 5.314 |
| 6 | 0.1943 | 4.935 |
| 7 | 0.1793 | 4.554 |
| 8 | 0.1644 | 4.176 |
| 9 | 0.1495 | 3.800 |
| 10 | 0.1345 | 3.416 |
| 11 | 0.1196 | 3.038 |
| 12 | 0.1046 | 2.657 |
| 13 | 0.0897 | 2.278 |
| 14 | 0.0747 | 1.900 |
| 15 | 0.0673 | 1.709 |
| 16 | 0.0598 | 1.519 |
| 17 | 0.0538 | 1.366 |
| 18 | 0.0478 | 1.214 |
| 19 | 0.0418 | 1.062 |
| 20 | 0.0359 | 0.912 |
| 21 | 0.0329 | 0.836 |
| 22 | 0.0299 | 0.759 |
| 23 | 0.0269 | 0.660 |
| 24 | 0.0239 | 0.607 |
| 25 | 0.0209 | 0.531 |
| 26 | 0.0179 | 0.455 |
| 27 | 0.0164 | 0.417 |
| 28 | 0.0149 | 0.378 |

Note: Table D1 is for information only. This product is commonly specified to decimal thickness, not to gage number.

| Table D2 Gage Numbers and Equivalent Thicknesses Galvanized Sheet | | |
|---|---|---|
| Galvanized Sheet Gage Number | Thickness Equivalent in. | mm |
| 8 | 0.1681 | 4.270 |
| 9 | 0.1532 | 3.891 |
| 10 | 0.1382 | 3.510 |
| 11 | 0.1233 | 3.132 |
| 12 | 0.1084 | 2.753 |
| 13 | 0.0934 | 2.372 |
| 14 | 0.0785 | 1.993 |
| 15 | 0.0710 | 1.803 |
| 16 | 0.0635 | 1.613 |
| 17 | 0.0575 | 1.460 |
| 18 | 0.0516 | 1.311 |
| 19 | 0.0456 | 1.158 |
| 20 | 0.0396 | 1.006 |
| 21 | 0.0366 | 0.930 |
| 22 | 0.0336 | 0.853 |
| 23 | 0.0306 | 0.777 |
| 24 | 0.0276 | 0.701 |
| 25 | 0.0247 | 0.627 |
| 26 | 0.0217 | 0.551 |
| 27 | 0.0202 | 0.513 |
| 28 | 0.0187 | 0.475 |
| 29 | 0.0172 | 0.437 |
| 30 | 0.0157 | 0.399 |
| 31 | 0.0142 | 0.361 |
| 32 | 0.0134 | 0.340 |

Note: Table D2 is for information only. This product is commonly specified to decimal thickness, not to gage number.

# Welding Symbols

## Standard Location of Elements of a Welding Symbol

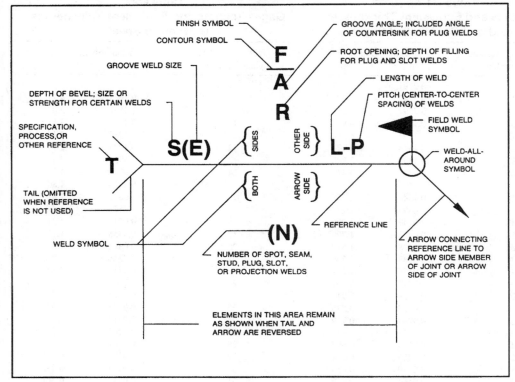

## Supplementary Symbols

| WELD ALL AROUND | FIELD WELD | MELT THROUGH | CONSUMABLE INSERT (SQUARE) | BACKING OR SPACER (RECTANGLE) | CONTOUR | | |
|---|---|---|---|---|---|---|---|
| | | | | | FLUSH OR FLAT | CONVEX | CONCAVE |

# Welding Symbols

| GROOVE | | | | | | | |
|---|---|---|---|---|---|---|---|
| SQUARE | SCARF | V | BEVEL | U | J | FLARE-V | FLARE-BEVEL |
| | | | | | | | |

| FILLET | PLUG OR SLOT | STUD | SPOT OR PROJECTION | SEAM | BACK OR BACKING | SURFACING | FLANGE | |
|---|---|---|---|---|---|---|---|---|
| | | | | | | | EDGE | CORNER |
| | | | | | | | | |

NOTE: THE REFERENCE LINE IS SHOWN DASHED FOR ILLUSTRATIVE PURPOSES.

# American Welding Society

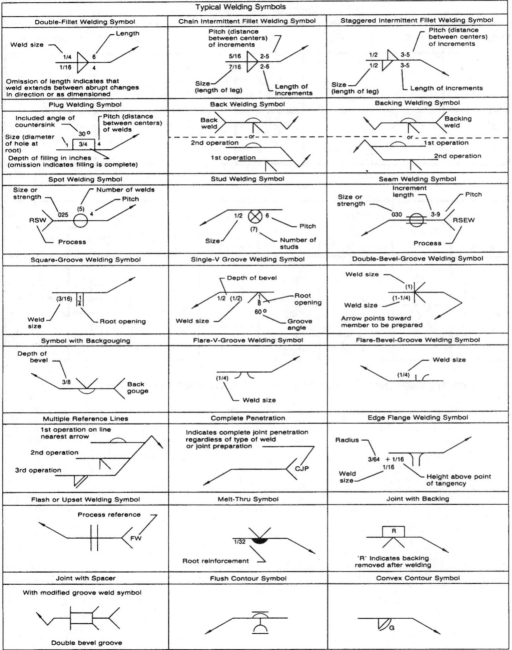

Typical Welding Symbols

| Double-Fillet Welding Symbol | Chain Intermittent Fillet Welding Symbol | Staggered Intermittent Fillet Welding Symbol |
|---|---|---|
| Weld size, Length, 1/4 6, 1/16 4. Omission of length indicates that weld extends between abrupt changes in direction or as dimensioned | Pitch (distance between centers) of increments, 5/16 2-5, 7/16 2-6, Size (length of leg), Length of increments | Pitch (distance between centers) of increments, 1/2 3-5, 1/2 3-5, Size (length of leg), Length of increments |
| **Plug Welding Symbol** | **Back Welding Symbol** | **Backing Welding Symbol** |
| Included angle of countersink, Pitch (distance between centers) of welds, 30°, Size (diameter of hole at root), 3/4 4, Depth of filling in inches (omission indicates filling is complete) | Back weld, — or —, 2nd operation, 1st operation | Backing weld, — or —, 1st operation, 2nd operation |
| **Spot Welding Symbol** | **Stud Welding Symbol** | **Seam Welding Symbol** |
| Size or strength, Number of welds, Pitch, 025 4, (5), RSW, Process | 1/2 6, (7), Size, Pitch, Number of studs | Size or strength, Increment length, Pitch, 030 3-9, RSEW, Process |
| **Square-Groove Welding Symbol** | **Single-V Groove Welding Symbol** | **Double-Bevel-Groove Welding Symbol** |
| (3/16) 1/4, Weld size, Root opening | Depth of bevel, 1/2 (1/2), Root opening, 1/8, 60°, Weld size, Groove angle | Weld size, (1), (1-1/4), Weld size, Arrow points toward member to be prepared |
| **Symbol with Backgouging** | **Flare-V-Groove Welding Symbol** | **Flare-Bevel-Groove Welding Symbol** |
| Depth of bevel, 3/8, Back gouge | (1/4), Weld size | Weld size, (1/4) |
| **Multiple Reference Lines** | **Complete Penetration** | **Edge Flange Welding Symbol** |
| 1st operation on line nearest arrow, 2nd operation, 3rd operation | Indicates complete joint penetration regardless of type of weld or joint preparation, CJP | Radius, 3/64 + 1/16, 1/16, Weld size, Height above point of tangency |
| **Flash or Upset Welding Symbol** | **Melt-Thru Symbol** | **Joint with Backing** |
| Process reference, FW | 1/32, Root reinforcement | R, 'R' Indicates backing removed after welding |
| **Joint with Spacer** | **Flush Contour Symbol** | **Convex Contour Symbol** |
| With modified groove weld symbol, Double bevel groove | | G |

\* It should be understood that these charts are intended only as shop aids. The only complete and official presentation of the standard welding symbols is in A2.4.

# American Welling Society

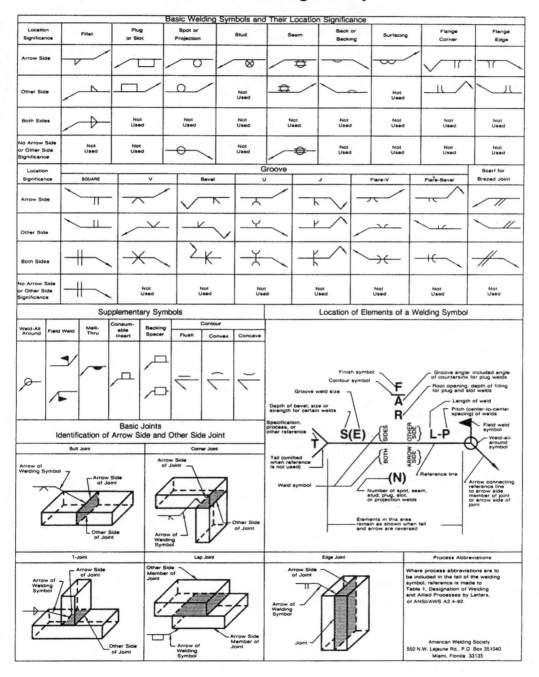

## Basic Welding Symbols and Their Location Significance

| Location Significance | Fillet | Plug or Slot | Spot or Projection | Stud | Seam | Back or Backing | Surfacing | Flange Corner | Flange Edge |
|---|---|---|---|---|---|---|---|---|---|
| Arrow Side | | | | | | | | | |
| Other Side | | | | Not Used | | | Not Used | | |
| Both Sides | | Not Used | Not Used | Not Used | Not Used | Not Used | Not Used | Not Used | Not Used |
| No Arrow Side or Other Side Significance | Not Used | Not Used | | Not Used | | Not Used | Not Used | Not Used | Not Used |

## Groove

| Location Significance | SQUARE | V | Bevel | U | J | Flare-V | Flare-Bevel | Scarf for Brazed Joint |
|---|---|---|---|---|---|---|---|---|
| Arrow Side | | | | | | | | |
| Other Side | | | | | | | | |
| Both Sides | | | | | | | | |
| No Arrow Side or Other Side Significance | | Not Used | Not Used | Not Used | Not Used | Not Used | Not Used | Not Used |

## Supplementary Symbols

| Weld-All Around | Field Weld | Melt-Thru | Consumable Insert | Backing Spacer | Contour — Flush | Convex | Concave |
|---|---|---|---|---|---|---|---|

## Location of Elements of a Welding Symbol

Finish symbol
Contour symbol
Groove weld size
Depth of bevel; size or strength for certain welds
Specification, process, or other reference
Tail (omitted when reference is not used)
Weld symbol

Groove angle: included angle of countersink for plug welds
Root opening: depth of filling for plug and slot welds
Length of weld
Pitch (center-to-center spacing) of welds
Field weld symbol
Weld-all-around symbol
Reference line
Arrow connecting reference line to arrow side member of joint or arrow side of joint
Number of spot, seam, stud, plug, slot, or projection welds
Elements in this area remain as shown when tail and arrow are reversed

F
A
R
S(E)
L-P
(N)
T

SIDES  OTHER SIDE  BOTH  ARROW SIDE

## Basic Joints
### Identification of Arrow Side and Other Side Joint

**Butt Joint**
Arrow of Welding Symbol
Arrow Side of Joint
Other Side of Joint

**Corner Joint**
Arrow Side of Joint
Other Side of Joint
Arrow of Welding Symbol

**T-Joint**
Arrow Side of Joint
Arrow of Welding Symbol
Other Side of Joint

**Lap Joint**
Other Side Member of Joint
Arrow Side Member of Joint
Arrow of Welding Symbol

**Edge Joint**
Arrow Side of Joint
Arrow of Welding Symbol
Joint

**Process Abbreviations**
Where process abbreviations are to be included in the tail of the welding symbol, reference is made to Table 1, Designation of Welding and Allied Processes by Letters, of ANSI/AWS A2.4-92.

American Welding Society
550 N.W. Lejeune Rd., P.O. Box 351040
Miami, Florida 33135

# B.  Structural Steel

## Allowable Roughness in Oxygen Cut Surfaces

D1.1, 3.2.2

D1.1, . . . Notches or gouges exceeding ³/₁₆ in. deep
may be repaired by grinding if the nominal cross
sectional area is not reduced by more than 2%. . . .

UBC, 2-604, M2.2

*Ground* or machined surfaces shall be faired to the
original surface with a slope not exceeding 1:10. . . .

*In thermal* cut surfaces, occasional notches or
gouges may, with the approval of the engineer, be
repaired by welding.

## Alternate Connections

UBC, 2-361, 2211.7.1.3

## Ambient Temperature

D1.1, 3.1.4

D1.1, . . . Welding shall not be done when the ambient temperature is lower than 0 deg. F (− 18
deg. C) (see 4.2), when surfaces are wet or exposed
to rain, snow, or high wind velocities, or when welding personnel are exposed to inclement conditions.

D1.1 Comm., 3.1.3

## Anchor Bolts

UBC, 2-358, 2210

UBC, . . . The protrusion of the threaded ends
through the connected material shall be sufficient
to fully engage the threads of the nuts, but shall not
be greater than the length of threads on the bolts.
Base plate holes for anchor bolts may be oversized
as follows:

AISC, 5-79, J10
AISC, 4-4, Table 1-C

| Bolt Size, inches | Hole Size, inches |
|---|---|
| $3/4$ | $5/16$ oversized |
| $7/8$ | $5/16$ oversized |
| $1 < 2$ | $1/2$ oversized |
| $> 2$ | $1 >$ bolt diameter |

**Anchor Bolt and Threaded Rod Material**

| ASTM Specification | Diameter in inches | Material |
|---|---|---|
| A307 | 4 | Carbon |
| A325 | 1/2 to 1-1/2 | Carbon, Quenched and Tempered |
| A354 Grades B,C & D | 1/4 to 4 | Alloy, Quenched and Tempered |
| A449 | 1/4 ot 3 | Carbon, Quenched and Tempered |
| A490 | 1/2 to 1-1/2 | Alloy, Quenched and Tempered |
| A687 | 5/8 to 3 | Alloy, Quenched and Tempered, Notch Tough |
| A36 | 8 | Carbon |
| A572 Grade 50 | 2 | High-Strength Low Alloy |
| A572 Grade 42 | 6 | High-Strength Low Alloy |
| A588 | 4 to 8 | High-Strength Low Alloy, Weather Resistant |

Quenched and Tempered Anchor Bolts should not be heated or welded without the expressed approval of the Engineer.

## Approval                                                           D1.1, 1.1.2

## Arc Shields (Ferrules)                                             D1.1, 7.2.2

D1.1, . . . An arc shield of heat resistant ceramic    D1.1, 7.4.4
or other suitable material shall be furnished with
each stud. . . .

*Any* arc shields that show signs of surface moisture
from dew or rain shall be oven dried at
250 deg. F (120 deg. C) for 2 hours before use.

## Arc Strikes                                                        D1.1, 3.10

D1.1, . . . Arc strikes outside the area of perma-     D1.1 Comm., 3.10
nent welds should be avoided on any base metal.
Cracks or blemishes caused by arc strikes shall be

ground to a smooth contour and checked to ensure
soundness.

## Arc Welds (Thickness Limitations)  D1.1, 1.2.3

D1.1, . . . The provisions of this code are not in-   UBC, 2-518, E2
tended to apply to welding base metals less than
1/8 in. thick.

## Areas to which Studs are to be Welded  D1.1, 7.4.3

D1.1, . . . The areas to which studs are to be
welded shall be free of scale, rust, moisture, or
other injurious material to the extent necessary to
obtain satisfactory welds. These areas may be
cleaned by wire brushing, scaling, prick-punching,
or grinding.

## Backing Bars

Structural  D1.1, 3.13.3

D1.1, . . . The suggested minimum nominal thick-   D1.1 Comm., 3.13 - 3.13.5
ness of backing bars, provided that they shall be of
sufficient thickness to prevent melt-through, is
shown in the following table:

| Process | Thickness (inches) |
|---|---|
| SMAW | 3/16 |
| GMAW | 1/4 |
| FCAW-SS | 1/4 |
| FCAW-G | 3/8 |
| SAW | 3/8 |

Tubular  D1.1, 10.2.4.2

## Base Metals

Reinforcing  D1.4, 1.2.1

Sheet Steel  D1.3, 1.2

Structural  D1.1, 1.2
D1.1, 8.2.1
D1.1 Comm., 1.2

Tubular  D1.1, 10.2

## Base Metal Preparation

D1.1, 3.2

D1.1, . . . Surfaces on which weld metal is to be deposited shall be free from fins, tears, cracks, and other discontinuities that would adversely affect the quality or strength of the weld. . . .

D1.1 Comm., 3.2.1

*For girders* in dynamically loaded structures, all mill scale shall be removed from the surfaces on which flange-to-web welds are to be made by the submerged arc welding or by shielded metal arc welding with low hydrogen electrodes.

D1.1, 7.5.5.1

For Studs

D1.1, . . . Surfaces to be welded and surfaces adjacent to a weld shall be free from loose scale, slag, rust, moisture, grease, and other foreign material that would prevent proper welding or produce objectionable fumes.

## Base Plates

AISC, 3-111

Oversized Holes for *(See Anchor Bolts)*

UBC, 2-358, 2210

## Bearing Devices (Base Plates)

AISC, 5-235, 7.6

It is the responsibility of the owner to set all leveling nuts, leveling plates, and loose bearing plates that can be set without the use of heavy equipment, to proper line and grade. All other bearing members are set by the erector to lines and grades established by the owner.

*Acceptance of* the final location and the proper grouting of base and bearing plates is the responsibility of the owner.
*The elevation* tolerance on bearing devices is plus or minus $1/8$ in.

## Bearing Stiffeners

AISC, 5-82, K1.8

D1.1, . . . The bearing ends of bearing stiffeners shall be square with the web and shall have at least 75% of the stiffener bearing cross sectional

D1.1, 3.5.2 - 3.5.3.3

area in contact with the inner surface of the flanges. . . .

*Where* tight fit of intermediate stiffeners is specified, it shall be defined as allowing a gap of up to $1/16$ in. between the stiffener and the flange.

## Bend Tests (Studs)                     D1.1, 7.7.1.4 - 7.7.1.5

D1.1, . . . In addition to visual examination, the test shall consist of bending the studs after they are allowed to cool, to an angle of approximately 30 deg. from their original axes by either striking the studs with a hammer or placing a pipe or other suitable hollow device over the studs and manually or mechanically bending the stud. . . .

*At temperatures* below 50 deg. F, bending shall preferably be done by continuous slow application of load.

### Amount to be Tested                    D1.1, 7.7.1.1 - 7.7.1.5

D1.1, . . . Before production welding with a particular set-up and with a given size and type of stud, and at the beginning of each day's or shift's production, testing shall be performed on the first two studs that are welded.

### Repaired Studs                         D1.1, 7.8.1

D1.1, . . . If a visual inspection reveals any stud that does not show a full 360 deg. flash or any stud that has been repaired by welding, such stud shall be bent to an angle of *approximately* 15 deg. from its original axis. Threaded studs shall be torque tested. The bent stud shear connectors (Type B) and *other studs to be embedded in concrete* (Type A) that show no sign of failure shall be acceptable for use and left in the bent position.

### Without Heat                           D1.1, 7.8.3

D1.1, . . . All bending and straightening when required shall be done without heating, before completion of the production stud welding operation, except as otherwise provided in the contract.

## Bolted Parts (Slope of)

UBC, 2-446, 2222 - 2222.3

UBC, . . . All material within the grip of the bolt shall be steel. There shall be no compressible material such as gaskets or insulation within the grip. Bolted steel parts shall fit solidly together after the bolts are tightened and may be coated or uncoated. The slope of the surfaces of parts in contact with the bolt head and nut shall not exceed 1:20 with respect to a plane normal to the bolt axis.

AISC, 5-267, 3 (a)
AISC, 5-88, M2.5

## Bolt Holes

UBC, 2-446, 2222.3

UBC, . . . When approved by the building official, oversize, short-slotted holes or long-slotted holes may be used, subject to the following joint detail requirements:

UBC, 2-448, 2226.2
UBC, 2-594, J3.7
UBC, 2-694, J3.2

Enlargement of

AISC, 5-88, M2.5

During erection, the use of drift pins to align bolt holes in members is allowed, provided their use does not distort, oblong, or otherwise enlarge the holes.

*Improper* or poor matching of bolt holes shall be cause for rejection of the member.

For Rivets and Bolts

UBC, 2-605, M2.2.5
AISC, 5-88, M2.5

Misalignment of

AISC, 5-88, M2.5

*Improper* or poor matching of bolt holes shall be cause for rejection of the member.

Nominal Dimensions of

AISC, 5-268, Table 1

Hole sizes for different diameter bolts can be found in this specification. For convenience, these sizes are provided in the following table.

AISC, 5-72, J3.2.a

## Bolt Hole Dimensions in Inches

| Bolt Diameter | 1/2 in. | 5/8 in. | 3/4 in. | 7/8 in | 1 in |
|---|---|---|---|---|---|
| Standard | 9/16 | 11/16 | 13/16 | 15/16 | 1-1/16 |
| Oversize | 5/8 | 13/16 | 15/16 | 1-1/16 | 1-1/4 |
| Short Slot | 9/16 x 1 1/4 | 11/16 x 7/8 | 13/16 x 1 | 15/16 x 1-1/8 | 1-1/16 x 2-1/2 |
| Long Slot | 9/16 x 1-1/4 | 11/16 x 1-9/16 | 13/16 x 1-7/8 | 15/16 x 2-3/16 | 1-1/16 x 2-1/2 |

| Bolt Diameter | 1-1/8 | 1-1/4 | 1-3/8 | 1-1/2 |
|---|---|---|---|---|
| Standard | 1-3/16 | 1-5/16 | 1-7/16 | 1-9/16 |
| Oversize | 1-7/16 | 1-9/16 | 1-11/16 | 1-13/16 |
| Short Slot | 1-3/16 x 1-1/2 | 1-5/16 x 1-5/8 | 1-7/16 x 1-3/4 | 1-9/16 x 1-7/8 |
| Long Slot | 1-3/16 x 2-13/16 | 1-5/16 x 3-1/8 | 1-7/17 x 3-7/16 | 1-9-16 x 3-3/4 |

*Bear in* mind that base plate holes for anchor bolts may be oversized as follows: (also, see Anchor Bolts in Section A)

| Bolt Size, inches (mm) | Hole Size, inches (mm) |
|---|---|
| $3/4$ (19.1) | $5/16$ (7.9) oversized |
| $7/8$ (22.2) | $5/16$ (7.9) oversized |
| 1 < 2 (25.4 < 50.8) | $1/2$ (12.7) oversized |
| >2 (> 50.8) | 1 (25.4) > bolt diameter |

Oversized and Slotted                                    UBC, 2-448, 2226.2

UBC, . . . The maximum size of holes for rivets          UBC, 2-594, J3.7
and bolts are given in Table J3.5, except that           UBC, 2-694, J3.2
larger holes, required for tolerance on location of      UBC, 2-698, J3.8b
anchor bolts in concrete foundations, may be             AISC,5-271, para. 7 (b)
used in column base details. . . .

*Oversized holes* may be used in any or all plies of     AISC, 5-268, Table 1
slip-critical connections, but they shall not be         AISC, 5-71, J3.2.a - 2.e
used in bearing-type connections. Hardened               AISC, 5-71, Table J3.1
washers shall be installed over oversized holes
in an outer ply. . . .

*(Continued)*

For Rivets and Bolts, Oversize and Slotted
*(Continued)*

*Short-slotted* holes may be used in any or all plies of slip-critical or bearing-type connections. . . .

*Washers* shall be installed over short-slotted holes in an outer ply; when high strength bolts are used, such washers shall be hardened. . . .

*Long-slotted* holes may be used in only one of the connected parts of either a slip-critical or bearing-type connection at an individual faying surface. . . .

*Plate washers* or continuous bar washers with a minimum thickness of 5/16 in. are required when long-slotted holes are used in an outer ply. The plate or bar washer must be of a structural grade and need not be hardened.

*If a slotted* or oversized hole is used in an outer ply of the connection, and ASTM A490 bolts over 1 in. in diameter are used, a 5/16 in. minimum thickness F436 washer must be used.

*The distance* between bolt holes shall not be less than one bolt diameter.

*This specification* allows bolt holes to be 1/16 in. greater than the diameter of the installed bolt. The engineer shall approve all usage of oversized holes.

AISC, 5-294, C3, para. 2
UBC, 2-694, J3.2
AISC, 5-33, B2

Sizes of

The diameter of a hole for a rivet or bolt shall be 1/16 in. greater than the nominal diameter of the rivet or bolt to be installed.

### Bolt Hole Dimensions in Inches

| Bolt Diameter | 1/2 in. | 5/8 in. | 3/4 in. | 7/8 in | 1 in |
|---|---|---|---|---|---|
| Standard | 9/16 | 11/16 | 13/16 | 15/16 | 1-1/16 |
| Oversize | 5/8 | 13/16 | 15/16 | 1-1/16 | 1-1/4 |
| Short Slot | 9/16 x 1 1/4 | 11/16 x 7/8 | 13/16 x 1 | 15/16 x 1-1/8 | 1-1/16 x 2-1/2 |
| Long Slot | 9/16 x 1-1/4 | 11/16 x 1-9/16 | 13/16 x 1-7/8 | 15/16 x 2-3/16 | 1-1/16 x 2-1/2 |

| Bolt Diameter | 1-1/8 | 1-1/4 | 1-3/8 | 1-1/2 |
|---|---|---|---|---|
| Standard | 1-3/16 | 1-5/16 | 1-7/16 | 1-9/16 |
| Oversize | 1-7/16 | 1-9/16 | 1-11/16 | 1-13/16 |
| Short Slot | 1-3/16 x 1-1/2 | 1-5/16 x 1-5/8 | 1-7/16 x 1-3/4 | 1-9/16 x 1-7/8 |
| Long Slot | 1-3/16 x 2-13/16 | 1-5/16 x 3-1/8 | 1-7/17 x 3-7/16 | 1-9-16 x 3-3/4 |

# Bolts

A325 and A490                                   AISC, 5-265

A449                                            AISC, 5-27, A3.4

ASTM A449 bolts are permitted only in non-slip-critical connections requiring bolt diameters over 1 ½ in. in diameter. Any diameter of ASTM A449 may be used in an anchor bolt or threaded rod application.

Application (F, N, X)                           AISC, 4-9

The specification allows for the following three different types of applications for high-strength bolts.

SC   Slip-Critical Connections

N    Bearing-Type Connection/Threads may be included in the shear plane

X    Bearing-Type Connection/Threads must be excluded from the shear plane

It is the responsibility of the design engineer to provide this information on the drawings.

In Combination with Rivets                      AISC, 5-62, J1.10 - J1.11

D1.1, . . . Rivets or bolts used in bearing type connections shall not be considered as sharing the stress in combination with welds. Welds, if used, shall be provided to carry the entire stress in the connection. However, connections that are welded to one member and riveted or bolted to the other member are permitted. High-strength

D1.1, 8.7
D1.1, 9.14

*(Continued)*

Bolts, Installation of *(Continued)*

bolts properly installed as a friction type connection prior to welding may be considered as sharing the stress with the welds.

Inspection of                                             AISC, 5-276, para. 9 a - c

Installation of                                           UBC, 2-450, 2227.4.1/4.2

UBC, . . . In connections requiring full preten-          UBC, 2-448, 2226.3
sion, fasteners, together with washers of size and        UBC, 2-451, 2227.4.5
quality specified. . . .                                  UBC,4-451, 2227.5
*shall be* installed in properly aligned holes and
tightened by one of the methods described in
Section 2227.4, items 2 through 5 to at least the
minimum tension specified in Table 22-lV-D
when all the fasteners are tight. . . .

*Tightening* may be done by turning the bolt
while the nut is prevented from rotating when it
is impractical to turn the nut.

*If impact* wrenches are used to tighten the bolts
in the connection, the specification dictates they
be capable of tightening the bolts in approximately 10 seconds.

*When turn-of-nut* tightening is used, hardened
washers are not required except as may be specified in Section 2226.3.

Nuts and Washers                                          UBC, 2-445, 2221

UBC, . . . Flat circular washers and square or
rectangular washers shall conform to the requirements of Tables 22-IV-K, 22-IV-L, 22-IV-M,
and 22-IV-N. Nuts shall conform to the chemical
and mechanical requirements of Tables 22-IV-G,
22-IV-H, and 22-IV-I. The grade and surface finish of nuts for each bolt type shall be as follows:

### TABLE 22-IV-L—HARDENED CIRCULAR AND CLIPPED CIRCULAR WASHERS[1]

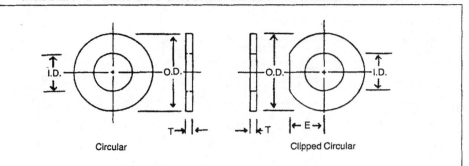

Circular                  Clipped Circular

| BOLT SIZE (inches) | CIRCULAR AND CLIPPED CIRCULAR | | | | CLIPPED |
| | Nominal Outside Diameter (O.D.) (inches) | Normal Inside Diameter (I.D.) (inches) | Thickness (T) (inches) | | Minimum Edge Distance (E)[2] (inches) |
| | | | min. | max. | |
| | | × 25.4 for mm | | | |
| $1/4$ | $5/8$ | $9/32$ | 0.051 | 0.080 | $7/32$ |
| $5/16$ | $11/16$ | $11/32$ | 0.051 | 0.080 | $9/32$ |
| $3/8$ | $13/16$ | $13/32$ | 0.051 | 0.080 | $11/32$ |
| $7/16$ | $59/64$ | $15/32$ | 0.051 | 0.080 | $13/32$ |
| $1/2$ | $1 1/16$ | $17/32$ | 0.097 | 0.177 | $7/16$ |
| $5/8$ | $1 5/16$ | $11/16$ | 0.122 | 0.177 | $9/16$ |
| $3/4$ | $1 15/32$ | $13/16$ | 0.122 | 0.177 | $21/32$ |
| $7/8$ | $1 3/4$ | $15/16$ | 0.136 | 0.177 | $25/32$ |
| 1 | 2 | $1 1/8$ | 0.136 | 0.177 | $7/8$ |
| $1 1/8$ | $2 1/4$ | $1 1/4$ | 0.136 | 0.177 | 1 |
| $1 1/4$ | $2 1/2$ | $1 3/8$ | 0.136 | 0.177 | $1 3/32$ |
| $1 3/8$ | $2 3/4$ | $1 1/2$ | 0.136 | 0.177 | $1 7/32$ |
| $1 1/2$ | 3 | $1 5/8$ | 0.136 | 0.177 | $1 5/16$ |
| $1 3/4$ | $3 3/8$ | $1 7/8$ | 0.178[3] | 0.28[3] | $1 17/32$ |
| 2 | $3 3/4$ | $2 1/8$ | 0.178[3] | 0.28[3] | $1 3/4$ |
| $2 1/4$ | 4 | $2 3/8$ | 0.24[4] | 0.34[4] | 2 |
| $2 1/2$ | $4 1/2$ | $2 5/8$ | 0.24[4] | 0.34[4] | $2 3/16$ |
| $2 3/4$ | 5 | $2 7/8$ | 0.24[4] | 0.34[4] | $2 13/32$ |
| 3 | $5 1/2$ | $3 1/8$ | 0.24[4] | 0.34[4] | $2 5/8$ |
| $3 1/4$ | 6 | $3 3/8$ | 0.24[4] | 0.34[4] | $2 7/8$ |
| $3 1/2$ | $6 1/2$ | $3 5/8$ | 0.24[4] | 0.34[4] | $3 1/16$ |
| $3 3/4$ | 7 | $3 7/8$ | 0.24[4] | 0.34[4] | $3 5/16$ |
| 4 | $7 1/2$ | $4 1/8$ | 0.24[4] | 0.34[4] | $3 1/2$ |

[1] Tolerances are as noted in Table 22-IV-N.
[2] Clipped edge, $E$, shall not be closer than $7/8$ of the bolt diameter from the center of the washer.
[3] $3/16$ inch (4.8 mm) nominal.
[4] $1/4$ inch (6.4 mm) nominal.

*(Continued)*

## Bolts, Installation of *(Continued)*

### Nuts for A325 and A490 Bolts

| A325 Bolts | Grade of Nuts | | | | | | |
|---|---|---|---|---|---|---|---|
| | A563 | | | | | A194 | |
| Type 1 Plain | C | C3 | D | DH | DH3 | 2 | 2H |
| Type 1 Galvanized | | | | DH[g] | | | 2H[g] |
| Type 2 Discontinued | | | | | | | |
| Type 3 Plain | | C3 | | | DH3 | | |

| A490 Bolts | Grade of Nuts | | | | |
|---|---|---|---|---|---|
| | A563 | | | | A194 |
| Type 1 Plain | | | DH | DH3 | 2H |
| Type 2 Plain | | | DH | DH3 | 2H |
| Type 3 Plain | | | | DH3 | |

### TABLE 22-IV-M—HARDENED BEVELED WASHERS[1]

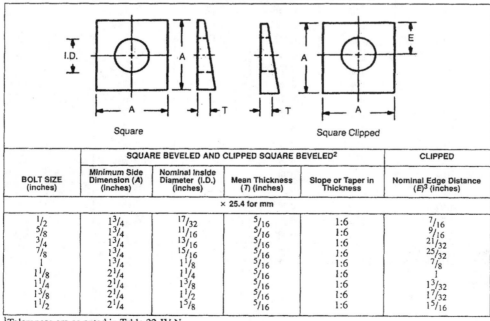

Square                     Square Clipped

| BOLT SIZE (inches) | SQUARE BEVELED AND CLIPPED SQUARE BEVELED[2] | | | | CLIPPED |
|---|---|---|---|---|---|
| | Minimum Side Dimension (A) (inches) | Nominal Inside Diameter (I.D.) (inches) | Mean Thickness (T) (inches) | Slope or Taper in Thickness | Nominal Edge Distance (E)[3] (inches) |
| | | | × 25.4 for mm | | |
| $1/2$ | $1^3/4$ | $17/32$ | $5/16$ | 1:6 | $7/16$ |
| $5/8$ | $1^3/4$ | $11/16$ | $5/16$ | 1:6 | $9/16$ |
| $3/4$ | $1^3/4$ | $13/16$ | $5/16$ | 1:6 | $21/32$ |
| $7/8$ | $1^3/4$ | $15/16$ | $5/16$ | 1:6 | $25/32$ |
| 1 | $1^3/4$ | $1^1/8$ | $5/16$ | 1:6 | $7/8$ |
| $1^1/8$ | $2^1/4$ | $1^1/4$ | $5/16$ | 1:6 | 1 |
| $1^1/4$ | $2^1/4$ | $1^3/8$ | $5/16$ | 1:6 | $1^3/32$ |
| $1^3/8$ | $2^1/4$ | $1^1/2$ | $5/16$ | 1:6 | $1^7/32$ |
| $1^1/2$ | $2^1/4$ | $1^5/8$ | $5/16$ | 1:6 | $1^5/16$ |

[1]Tolerances are as noted in Table 22-IV-N.
[2]Rectangular beveled washers shall conform to the dimensions shown above, except that one side may be longer than that shown for the A dimension.
[3]Clipped edge E shall not be closer than $7/8$ of the bolt diameter from the center of the washer.

## TABLE 22-IV-D—FASTENER TENSION REQUIRED FOR SLIP-CRITICAL CONNECTIONS AND CONNECTIONS SUBJECT TO DIRECT TENSION

| NOMINAL BOLT SIZE, INCHES | MINIMUM TENSION[1] IN 1000's OF POUNDS (kips) | |
|---|---|---|
| | × 4448 for N | |
| × 25.4 for mm | A 325 Bolts | A 490 Bolts |
| $1/2$ | 12 | 15 |
| $5/8$ | 19 | 24 |
| $3/4$ | 28 | 35 |
| $7/8$ | 39 | 49 |
| 1 | 51 | 64 |
| $1 1/8$ | 56 | 80 |
| $1 1/4$ | 71 | 102 |
| $1 3/8$ | 85 | 121 |
| $1 1/2$ | 103 | 148 |

[1]Equal to 70 percent of specified minimum tensile strength of bolts (as specified for tests of full-size A 325 and A 490 bolts with UNC threads loaded in axial tension) rounded to the nearest kip.

## TABLE 22-IV-E—NUT ROTATION FROM SNUG-TIGHT CONDITION[1,2]

| BOLT LENGTH (Underside of head to end of bolt) | DISPOSITION OF OUTER FACE OF BOLTED PARTS | | |
|---|---|---|---|
| | Both faces normal to bolt axis | One face normal to bolt axis and other sloped not more than 1:20 (beveled washer not used) | Both faces sloped not more than 1:20 from normal to the bolt axis (beveled washer not used) |
| Up to and including 4 diameters | $1/3$ turn | $1/2$ turn | $2/3$ turn |
| Over 4 diameters but not exceeding 8 diameters | $1/2$ turn | $2/3$ turn | $5/6$ turn |
| Over 8 diameters but not exceeding 12 diameters[3] | $2/3$ turn | $5/6$ turn | 1 turn |

[1]For bolts installed by $1/2$ turn and less, the tolerance shall be plus or minus 30 degrees; for bolts installed by $2/3$ turn and more, the tolerance shall be plus or minus 45 degrees.
[2]Applicable only to connections in which all material within the grip of the bolt is steel.
[3]The rotation shall be determined by actual test in a suitable tension measuring device which simulates conditions of solidly fitted steel.

Reuse

UBC, . . . A490 and galvanized A325 bolts shall not be reused. Other A325 bolts may be reused if approved by the building official. Touching up or retightening previously tightened bolts that may have been loosened by tightening of adjacent bolts shall not be considered as reuse provided the snugging up continues from the initial position and does not require greater rotation, including the tolerance, than that required by Table 22-IV-E.

UBC, 2-451, 2227.5

AISC, 5-276, para. 8 (e)
AISC, 5-276, (e)

Tension Table

UBC, 2-454, Table 22-IV-D
UBC, 2-591, Table J3.1

Tightening (see *Tightening,* Section B)

UBC, 2-450, 2227.4.1/4.2
UBC, 2-448, 2226.3
UBC, 2-451, 2227.4.5
and 2227.5

*(Continued)*

## Box Beams

UBC, 2-678, F6

## Boxing

D1.1, 8.8.6

D1.1, . . . Side or end fillet welds terminating at
ends or sides of header angles, brackets, beam
seats, and similar connections shall be returned
continuously around the corners for a distance at
least twice the nominal size of the weld, except as
provided in 8.8.5. . . .

*Fillet* welds deposited on opposite sides of a common plane of contact between two parts shall be interrupted at a corner common to both welds (see
Figure 8.2).

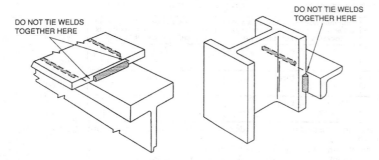

Figure 8.2 — Fillet Welds on Opposite Side of a Common Plane of Contact (see 8.8.5)

## Built-Up Members

UBC, 2-678, F6
AISC, 5-50, F6

## Buttering

D1.1, 3.3.4.1
D1.1, Comm., 3.3.4.1

D1.1, . . . Root openings greater than those permitted in 3.3.4, but not greater than twice the
thickness of the thinner part or $3/4$ in., whichever is
less, may be corrected by welding to acceptable dimensions prior to joining the parts by welding.

## Butt Joints

D1.1, 3.6.3

D1.1, . . . Surfaces of butt joint welds required to
be flush shall be finished so as not to reduce the
thickness of the thinner base metal or weld metal
by more than $1/32$ in. or 5% of the thickness,
whichever is smaller, nor leave reinforcement that
exceeds $1/32$ in. However, all reinforcement must be
removed where the weld forms part of a faying or
contact surface. Any reinforcement must blend
smoothly into the plate surfaces with transition
areas free from undercut.

Alignment of                                    D1.1, 3.3.3

D1.1, . . . Parts to be joined at butt joints shall
be carefully aligned. Where the parts are effectively restrained against bending due to eccentricity in alignment, an offset not exceeding 10% of
the thickness of the thinner part joined, but in no
case more than $1/8$ in., shall be permitted as a departure from the theoretical alignment. . . .

Variance in                                    D1.1, 3.3.4

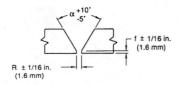

(A) GROOVE WELD WITHOUT BACKING -
    ROOT NOT BACKGOUGED

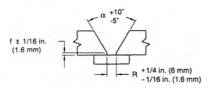

(B) GROOVE WELD WITH BACKING -
    ROOT NOT BACKGOUGED

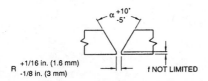

(C) GROOVE WELD WITHOUT BACKING -
    ROOT BACKGOUGED

|  | Root not back gouged* | | Root back gouged | |
|---|---|---|---|---|
|  | in. | mm | in. | mm |
| (1) Root face of joint | ±1/16 | 1.6 | Not limited | |
| (2) Root opening of joints without backing | ±1/16 | 1.6 | +1/16<br>−1/8 | 1.6<br>3 |
| Root opening of joints with backing | +1/4<br>−1/16 | 6<br>1.6 | Not applicable | |
| (3) Groove angle of joint | +10°<br>−5° | | +10°<br>−5° | |

*see 10.13.1 for tolerances for complete joint penetration
tubular groove welds made from one side without backing.

**Figure 3.3 — Workmanship Tolerances in
Assembly of Groove Welded Joints (see 3.3.4)**

## Calibrated Wrench Method

AISC, 5-305, para. 3 - 5

The calibrated wrench method of tightening, when used, should be performed with regard to any variables that affect torque. The most commonly found variables are:

UBC, 2-450, 2227.4.3

1.  The finish and tolerance of the nut and bolt threads.
2.  Different manufacturers of nuts vs. bolts.
3.  Amount of, and type of lubrication.
4.  Storage conditions at the job site.
5.  Dirt or burrs on the threads.
6.  Condition and capacity of the impact wrench.
7.  Air supply, hose diameter, etc.

*When the calibrated* wrench method of tightening is used, hard washers must be installed, fasteners must be protected from dirt, and the elements and wrenches must be calibrated daily.

## Cambering

D1.1, 3.2.8, Tables 3.2 and 3.3

D1.1, . . . Correction of errors in camber of quenched and tempered steel shall be given prior approval by the engineer. . . .

AISC, 5-87, M2.1
D1.1, 3.7.3

*Members distorted* by welding shall be straightened by mechanical means or by application of a limited amount of localized heat. The temperature of heated areas as measured by approved methods shall not exceed 1100 deg. F (590 deg. C) for quenched and tempered steel nor 1200 deg. F (650 deg. C) for other steels. The part to be heated for straightening shall be substantially free of stress and from external forces, except those forces resulting from the mechanical straightening method used in conjunction with the application of heat.

## Caulking

D1.1, 3.9

(Plastic deformation of weld and base metal surfaces by mechanical means to seal or obscure discontinuities)

D1.1, . . . Caulking of welds shall not be permitted.

## Cold Formed Steel Construction

SBCCI, 1503

## Column Base Plate Finishing

UBC, 2-707, M2.8

UBC, . . . Column bases and base plates shall be finished in accordance with the following requirements:

UBC, 2-605, M2.2.6
AISC, 3-111

1. Rolled steel bearing plates 2 in. or less in thickness are permitted without milling, provided a satisfactory contact bearing is obtained; rolled steel bearing plates over 2 in. but not over 4 in. in thickness may be straightened by pressing, or if presses are not available, by milling for all bearing surfaces (except as noted in Subparagraphs 3 and 4 of this section), to obtain a satisfactory contact bearing; rolled steel bearing plates over 4 in. thick shall be milled for all bearing surfaces (except as noted. . . .).

2. Column bases other than rolled steel bearing plates shall be milled for all bearing surfaces (except as noted. . . .)

3. The bottom surfaces of bearing plates and column bases that are grouted to ensure full bearing contact on foundations need not be milled.

4. The top surfaces of base plates with columns full-penetration welded need not be pressed or milled.

## Column Compression Joints

AISC, 5-90, M4.4

A gap of $1/16$ in. is allowed in column compression joints. If the gap is greater than 1/16 in. but is less than $1/4$ in., and with engineering approval, the gap can be packed with non-tapered steel shims. These shims may be mild steel, regardless of the grade of the main member.

## Columns (Determining Plumb)

AISC, 5-238, 7.11.3.1

Subject to the limitations set forth in the specification, individual columns are considered plumb if the working line does not deviate from the plumb line by more than 1 in. in 500 in. or about 1 in. in 41 ft.

## Column Splices

AISC, 4-132
UBC, 2-360, 2211.5.2

## Complete Joint Penetration Required

D1.1, 2.1.3.1

D1.1, . . . The welding symbol without dimensions designates a complete joint penetration weld. . . .

A2.4, 9.2.2

## Composite Beams with Formed Steel Deck

UBC, 2-686, I5

UBC, . . .

AISC, 5-60, I5 - I5.3

1. Section I5 is applicable to decks with nominal rib height not greater than 3 in.

AISC, 5-60, I5
AISC, 5-159, Fig. C-15.1

2. The average width of concrete rib or haunch shall not be less than 2 in. . . .

3. The concrete slab shall be connected to the steel beam or girder with welded stud shear connectors $3/4$ in. or less in diameter. Studs may be welded through the deck or directly to the steel member.

4. Stud shear connectors shall extend not less than $1\frac{1}{2}$ in. above the top of the steel deck after installation.

5. The slab thickness above the steel deck shall not be less than 2 in.

*The maximum* center-to-center spacing of shear connectors on a beam or girder shall not be greater than 36 in.

*Decking shall* be anchored, by welds or shear connectors, at a maximum center-to-center spacing of 16 in. Welds and studs are used as deck anchors to help resist uplifting of the deck.

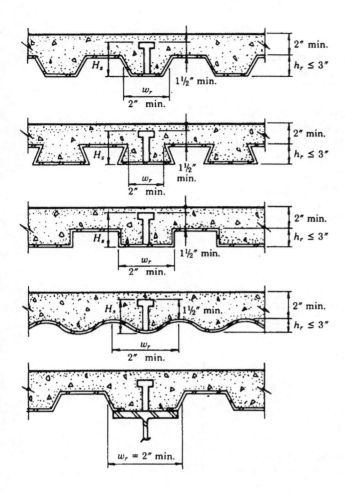

## Composite Construction

UBC, 2-682, Chapter I

UBC, . . . This chapter applies to steel beams supporting a reinforced concrete slab so interconnected that the beams and the slab act together to resist bending.

AISC, 5-155, J1
SBCCI, 1511

## Compression Members

AISC, 5-43, E4

D1.1, . . . The maximum longitudinal spacing of stitch welds connecting two or more rolled shapes in contact with one another shall not exceed 24 in.

AISC, 5-62, J1.4
UBC, 2-687, J1.4
UBC, 2-708, M4.4

*(Continued)*

Compression Members *(Continued)*

*Connections* or splices of tension or compression members made by groove welds shall have complete joint penetration welds. . . .

UBC, 2-672, E4
D1.1, 9.17

UBC, . . . Lack of contact bearing not exceeding a gap of $^1/_{16}$ in., regardless of the type of splice used (partial-penetration, groove-welded, or bolted), shall be acceptable. If the gap exceeds $^1/_{16}$ in., but is less than $^1/_4$ in. and if an engineering investigation shows sufficient contact area does not exist, the gap shall be packed with non-tapered steel shims. Shims may be of mild steel, regardless of the grade of the main material.

D1.1, 8.12.1

# Concavity (Excessive)

D1.1, 3.7.2.2

D1.1, . . . The surfaces shall be prepared and additional weld metal deposited.

D1.1, Fig. 3.4

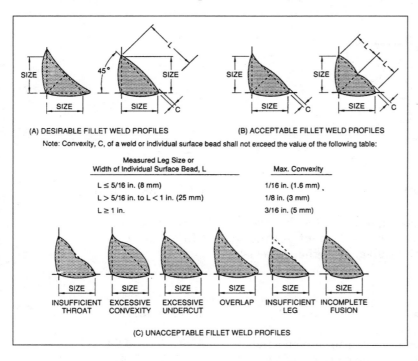

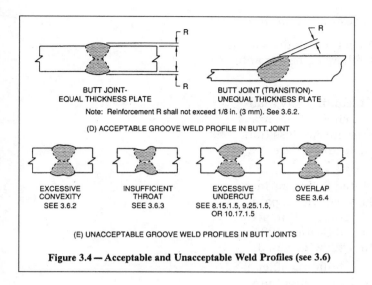

BUTT JOINT-
EQUAL THICKNESS PLATE

BUTT JOINT (TRANSITION)-
UNEQUAL THICKNESS PLATE

Note: Reinforcement R shall not exceed 1/8 in. (3 mm). See 3.6.2.

(D) ACCEPTABLE GROOVE WELD PROFILE IN BUTT JOINT

| EXCESSIVE CONVEXITY SEE 3.6.2 | INSUFFICIENT THROAT SEE 3.6.3 | EXCESSIVE UNDERCUT SEE 8.15.1.5, 9.25.1.5, OR 10.17.1.5 | OVERLAP SEE 3.6.4 |

(E) UNACCEPTABLE GROOVE WELD PROFILES IN BUTT JOINTS

**Figure 3.4 — Acceptable and Unacceptable Weld Profiles (see 3.6)**

## Connections (Stiffeners, Bolts, and Welds)

AISC, 5-96, N8

## Connection Slip

AISC, 5-297, para.1

## Connection Strength

UBC, 2-361, 2211.7.1.2

UBC, . . . The girder-to-column connection may be considered to be adequate to develop the flexural strength of the girder if it conforms to the following:

1. The flanges have full penetration butt welds to the columns.

2. The girder web-to-column connection shall be capable of resisting the girder shear determined for the combination of gravity loads and the seismic shear forces that result from compliance with Section 2211.7.2.1. . . .

## Contact Surfaces (Paint in, Classes of)

AISC, 5-89, M3.3

UBC, . . . Paint is permitted on the faying surfaces unconditionally in connections, except in slip-critical connections as defined in Section 2224.1.

UBC, 2-446, 2222.2

*(Continued)*

Contact Surfaces (Paint in, Classes of) *(Continued)*

The faying surfaces of slip-critical connections shall meet the requirements of the following paragraphs, as applicable:

1. In noncoated joints, paint, including any inadvertent overspray, shall be excluded from areas closer than one bolt diameter but not less than 1 in. from the edge of any hole and all areas within the bolt pattern.

2. Joints specified to have painted faying surfaces shall be blast cleaned and coated with a paint that has been qualified as Class A or B in accordance with the requirements of Section 2221.7, except as provided in Section 2222.2, Item 3.

3. Subject to the approval of the building official, coatings providing a slip coefficient less than 0.33 may be used, provided. . . .

4. Coated joints shall not be assembled before the coatings have cured for the minimum time used in the qualifying test.

5. Galvanized faying surfaces shall be hot dip galvanized in accordance with Table 22-IV-O and shall be roughened by means of hand wire brushing. Power wire brushing is not permitted.

## Continuity Plates                                        UBC, 2-362, 2211.7.4

## Contractor Compliance                                    D1.1, 6.6.2

D1.1, . . . The contractor shall comply with all requests of the inspector(s) to correct deficiencies in materials and workmanship as provided in the contract documents. . . .

## Control of Shrinkage                                      D1.1, 3.4.1

D1.1, . . . In assembling and joining parts of a       D1.1, 3.4.5
structure or of built-up members and in welding re-    D1.1 Comm., 3.3.4.1
inforcing parts to members, the procedure and sequence shall be such as will minimize distortion and shrinkage. . . .

*In assemblies,* joints expected to have significant shrinkage should usually be welded before joints expected to have lesser shrinkage. They should also be welded with as little restraint as possible.

## Convexity (Excessive)

D1.1, 3.7.2.1

D1.1, . . . The contractor has the option of either repairing an unacceptable weld or removing and replacing the entire weld, except as modified by 3.7.4. . . .

D1.1, Fig., 3.4

*Excessive* weld metal shall be removed.

## Copes and Blocks

AISC, 4-175

D1.1, . . . Radii of beam copes and weld access holes shall provide a smooth transition free of notches or cutting past the points of tangency between adjacent surfaces. . . .

D1.1, 3.2.5

*All weld access* holes required to facilitate welding operations shall have a length from the toe of the weld preparation, not less than $1\frac{1}{2}$ times the thickness of the material in which the hole is made. The height of the access hole shall be adequate for deposition of sound weld metal in the adjacent plates and provide clearance for weld tabs for the weld in the material in which the hole is made, but not less than the thickness of the material.

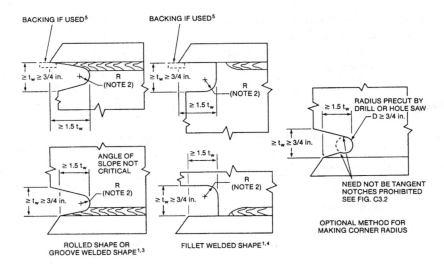

Notes:
1. For ASTM A6 Group 4 and 5 shapes and welded built-up shapes with web thickness more than 1-1/2 in. (38.1 mm), preheat to 150°F (66°C) prior to thermal cutting, grind and inspect thermally cut edges of access hole using magnetic particle or dye penetration methods prior to making web and flange splice groove welds.
2. Radius shall provide smooth notch-free transition; R ≥ 3/8 in. (9 mm) [Typical 1/2 in. (13 mm)].
3. Access opening made after welding web to flange.
4. Access opening made before welding web to flange. Weld not returned through opening.
5. These are typical details for joints welded from one side against steel backing. Alternative joint designs should be considered.

**Figure 3.2 — Weld Access Hole Geometry (see 3.2.2, 3.2.2.1, 3.2.5 and 3.2.5.1)**

# Correction of Errors

D1.1, 3.7

D1.1, . . . Oxyfuel gas gouging shall not be used in quenched and tempered steel. Unacceptable portions of the weld shall be removed without substantial removal of the base metal. . . .

D1.1 Comm., 3.7

*The contractor* has the option of either repairing an unacceptable weld or removing and replacing the entire weld, except as modified by 3.7.4. . . .

*The crack* and sound metal 2 in. beyond each end of the crack shall be removed and rewelded. . . .

*The temperature* of heated areas as measured by approved methods shall not exceed 1100 deg. F for quenched and tempered steel nor 1200 deg. F for other steels. . . .

*Prior approval* of the engineer shall be obtained for repairs to the base metal (other than those required by 3.2) and repair of major or delayed cracks. . . .

*The engineer* shall be notified before improperly fitted and welded members are cut apart. . . .

## Correction of Welds

D1.1 Comm., 3.7.2

D1.1 Comm . . . The code permits the contractors, at their option, to either repair or remove and replace an unacceptable weld. It is not the intent of the code to give the Inspector authority to specify the mode of correction.

## Cover Plates

AISC, 5-37, B10

The cross sectional area of a cover plate on bolted or riveted girders shall not be greater than 70% of the flange area.

D1.1, 9.21.6 - 9.21.6.3
UBC, 2-669,  B10

D1.1, . . . Cover plates shall preferably be limited to one on any flange. The maximum thickness of cover plates on a flange (total thickness of all cover plates if more than one is used) shall not be greater than $1\frac{1}{2}$ times the thickness of the flange to which the cover plate is attached. . . .

## Cracks

D1.1, 3.7.2.4

D1.1, . . . The crack and sound metal 2 in. beyond each end of the crack shall be removed and rewelded.

Development of

D1.1 Comm., 4.7.7

D1.1 Comm . . . Solidification of molten weld metal due to the quenching effect of the base metal starts along the sides of the weld metal and progresses inward until completed. The last liquid metal to solidify lies in a plane through the centerline of the weld. If the weld depth is greater than the width of the face, the weld surface may solidify prior to center solidification. When this occurs, the shrinkage forces acting on the still hot, semi-liquid center or core of the weld may cause a centerline crack to develop,. . . .

## Curving

AISC, 5-87, M2.1

D1.1, . . . *Members distorted* by welding shall be
straightened by mechanical means or by application of a limited amount of localized heat. The temperature of heated areas as measured by approved
methods shall not exceed 1100 deg. F (590 deg. C)
for quenched and tempered steel nor 1200 deg. F
(650 deg. C) for other steels. The part to be heated
for straightening shall be substantially free of
stress and from external forces, except those forces
resulting from the mechanical straightening
method used in conjunction with the application of
heat.

D1.1, 3.7.3

## Cut Edge Discontinuities

D1.1, Table 3.1

### Table 3.1
### Limits on Acceptability and Repair of Mill Induced
### Laminar Discontinuities in Cut Surfaces (see 3.2.3)

| Description of Discontinuity | Repair Required |
| --- | --- |
| Any discontinuity 1 in. (25 mm) in length or less | None, need not be explored. |
| Any discontinuity over 1 in. (25 mm) in length and 1/8 in. (3 mm) maximum depth | None, but the depth should be explored.* |
| Any discontinuity over 1 in. (25 mm) in length with depth over 1/8 in. (3 mm) but not greater then 1/4 in. (6 mm) | Remove, need not weld. |
| Any discontinuity over 1 in. (25 mm) in length with depth over 1/4 in. (6 mm) but not greater than 1 in. | Completely remove and weld. |
| Any discontinuity over 1 in. (25 mm) in length with depth greater than 1 in. | See 3.2.3.2. |

*A spot check of 10% of the discontinuities on the cut surface in question should be explored by grinding to determine depth. If the depth of any one of the discontinuities explored exceeds 1/8 in. (3 mm), then all of the discontinuities over 1 in. (25 mm) in length remaining on that cut surface shall be explored by grinding to determine depth. If none of the discontinuities explored in the 10% spot check have a depth exceeding 1/8 in. (3 mm), then the remainder of the discontinuities on that cut surface need not be explored.

## Cutting (Thermal)

AISC, 5-87, M2.2

D1.1, . . . Roughness exceeding these values and
notches or gouges not more than 3/16 in. deep on otherwise satisfactory surfaces shall be removed by
machining or grinding. Notches or gouges exceeding
3/16 in. deep may be repaired by grinding if the nom-

D1.1, 3.2.2.1

inal cross-sectional area is not reduced by more than 2%. Ground or machined surfaces shall be faired to the original surface with a slope not exceeding 1:10. Cut surfaces and adjacent edges shall be left free of slag. In thermal cut surfaces, occasional notches or gouges may, with the approval of the engineer, be repaired by welding.

*Reentrant corners* of cut material shall be formed to provide a gradual transition with a radius of not less than 1 in. Adjacent surfaces shall meet without offset or cutting past the point of tangency. The reentrant corners may be formed by thermal cutting, followed by grinding, if necessary, to meet the surface requirements of 3.2.2.

D1.1, 3.2.4

*Radii of* beam copes and weld access holes shall provide a smooth transition free of notches or cutting past the points of tangency between adjacent surfaces. . . .

D1.1, 3.2.5

*All weld* access holes required to facilitate welding operations shall have a length from the toe of the weld preparation, not less than 1½ times the thickness of the material in which the hole is made. . . .

D1.1, 3.2.5.1

*Do not* return fillet welds through weld access holes. (See Figure 3.2.)

## Decking

AISC, 5-158, I5

Composite construction shall, where applicable, be limited to decking having a 3 in. maximum rib height and an average rib height of not less than 2 in.

AISC, 5-60, I5
AISC, 2-246
AISC, 2-255

*Welded shear* connectors shall be ¾ in. or less in diameter.

*After installation* by welding, shear connectors shall extend a minimum of 1 ½ in. above the top of the steel deck.

*The top* of the composite slab shall be a minimum of 2 in. above the top of the decking.

*The maximum* center-to-center spacing of shear connectors on a beam or girder shall not be greater than 36 in.

*(Continued)*

Decking *(Continued)*

*Decking shall* be anchored, by welds or shear connectors, at a maximum center-to-center spacing of 16 in.

*Welds and* studs are used as deck anchors to help resist uplifting of the deck.

*Stud welds* not located directly over the web of a beam tend to tear out of a thin flange before attaining their full shear-resistance capacity. To guard against this contingency, the size of a stud not located over the beam web is limited to 2 $\frac{1}{2}$ times the flange thickness.

## Deck Ribs

AISC, 5-60, I5.1 - I5.3

Composite construction shall, where applicable, be limited to decking having a 3 in. maximum rib height and an average rib height of not less than 2 in.

*Welded shear* connectors shall be $\frac{3}{4}$ in. or less in diameter.

*Studs* may be welded through the deck or directly to the steel member. . . .

*After installation* by welding, shear connectors shall extend a minimum of 1 $\frac{1}{2}$ in. above the top of the steel deck.

*The top* of the composite slab shall be a minimum of 2 in. above the top of the decking.

*The maximum* center-to-center spacing of shear connectors on a beam or girder shall not be greater than 36 in.

*Decking shall* be anchored, by welds or shear connectors, at a maximum center-to-center spacing of 16 in. Welds and studs are used as deck anchors to help resist uplifting of the deck.

Formed Steel Decking with Studs

AISC, 2-246

  *Composite construction* shall, where applicable, be limited to decking having a 3 in. maximum rib height and an average rib height of not less than 2 in.

AISC, 2-255

*Welded shear* connectors shall be $^3/_4$ in. or less in diameter.

*After installation* by welding, shear connectors shall extend a minimum of 1 $^1/_2$ in. above the top of the steel deck.

*The top* of the composite slab shall be a minimum of 2 in. above the top of the decking.

Parallel

AISC, 5-60, I5.3

The maximum diameter of a welded shear connector shall be 2 $^1/_2$ times the thickness of the flange to which it is welded. Larger studs may be used if welded directly over the web of the member.

AISC, 2-246 and 2-255
UBC, 2-687, I5.3

Perpendicular

AISC, 5-60, I5.2
AISC, 2-246 and 2-255
UBC, 2-686, I5.2
UBC, 2-358, 2211.2
D1.1 App. B
A3.0
D1.1, 3.7.6

## Definition of Terms

## Design Revisions

D1.1, . . . If, after an unacceptable weld has been made, work has been performed that has rendered that weld inaccessible or has created new conditions that make correction of the unacceptable weld dangerous or ineffectual, then the original conditions shall be restored by removing welds or members, or both, before the corrections are made. If this is not done, the deficiency shall be compensated for by additional work performed according to an approved revised design.

## Development of Cracks *(See Cracks)*

D1.1 Comm., 4.7.7

## Dew Point

D1.1, 4.13

D1.1, . . . A gas or gas mixture used for shielding in gas metal arc welding or flux cored arc welding shall be of a welding grade having a dew point of −40 deg. F or lower.

## Diameter of Electrodes (Maximum)

S.M.A.W.                                                 D1.1, 4.6.3

D1.1, . . . The maximum diameter of electrodes
shall be:

1. $5/16$ in. for all welds made in the flat position,
   except root passes.
2. $1/4$ in. for horizontal fillet welds.
3. $1/4$ in. for root passes of fillet welds made in
   the flat position and groove welds made in the
   flat position with backing and with a root
   opening of 1/4 in. or more.
4. $5/32$ in. for welds made with EXX14 and low
   hydrogen electrodes in the vertical and overhead positions.
5. $3/16$ in. for root passes of groove welds and for
   all other welds not included under 4.6.3(1),
   (2), (3), and (4).

G.M.A.W.                                                 D1.1, 4.14.1.2

D1.1, . . . The maximum diameter of welding
electrodes shall be:

1. $5/32$ in. for the flat and horizontal positions.
2. $3/32$ in. for the vertical position.
3. $5/64$ in. for the overhead position.

F.C.A.W.                                                 D1.1, 4.14.1.2

D1.1, . . . The maximum diameter of welding
electrodes shall be:

1. $5/32$ in. for the flat and horizontal positions.
2. $3/32$ in. for the vertical position.
3. $5/64$ in. for the overhead position.

S.A.W.                                                   D1.1, 4.7.3

D1.1, . . . The diameter of electrodes shall not
exceed $1/4$ in.

## Dimensions and Properties                    AISC, 1-9

## Distortion                                    D1.1, 3.4

D1.1, . . . In assembling and joining parts of a
structure or of built-up members and in welding reinforcing parts to members, the procedure and sequence shall be such as will minimize distortion
and shrinkage.

## Doubler Plates                    UBC, 2-362, 2211.7.2.3

UBC, . . . Doubler plates provided to reduce panel
zone shear stress or to reduce the web depth thickness ratio shall be placed not more than $1/16$ in. from
the column web and shall be welded across the
plate width top and bottom with at least a
$3/16$ in. fillet weld. They shall be either butt or
fillet welded to the column flanges to develop
the shear strength of the doubler plate. . . .

## Dryness of Ferrules                          D1.1, 7.4.4

D1.1, . . . The arc shields or ferrules shall be kept
dry. Any arc shields that show signs of surface
moisture from dew or rain shall be oven dried at
250 deg. F for 2 hours before use.

## Dye Penetrant                                D1.1, 6.7.7

D1.1, . . . For detecting discontinuities that are
open to the surface, dye penetrant inspection may
be used. . . .

## Edge Distance

Maximum                                UBC, 2-699, J3.10

When fasteners are used to connect a plate with     AISC, 5-76, J3.10
a shape, or two plates together, the center-to-
center spacing shall not exceed a maximum of
14 times the thickness of the thinner part, nor
7 in.

*(Continued)*

Edge Distance Maximum *(Continued)*

> *For unpainted* built-up members made of weather resistant steel, the maximum edge distance shall not be greater than 8 times the thickness of the thinner part, or 5 in.

> *The maximum* edge distance as measured from the center of a bolt hole to the nearest edge of the connection shall not be greater than 12 times the thickness of the connected part and shall not be greater than 6 in.

Minimum

> The minimum distance from the center of a standard size bolt hole to the edge of a connected part shall be as indicated in the following table:

UBC, 2-698, J3.9
UBC, 2-699, Table J3.5
AISC, 5-75, J3.9
AISC, 5-76, Table J3.5

### Minimum Edge Distance

| Bolt Diameter in inches | From a sheared Edge | From a Rolled Edge |
|---|---|---|
| 1/2 | 7/8 | 3/4 |
| 5/8 | 1-1/8 | 7/8 |
| 3/4 | 1-1/4 | 1 |
| 7/8 | 1-1/2 | 1-1/8 |
| 1 | 1-3/4 | 1-1/4 |
| 1-1/8 | 2 | 1-1/2 |
| 1-1/4 | 2-1/4 | 1-5/8 |
| 1-1/2 | 2-5/8 | 1-7/8 |

Edge distances are measured from the edge of the connected material to the centerline of the hole.

## Effective Areas of Welds

UBC, 2-690, J2.2a
AISC, 5-65, J2.1

Fillets

D1.1, 2.3.2 - 2.3.2.4

> D1.1, . . . The effective area of fillet welds shall be taken as the effective length times the effective throat thickness.

Grooves

D1.1, 2.3.1

> D1.1, . . . The effective area shall be the effective weld length multiplied by the weld size.

Plug and Slots

D1.1, 2.3.3

> D1.1, . . . The effective area shall be the nominal area of the hole or slot in the plane of the faying surface.

AISC, 5-68, J2.3.b

# Effective Length of Welds

UBC, 2-689, J2.1
UBC, 2-690, J2.2

Fillet

D1.1, 2.3.2.1 - 2.3.2.3

D1.1, . . . The effective length of a fillet weld shall be the overall length of the full size fillet, including boxing. No reduction in effective length shall be made for either the start or crater of the weld if the weld is full size throughout its length. . . .

*The effective* length of a curved fillet weld shall be measured along the centerline of the effective throat. . . .

*The minimum* effective length of a fillet weld shall be at least four times the nominal size, or the size of the weld shall be considered not to exceed 25% of its effective length.

Groove

D1.1, 2.3.1.1

D1.1, . . . The effective weld length for any groove weld, square or skewed, shall be the width of the part joined, perpendicular to the direction of stress.

# Effective Size of Welds

UBC, 2-689, J2
UBC, 2-691, Table J2.4
D1.1, 2.3
UBC, 2-691, J2.2.b

Maximum Effective Size of Fillet Welds

UBC, . . . The maximum size of fillet welds that is permitted along edges of connected parts shall be:

UBC, 2-691, Table J2.4
D1.1, 2.7.1.2, Table 2.2
AISC, 5-67, J2.2.b
AISC, 5-65, J2

- Material less than $\frac{1}{4}$ in. thick, not greater than the thickness of the material.

- Material $\frac{1}{4}$ in. or more in thickness, not greater than the thickness of the material minus $\frac{1}{16}$ in., unless the weld is especially designated on the drawings to be built out to obtain full-throat thickness.

*(Continued)*

Effective Size of Welds, Maximum Effective Size of Fillet Welds *(Continued)*

## TABLE J2.4
## Minimum Size of Fillet Welds

| Material Thickness of Thicker Part Joined (in.) | Minimum Size of Fillet Weld[a] (in.) |
|---|---|
| × 25.4 for mm | |
| To $1/4$ inclusive | $1/8$ |
| Over $1/4$ to $1/2$ | $3/16$ |
| Over $1/2$ to $3/4$ | $1/4$ |
| Over $3/4$ | $5/16$ |

[a]Leg dimension of fillet welds. Single-pass welds must be used.

Minimum Size of Fillet Welds and Partial Penetration Welds

UBC, . . . The minimum size of fillet welds shall be shown in Table J2.4. Minimum weld size is dependent upon the thicker of the two parts joined, except that the weld size need not exceed the thickness of the thinner part. . . .

D1.1, . . . The minimum fillet weld size, except for fillet welds used to reinforce groove welds, shall be as shown in Table 2.2. . . .

Length of Fillet Welds

UBC, . . . The minimum effective length of fillet welds designed on the basis of strength shall be not less than four times the nominal size, or else the size of the weld shall be considered not to exceed $1/4$ of its effective length. . . .

Intermittent Fillet Welds

UBC, . . . The effective length of any segment of intermittent fillet welding shall not be less than four times the weld size, with a minimum of $1 1/2$ in.

D1.1, . . . The minimum length of an intermittent fillet weld shall be $1 1/2$ in.

UBC, 2-691, J2.2.b

D1.1, 2.7.1.1
AISC, J2.2b

UBC, 2-692, J2.2.b

D1.1, 2.3.2.1 - 2.3.2.3
AISC, 5-67, J2.2.a

UBC, 2-692, J2.2.b

D1.1, 2.7.1.4

# Effective Throat of Partial Penetration Welds (Minimum)

AISC, 5-66, J2.1.b

D1.1, . . . The minimum weld size of a partial joint penetration groove weld shall be as specified in Table 2.3.

AISC, Table J2.3
UBC, 2-691, Table J2.3

*The size* of the weld need not exceed the thickness of the thinner part.

D1.1, 2.10
D1.1, Table 2.3
D1.1, 2.3.1.2
D1.1, 2.3.1.3
D1.1, 2.3.1.5

### TABLE J2.3
### Minimum Effective Throat Thickness of
### Partial-penetration Groove Welds

| Material Thickness of Thicker Part Joined (in.) | Minimum Effective Throat Thickness[a] (in.) |
|---|---|
| × 25.4 for mm | |
| To $1/4$ inclusive | $1/8$ |
| Over $1/4$ to $1/2$ | $3/16$ |
| Over $1/2$ to $3/4$ | $1/4$ |
| Over $3/4$ to $1^1/2$ | $5/16$ |
| Over $1/2$ to $2^1/4$ | $3/8$ |
| Over $2^1/4$ to 6 | $1/2$ |
| Over 6 | $5/8$ |

[a]See Sect. J2.

# Electrodes

Low Hydrogen Storage

D1.1, 4.5.2

D1.1, . . . All electrodes having low hydrogen coverings conforming to ANSI/AWS A5.1 shall be purchased in hermetically sealed containers or shall be dried for at least 2 hours between 500 and 800 deg. F before they are used. Electrodes having a low hydrogen covering conforming to ANSI/AWS A5.5 shall be purchased in hermetically sealed containers or shall be dried at least 1 hour between 700 and 800 deg. F before being used. Electrodes shall be dried prior to use if the hermetically sealed container shows evidence of damage. Immediately after opening of the hermetically sealed container or removal of the electrodes from the drying ovens, electrodes shall be stored in ovens held at a temperature of at least 250 deg. F. . . .

D1.1, Table 4.6
D1.1 Comm., 4.5

*(Continued)*

Electrodes, Low Hydrogen Storage *(Continued)*

### Table 4.6
### Permissible Atmospheric Exposure of
### Low Hydrogen Electrodes (see 4.5.2.1)

| Electrode | Column A (hours) | Column B (hours) |
|---|---|---|
| **A5.1** | | |
| E70XX | 4 max | Over 4 to 10 max |
| E70XXR | 9 max | |
| E70XXHZR | 9 max | |
| E7018M | 9 max | |
| **A5.5** | | |
| E70XX-X | 4 max | Over 4 to 10 max |
| E80XX-X | 2 max | Over 2 to 10 max |
| E90XX-X | 1 max | Over 1 to 5 max |
| E100XX-X | 1/2 max | Over 1/2 to 4 max |
| E110XX-X | 1/2 max | Over 1/2 to 4 max |

Notes:
1. Column A: Electrodes exposed to atmosphere for longer periods than shown shall be redried before use.
2. Column B: Electrodes exposed to atmosphere for longer periods than those established by testing shall be redried before use.
3. Entire table: Electrodes shall be issued and held in quivers, or other small open containers. Heated containers are not mandatory.
4. The optional supplemental designator, R, designates a low hydrogen electrode which has been tested for covering moisture content after exposure to a moist environment for 9 hours and has met the maximum level permitted in ANSI/AWS A5.1-91, *Specification for Carbon Steel Electrodes for Shielded Metal Arc Welding.*

Maximum Diameter for:

S.M.A.W.                                        D1.1, 4.6.3

D1.1, . . . The maximum diameter of electrodes shall be:

1. 5/16 in. for all welds made in the flat position.

2. 1/4 in. for horizontal fillet welds.

3. 1/4 in. for root passes of fillet welds made in the flat position and groove welds made in the flat position with backing and with a root opening of 1/4 in. or more.

4. 5/32 in. for welds made with EXX14 and low hydrogen electrodes in the vertical and overhead positions.

5. $^3/_{16}$ in. for root passes of groove welds and for all other welds not included under 4.6.3(1), (2), (3), and (4).

**G.M.A.W.**                                                                   D1.1, 4.14.1.2

D1.1, . . . The maximum diameter of welding electrodes shall be:

1. $^5/_{32}$ in. for the flat and horizontal positions.
2. $^1/_2$ in. for the flat and vertical positions.
3. $^5/_{64}$ in. for the overhead position.

**F.C.A.W. (see G.M.A.W. above)**                                              D1.1, 4.14.1.2

**S.A.W.**                                                                     D1.1, 4.7.3

D1.1, . . . The diameter of electrodes shall not exceed $^1/_4$ in.

**Redrying**                                                                   D1.1, 4.5.4

D1.1, . . . Electrodes that conform to the requirements of 4.5.2 shall subsequently be redried no more than one time. Electrodes that have been wet shall not be used.

**Restrictions for A514 and A517**                                             D1.1, 4.5.3

D1.1, . . . When used for welding ASTM A514 or A517 steels, electrodes for any classification lower than E100XX-X, except for E7018M and E70XXH4R, shall be dried at least 1 hour at temperatures between 700 and 800 deg. F before being used, whether furnished in hermetically sealed containers or otherwise.

**Size and Type for Studs**                                                    D1.1, 7.5.5.6

D1.1, . . . S.M.A.W. welding shall be performed using low hydrogen electrodes $^5/_{32}$ or $^3/_{16}$ in. in diameter, except that a smaller diameter electrode may be used on studs $^7/_{16}$ in. or less in diameter for out-of-position welds.

**Size for Repair (see Maximum above)**                                        D1.1, 3.7.1

## Electrogas Welding (On Quenched and Tempered Steel)

D1.1, 4.15.2

D1.1, . . . Prior to use, the contractor shall prepare a welding procedure specification and qualify each procedure for each process to be used according to the requirements in Section 5,. . . .

*The electroslag* and electrogas processes shall not be used for welding quenched and tempered steel nor for welding dynamically loaded structural members subject to tensile stresses or reversal of stress.

## Electroslag Welding (On Quenched and Tempered Steel)

D1.1, 4.15.2

(See Electrogas Welding above)

## End Shear

UBC, 2-684, I3

## Erection

AISC, 5-90, M4.7

To protect against any stresses that might be caused by wind, dead loads, or the erection process itself, the work shall be securely bolted or welded as the work progresses.

Alignment

AISC, 5-90, M4.3

Until properly aligned, permanent bolting and/or welding of the structure is not permitted by this specification.

Fit of Column Compression Members

AISC, 5-90, M4.4

A gap of $1/16$ in. is allowed in column compression joints. If the gap is greater than $1/16$ in. but is less than $1/4$ in., and with engineering approval, the gap can be packed with non-tapered steel shims. These shims may be mild steel, regardless of the grade of the main member.

## Extension Bars

D1.1, 3.12.1 - 3.12.3

D1.1, . . . Welds shall be terminated at the end of a joint in a manner that will ensure sound welds.

D1.1, 6.10.3.1
D1.1 Comm., 3.12

Whenever necessary, this shall be done by the use of weld tabs aligned in such a manner to provide an extension of the joint preparation. . . .

*For statically* loaded structures, weld tabs need not be removed unless required by the engineer. . . .

*For dynamically* loaded structures, weld tabs shall be removed upon completion and cooling of the weld, and the ends of the weld shall be made smooth and flush with the edges of the abutting parts.

*Weld tabs* shall be removed prior to radiographic inspection unless other approved by the engineer.

AISC, 5-231, 6.3.2

## Fabrication (Conformance to)

UBC, 2-357, 2203

UBC, . . . The design, fabrication, and erection of structural steel shall be in accordance with the requirements of Division VIII (LRFD) or Division IX (ASD). Seismic design of structures, where required, shall comply with Section 2211 or 2212 for structures designed in accordance with Division IX (ASD) or approved national standards.

UBC, 2-555, A3.1
UBC, 2-706, M2
AISC, 4-175
BOCA, 201.0
D1.1, 3.2.4
D1.1 Comm., 3.2.4

## Fabrication Identification

UBC, 2-356, 2202.2

UBC, . . . Steel furnished for structural load carrying purposes shall be properly identified for conformity to the ordered grade. . . .

UBC, 2-555, A3.1
AISC, 5-92, M5.5
BOCA, 1802.5

*Steel* that is not readily identifiable as to grade from marking and test records shall be tested to determine conformity to such standards. . . .

*The fabricator* shall maintain identity of the material and shall maintain suitable procedures and records attesting that the specified grade has been furnished in conformity with the applicable standard. The fabricators' identification mark system shall be established and on record prior to fabrication. . . .

*When structural steel* is furnished to a specified minimum yield point greater than 36,000 psi, the ASTM or other specification designation shall be in-
*(Continued)*

Fabrication Identification *(Continued)*

cluded near the erection mark on each shipping assembly or important construction component over any shop coat of paint prior to shipment from the fabricator's plant. . . .

## Failure to Comply with Orders

BOCA, 3003.2

## Fatigue

UBC, 2-374, 2214

## Faying Surfaces

Gaps in

D1.1, 3.3.1

D1.1, . . . The parts to be joined by fillet welds shall be brought into as close contact as practicable. The root opening shall not exceed $3/16$ in. except in cases involving either shapes or plates 3 in. or greater in thickness if, after straightening and in assembly, the root opening cannot be closed sufficiently to meet this tolerance. In such cases, a maximum root opening of $5/16$ in. is acceptable, provided suitable backing is used. If the separation is greater than $1/16$ in., the leg of the fillet weld shall be increased by the amount of the root opening, or the contractor shall demonstrate that the required effective throat has been obtained. . . .

*The separation* between faying surfaces of plug and slot welds, and of butt joints landing on a backing, shall not exceed $1/16$ in. The use of fillers is prohibited except as specified on the drawings or as specially approved by the engineer and made in accordance with 2.4.

Paint in

AISC, 5-295

AISC, 5-293, para. 3

In slip-critical connections that have been designated by the engineer to be free of paint, even the smallest amount of overspray is prohibited and must be removed.

## Ferrules

D1.1, 7.2.2

D1.1, . . . An arc shield (ferrule) of heat resistant ceramic or other suitable material shall be furnished with each stud. . . .

D1.1, 7.4.4

*The arc shields* or ferrules shall be kept dry. Any arc shields that show signs of surface moisture from dew or rain shall be oven dried at 250 deg. F for 2 hours before use.

## Field Coat

BOCA, 1802.3.2

## Field Welds (Surfaces Adjacent to)

AISC, 5-90, M3.5

D1.1, . . . Surfaces on which weld metal is to be deposited shall be smooth, uniform, and free from fins, tears, cracks, and other discontinuities that would adversely affect the quality or strength of the weld. Surfaces to be welded, and surfaces adjacent to a weld, shall also be free from loose or thick scale, slag, rust, moisture, grease, and other foreign material that would prevent proper welding or produce objectionable fumes. Mill scale that can withstand vigorous wire brushing, a thin rust-inhibitive coating, or antispatter compound may remain with the following exception: for girders in dynamically loaded structures, all mill scale shall be removed from the surfaces on which flange-to-web welds are to be made by submerged arc welding or by shielded arc welding with low hydrogen electrodes.

D1.1, 3.2.1

## Filler Metal Requirements

AISC, 5-28, A3.6
D1.1, 4.1
SBCCI, 1507
D1.1, 4.1.6

A242 and A588

D1.1, . . . For exposed, bare, unpainted applications of ASTM A242 and A588 steel requiring weld metal with atmospheric corrosion resistance and coloring characteristics similar to that of the base metal, the electrode-flux combination shall be in accordance with 4.16, and the chemical composition shall conform to one of the filler metals listed in Table 4.2.

D1.1, Table 4.2

*(Continued)*

Filler Metal Requirements, A 242 and A 588
*(Continued)*

### Table 4.2
### Filler Metal Requirements for Exposed Bare Applications of
### ASTM A242 and A588 Steel (see 4.1.4)

| Welding Processes | | | |
|---|---|---|---|
| Shielded Metal Arc | Submerged Arc | Gas Metal Arc or Gas Tungsten Arc | Flux Cored Arc |
| A5.5 | A5.23[1,4] | A5.28[4] | A5.29 |
| E7018-W | F7AX-EXXX-W | | |
| E8018-W | | | E8XT1-W |
| E8016-C3 or E8018-C3 | F7AX-EXXX-Ni1[2] | ER80S-Ni1 | E8XTX-Ni1 |
| E8016-C1 or E8018-C1 | F7AX-EXXX-Ni4[2] | | |
| E8016-C2 or E8018-C2 | | | |
| E7016-C1L or E7018-C1L | F7AX-EXXX-Ni2[2] | ER80S-Ni2 | E8XTX-Ni2 |
| E7016-C2L or E7018-C2L | F7AX-EXXX-Ni3[2] | ER80S-Ni3 | E80T5-Ni3 |
| E8018-B2L[1] | | ER80S-B2L[1] | E80T5-B2L[1] |
| | | ER80S-G[1,3] | |
| | | | E71T8-Ni1 |
| | | | E71T8-Ni2 |
| | | | E7XTX-K2 |

A514 and A517                                      D1.1, 4.5.3

   D1.1, . . . When used for welding ASTM A514
or A517 steels, electrodes of any classification
lower than E100XX-X and E70XXH4R, shall be
dried at least 1 hour at temperatures between
700 and 800 deg. F before being used, whether
furnished in hermetically sealed containers or
otherwise.

S.M.A.W.                                           D1.1, 4.1

   D1.1, . . . Single-pass fillet welds up to $1/4$ in.    D1.1, Table 4.1
maximum and $1/4$ in. groove welds made with a
single pass or a single pass each side, may be
made by using an E70XX or E70XX-X low hydrogen electrode.

G.M.A.W.                                           D1.1, 4.1

   D1.1, . . . Single pass fillet welds up to $5/16$ in.    D1.1, Table 4.1
maximum and groove welds made with a single
pass or a single pass each side, may be made using an ER70S-X electrode.

| | |
|---|---|
| F.C.A.W. | D1.1, 4.1 |
| D1.1, . . . Single-pass fillet welds up to $^5/_{16}$ in. maximum and groove welds made with a single pass each side, may be made using an E70TX-X electrode. | D1.1, Table 4.1 |
| S.A.W. | D1.1, 4.1 |
| D1.1, . . . Single-pass fillet welds up to $^5/_{16}$ in. maximum and groove welds made with a single pass or a single pass each side, may be made using an F7X-EXXX or F7X-EXX-XX electrode-flux combination. | D1.1, Table 4.1 |

See turned table on pages 128 through 131.

Reinforcing Steel

D1.4, . . . When joining different grades of steels, the filler metal shall be selected for the lower strength base metal.

See turned table on pages 132 through 133.

*(Continued)*

## Table 4.1
## Matching Filler Metal Requirements (see 4.1.1)

### Steel Specification Requirements

| Steel Specification[1,2] | | Minimum Yield Point/Strength (ksi) | (MPa) | Tensile Range (ksi) | (MPa) |
|---|---|---|---|---|---|
| ASTM A36[5] | | 36 | 250 | 58-80 | 400-550 |
| ASTM A53 | Grade B | 35 | 240 | 60 min | 415 min |
| ASTM A106 | Grade B | 35 | 240 | 60 min | 415 min |
| ASTM A131 | Grades A, B, CS, D, DS, E | 34 | 235 | 58-71 | 400-490 |
| ASTM A139 | Grade B | 35 | 241 | 60 min | 414 min |
| ASTM A381 | Grade Y35 | 35 | 240 | 60 min | 415 min |
| ASTM A500 | Grade A | 33 | 228 | 45 min | 310 min |
| | Grade B | 42 | 290 | 58 min | 400 min |
| ASTM A501 | | 36 | 250 | 58 min | 400 min |
| ASTM A516 | Grade 55 | 30 | 205 | 55-75 | 380-515 |
| | Grade 60 | 32 | 220 | 60-80 | 415-550 |
| ASTM A524 | Grade I | 35 | 240 | 60-85 | 415-586 |
| | Grade II | 30 | 205 | 55-80 | 380-550 |
| ASTM A529 | | 42 | 290 | 60-85 | 415-585 |
| ASTM A570 | Grade 30 | 30 | 205 | 49 min | 340 min |
| | Grade 33 | 33 | 230 | 52 min | 360 min |
| | Grade 36 | 36 | 250 | 53 min | 365 min |
| | Grade 40 | 40 | 275 | 55 min | 380 min |
| | Grade 45 | 45 | 310 | 60 min | 415 min |
| | Grade 50 | 50 | 345 | 65 min | 450 min |
| ASTM A573 | Grade 65 | 35 | 240 | 65-77 | 450-530 |
| | Grade 58 | 32 | 220 | 58-71 | 400-490 |
| ASTM A709 | Grade 36[5] | 36 | 250 | 58-80 | 400-550 |
| API 5L | Grade B | 35 | 240 | 60 | 415 |
| | Grade X42 | 42 | 290 | 60 | 415 |
| ABS | Grades A, B, D, CS, DS | | | 58-71 | 400-490 |
| | Grade E[6] | | | 58-71 | 400-490 |

(Group **I**)

### Filler Metal Requirements

| Electrode Specification[3,4] | Minimum Yield Point/Strength (ksi) | (MPa) | Tensile Strength Range (ksi) | (MPa) |
|---|---|---|---|---|
| SMAW | | | | |
| AWS A5.1 or A5.5[7,9] | | | | |
| E60XX | 50 | 345 | 62 min | 425 |
| E70XX | 60 | 415 | 72 min | 495 |
| E70XX-X | 57 | 390 | 70 min | 480 |
| SAW | | | | |
| AWS A5.17 or A5.23[7,9] | | | | |
| F6XX-EXXX | 48 | 330 | 60-80 | 415-550 |
| F7XX-EXXX or | 58 | 400 | 70-95 | 485-660 |
| F7XX-EXX-XX | | | | |
| GMAW, GTAW | | | | |
| AWS A5.18 | | | | |
| ER70S-X | 60 | 415 | 72 min | 495 |
| FCAW | | | | |
| AWS A5.20 | | | | |
| E6XT-X | 50 | 345 | 62 min | 425 |
| E7XT-X | 60 | 415 | 72 min | 495 |
| (Except -2, -3, -10, -GS) | | | | |

(continued)

**Table 4.1 (continued)**

| Group | Steel Specification[1,2] | Minimum Yield Point/Strength ksi | MPa | Tensile Range ksi | MPa | Electrode Specification[3,4] | Minimum Yield Point/Strength ksi | MPa | Tensile Strength Range ksi | MPa |
|---|---|---|---|---|---|---|---|---|---|---|
| | ASTM A131  Grades AH32, DH32, EH32 | 46 | 315 | 68-85 | 470-585 | SMAW | | | | |
| | Grades AH36, DH36, EH36 | 51 | 350 | 71-90 | 490-620 | AWS A5.1 or A5.5[7,9] | | | | |
| | ASTM A242[6] | 42-50 | 290-345 | 63-70 | 435-485 | E7015, E7016 | | | | |
| | ASTM A441 | 40-50 | 275-345 | 60-70 | 415-485 | E7018, E7028 | 60 | 415 | 72 min | 495 |
| | ASTM A516  Grade 65 | 35 | 240 | 65-85 | 450-585 | E7015-X, E7016-X | | | | |
| | Grade 70 | 38 | 260 | 70-90 | 485-620 | E7018-X | 57 | 390 | 70 min | 480 |
| | ASTM A537  Class 1 | 45-50 | 310-345 | 65-90 | 450-620 | | | | | |
| | ASTM A572  Grade 42 | 42 | 290 | 60 min | 415 min | SAW | | | | |
| | ASTM A572  Grade 50 | 50 | 345 | 65 min | 450 min | AWS A5.17 or A5.23[7,9] | 58 | 400 | 70-95 | 485-660 |
| | ASTM A588[6]  (4 in. and under) | 50 | 345. | 70 min | 485 min | F7XX-EXXX or F7XX-EXX-XX | | | | |
| | ASTM A595  Grade A | 55 | 380 | 65 min | 450 min | | | | | |
| | Grades B and C | 60 | 415 | 70 min | 480 min | | | | | |
| II | ASTM A606[6] | 45-50 | 310-340 | 65 min | 450 min | GMAW, GTAW | | | | |
| | ASTM A607  Grade 45 | 45 | 310 | 60 min | 410 min | AWS A5.18 | 60 | 415 | 72 min | 495 |
| | Grade 50 | 50 | 345 | 65 min | 450 min | ER70S-X | | | | |
| | Grade 55 | 55 | 380 | 70 min | 480 min | | | | | |
| | ASTM A618 | 46-50 | 315-345 | 65 min | 450 min | FCAW | | | | |
| | ASTM A633  Grade A | 42 | 290 | 63-83 | 430-570 | AWS A5.20 | 60 | 415 | 72 min | 495 |
| | Grades C, D  (2-1/2 in. and under) | 50 | 345 | 70-90 | 485-620 | E7XT-X | | | | |
| | ASTM A709  Grade 50 | 50 | 345 | 65 min | 450 min | (Except -2, -3, -10, -GS) | | | | |
| | Grade 50W | 50 | 345 | 70 min | 485 min | | | | | |
| | ASTM A710  Grade A. Class 2  >2 in. | 55 | 380 | 65 min | 450 min | | | | | |
| | ASTM A808  (2-1/2 in. and under) | 42 | 290 | 60 min | 415 min | | | | | |
| | API 2H[6]  Grade 42 | 42 | 290 | 62-80 | 430-550 | | | | | |
| | Grade 50 | 50 | 345 | 70 min | 485 min | | | | | |
| | API 5L  Grade X52 | 52 | 360 | 66-72 | 455-495 | | | | | |
| | ABS  Grades AH32, DH32, EH32 | 45.5 | 315 | 71-90 | 490-620 | | | | | |
| | Grades AH36, DH36, EH36[6] | 51 | 350 | 71-90 | 490-620 | | | | | |

(continued)

## Table 4.1 (continued)

| Group | Steel Specification Requirements | | | | | Filler Metal Requirements | | | | |
|---|---|---|---|---|---|---|---|---|---|---|
| | Steel Specification[1,2] | Minimum Yield Point/Strength ksi | MPa | Tensile Range ksi | MPa | Electrode Specification[3,4] | Minimum Yield Point/Strength ksi | MPa | Tensile Strength Range ksi | MPa |
| III | ASTM A572 Grade 60 | 60 | 415 | 75 min | 515 min | SMAW AWS A5.5[7,9] E8015-X, E8016-X E8018-X | 67 | 460 | 80 min | 550 |
| | Grade 65 | 65 | 450 | 80 min | 550 min | SAW AWS A5.23[7,9] F8XX-EXX-XX | 68 | 470 | 80-100 | 550-690 |
| | ASTM A537 Class 2[6] | 46-60 | 315-415 | 80-100 | 550-690 | GMAW, GTAW AWS A5.28[7,9] ER80S-X | 68 | 470 | 80 min | 550 |
| | ASTM A633 Grade E[6] | 55-60 | 380-415 | 75-100 | 515-690 | FCAW AWS A5.29[7,9] E8XTX-X | 68 | 470 | 80-100 | 550-690 |
| | ASTM A710 Grade A. Class 2 ≤2 in. | 60-65 | 415-450 | 72 min | 495 min | | | | | |
| | ASTM A710 Grade A. Class 3 >2 in. | 60-65 | 415-450 | 70 min | 485 min | | | | | |
| IV | ASTM A514 Over 2-1/2 in. (63.5 mm) Grades 100, 100W ASTM A709 2-1/2 in. to 4 in. (63.5 to 102 mm) | 90 | 620 | 100-130 | 690-895 | SMAW AWS A5.5[7] E10015-X, E10016-X E10018-X | 87 | 600 | 100 min | 690 |
| | | | | | | SAW AWS A5.23[7] F10XX-EXX-XX | 88 | 610 | 100-120 | 690-830 |
| | ASTM A710 Grade A. Class 1 ≤3/4 in. | 90 | 620 | 100-130 | 690-895 | GMAW, GTAW AWS A5.28[7] ER100S-X | 88-102 | 610-700 | 100 min | 690 |
| | | 80 | 550 | 90 min | 620 min | | | | | |
| | ASTM A710 Grade A. Class 3 ≤2 in. | 75 | 515 | 85 min | 585 min | FCAW AWS A5.29[7] E10XTX-X | 88 | 605 | 100-120 | 690-830 |

(continued)

## Table 4.1 (continued)

| Group | Steel Specification Requirements | | | | | Filler Metal Requirements | | | | |
|---|---|---|---|---|---|---|---|---|---|---|
| | Steel Specification[1,2] | Minimum Yield Point/Strength ksi | MPa | Tensile Range ksi | MPa | Electrode Specification[3,4] | Minimum Yield Point/Strength ksi | MPa | Tensile Strength Range ksi | MPa |
| V | ASTM A514   2-1/2 in. (63.5 mm) and under | 100 | 690 | 110-130 | 760-895 | SMAW AWS A5.5[7] E11015-X, E11016-X E11018-X | 97 | 670 | 110 min | 760 |
| | ASTM A517 | 90-100 | 620-690 | 105-135 | 725-930 | SAW AWS A5.23[7] F11XX-EXX-XX | 98 | 680 | 110-130 | 760-895 |
| | ASTM A709   2-1/2 in. (63.5 mm) and under | 100 | 690 | 110-130 | 760-895 | GMAW, GTAW AWS A5.28[7] ER110S-X | 95-107 | 660-740 | 110 min | 760 |
| | | | | | | FCAW AWS A5.29[7] E11XTX-X | 98 | 675 | 110-130 | 760-900 |

Notes:

1. In joints involving base metals of different groups, low-hydrogen filler metal requirements applicable to the lower strength group may be used. The low-hydrogen processes shall be subject to the technique requirements applicable to the higher strength group.

2. Match API Standard 2B (fabricated tubes) according to steel used.

3. When welds are to be stress-relieved, the deposited weld metal shall not exceed 0.05 percent vanadium.

4. See 4.16 for electrogas and electroslag weld metal requirements.

5. Only low hydrogen electrodes shall be used when welding A36 or A709 Grade 36 steel more than 1 in. (25.4 mm) thick for dynamically loaded structures.

6. Special welding materials and procedures (e.g., E80XX-X low alloy electrodes) may be required to match the notch toughness of base metal (for applications involving impact loading or low temperature), or for atmospheric corrosion and weathering characteristics (see 4.1.4).

7. Deposited weld metal shall have a minimum impact strength of 20 ft • lbs (27.1 J) at 0° F (-18° C) when Charpy V-notch specimens are required.

8. The designation of ER70S-1B has been reclassified as ER80S-D2 in A5.28-79. Prequalified joint welding procedures prepared prior to 1981 and specifying AWS A5.18, ER70S-1B, may now use AWS A5.28-79 ER80S-D2 when welding steels in Groups I and II.

9. Filler metals of alloy groups B3, B3L, B4, B4L, B5, B5L, B6, B6L, B7, B7L, B8, B8L, or B9, in ANSI/AWS A5.5, A5.23, A5.28, or A5.29, are not prequalified for use in the as-welded condition.

## Table 5.1
## Matching Filler Metals Requirements
### (See 5.1)

| Group | Steel Specification | Grade | Min Yield Point/Strength ksi | MPa | Min Tensile Strength ksi | MPa | Electrode Specification[4] | Yield Point/Strength[1] ksi | MPa | Tensile Strength[1] ksi | MPa |
|---|---|---|---|---|---|---|---|---|---|---|---|
| I | ASTM A615 | Grade 40 | 40 | — | 70 | — | SMAW AWS A5.1 and A5.5 E7015, E7016, E7018, E7028, E7015-X, E7016-X, E7018-X | 60 | 415 | 72 | 495 |
| | ASTM A615M | Grade 300 | — | 300 | — | 500 | | 57 | 390 | 70 | 480 |
| | ASTM A617 | Grade 40 | 40 | — | 70 | — | GMAW AWS A5.18 ER70S-X | 60 | 415 | 72 | 495 |
| | ASTM A617M | Grade 300 | — | 300 | — | 500 | FCAW AWS A5.20 E7XT-X (Except -2, -3, -10, -GS) | 60 | 415 | 72 | 495 |
| II | ASTM A616 | Grade 50 | 50 | — | 80 | — | SMAW AWS A5.5 E8015-X, E8016-X, E8018-X | 67 | 460 | 80 | 550 |
| | ASTM A616M | Grade 350 | — | 350 | — | 550 | | | | | |
| | ASTM A706 | Grade 60 | 60 | — | 80 | — | GMAW AWS A5.28 ER80S-X | 68 | 470 | 80 | 550 |
| | ASTM A706M | Grade 400 | — | 400 | — | 550 | FCAW AWS A5.29 E8XTX-X | 68 | 470 | 80-100 | 550-690 |
| III | ASTM A615 | Grade 60 | 60 | — | 90 | — | SMAW AWS A5.5 E9015-X, E9016-X, E9018-X | 77 | 530 | 90 | 620 |
| | ASTM A615M | Grade 400 | — | 400 | — | 600 | | | | | |
| | ASTM A616 | Grade 60 | 60 | — | 90 | — | GMAW AWS A5.28 ER90S-X | 78 | 540 | 90 | 620 |
| | ASTM A616M | Grade 400 | — | 400 | — | 600 | | | | | |
| | ASTM A617 | Grade 60 | 60 | — | 90 | — | FCAW AWS A5.29 E9XTX-X | 78 | 540 | 90-110 | 620-760 |
| | ASTM A617M | Grade 400 | — | 400 | — | 600 | | | | | |

Steel Specification Requirements — Filler Metal Requirements

(Continued)

## Table 5.1
## (Continued)

| Group | Steel Specification | Minimum Yield Point/Strength ksi | MPa | Minimum Tensile Strength ksi | MPa | Electrode Specification[4] | Yield Point/Strength[1] ksi | MPa | Tensile Strength[1] ksi | MPa |
|---|---|---|---|---|---|---|---|---|---|---|
| IV | | | | | | SMAW AWS A5.5 | | | | |
| | | | | | | E10015-X, E10016-X, E10018-X | 87 | 600 | 100 | 690 |
| | | | | | | E10018-M | 88-100 | 610-690 | 100 | 690 |
| | ASTM A615   Grade 75[2] | 75 | — | 100 | — | GMAW AWS A5.28 | | | | |
| | | | | | | ER100S-X | 88-102 | 610-700 | 100 | 690 |
| | ASTM A615M   Grade 500[3] | — | 500 | — | 700 | FCAW AWS A5.29 | | | | |
| | | | | | | E10XTX-X | 88 | 610 | 100-120 | 690-830 |

**Steel Specification Requirements** — columns under Minimum Yield Point/Strength and Minimum Tensile Strength.
**Filler Metal Requirements** — columns under Yield Point/Strength[1] and Tensile Strength[1].

Notes:

1. This table is based on filler metal as-welded properties. Single values are minimums. Hyphenated values indicated minimum and maximum.
2. Applicable to bar sizes Nos. 11, 14, and 18.
3. Applicable to bar sizes Nos. 35, 45, and 55.
4. Filler metals classified in the postweld heat treated (PWHT) condition by the AWS filler metal specification may be used when given prior approval by the Engineer. Consideration shall be made of the differences in tensile strength, ductility and hardness between the PWHT versus as-welded condition.

Filler Metal Requirements *(Continued)*

Structural Steel                                          D1.1, Table 4.1
                                                          AISC, 5-28, A3.6

# Fillers                                                 UBC, 2-700, J6

D1.1, . . . A filler less than ¼ in. thick shall not be     AISC, 5-78, J6
used to transfer stress, but shall be kept flush with      D1.1, 2.4.2 - 2.4.3
the welded edges of the stress-carrying part. The          D1.1, Fig. 2.1 and 2.2
sizes of welds along such edges shall be increased
over the required sizes by an amount equal to the
thickness of the filler (see Figure 2.1). . . .

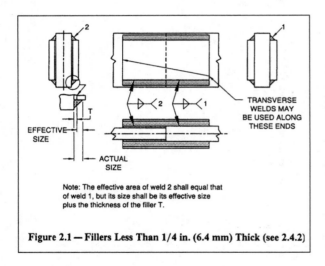

EFFECTIVE SIZE

ACTUAL SIZE

TRANSVERSE WELDS MAY BE USED ALONG THESE ENDS

Note: The effective area of weld 2 shall equal that of weld 1, but its size shall be its effective size plus the thickness of the filler T.

Figure 2.1 — Fillers Less Than 1/4 in. (6.4 mm) Thick (see 2.4.2)

*Any filler* ¼ in. or more in thickness shall extend
beyond the edges of the splice plate or connection
material. It shall be welded to the part on which it
is fitted, and the joint shall be of sufficient strength
to transmit the splice plate or connection material
stress applied at the surface of the filler as an ec-
centric load . . . (see Figure 2.1).

# Fillet Welds

Details of

    D1.1, . . . The transverse spacing of longitudi-      D1.1, 8.8
nal fillet welds used in end connections shall not    D1.1 Comm., 3.6.1
exceed 8 in. unless end transverse welds or inter-
mediate plug or slot welds are used. . . .

    *For lap* joints, the minimum amount of lap shall
be five times the thickness of the thinner part
joined but not less than 1 in. . . .

    *Fillet welds* deposited on the opposite sides of a
common plane of contact between two parts
shall be interrupted at a corner common to both
welds. . . .

    *Side* or end fillet welds terminating at ends or
sides of header angles, brackets, beam seats, and
similar connections shall be returned continuously around the corners for a distance at least
twice the nominal size of the weld, except as provided in 8.8.5. . . .

    *Boxing* shall be indicated on the drawings.

Effective Areas of                          UBC, 2-692, J2.3
                                            D1.1, 2.3.2 - 2.3.2.4
                                            AISC, 5-67, J2.2.a
    Fillet Welds                             UBC, 2-690, J2.2a

    UBC, . . . The effective area of fillet welds      D1.1, 2.3.2 - 2.3.2.4
shall be taken as the effective length times the    AISC, 5-67, J2.2.a
effective throat thickness. . . .

    *The* effective length of fillet welds, except
fillet welds in holes and slots, shall be the
overall length of full size fillets, including
returns. . . .

    *The* effective throat thickness of a fillet weld
shall be the shortest distance from the root of
the joint to the face of the diagrammatic
weld,. . . .

    *For* fillet welds in holes and slots, the effective
length shall be the length of the centerline of

*(Continued)*

Fillet Welds *(Continued)*

the weld along the center of the plane through the throat. . . .

Plug and Slot Welds

D1.1, . . . The effective area shall be the nominal area of the hole or slot in the plane of the faying surface.

Complete Penetration Groove Welds

UBC, . . . The effective area of groove welds shall be considered as the effective length of the weld times the effective throat thickness. . . .

*The* effective length of a groove weld shall be the width of the part joined. . . .

*The* effective throat thickness of a complete-penetration groove weld shall be the thickness of the thinner part joined.

Partial Joint Penetration Groove Welds

D1.1, . . . The effective area shall be the effective weld length multiplied by the weld size.

*The minimum* weld size of a partial joint penetration groove weld shall be as specified in Table 2.3.

*The weld* size of a partial joint penetration groove weld shall be the depth of bevel less $\frac{1}{8}$ in. for grooves having a groove angle less than 60 deg., but not less than 45 deg., at the root of the groove, when made by shielded metal arc or submerged arc welding or when made in the vertical or overhead welding positions by gas metal arc, flux cored arc, or gas tungston arc welding. . . .

Flare Groove Welds

D1.1, . . . The effective weld size for flare groove welds when filled flush to the surface of a bar, a 90 deg. bend in a formed section, or a rectangular tube, shall be as shown in

---

Reference column:

Plug and Slot Welds — D1.1, 2.3.3

D1.1, . . . The effective area — AISC, 5-68, J2.3 / UBC, 2-692, J2.3

Complete Penetration Groove Welds — UBC, 2-689, J2

UBC, . . . — AISC, 5-65, J2.1

Partial Joint Penetration Groove Welds — UBC, 2-690, J2.2a

D1.1, . . . — AISC, 5-65, J2.2a / D1.1, 2.3 / D1.1, 2.10

Flare Groove Welds — AISC, 5-65, J2.1.a

D1.1, . . . — AISC, 5-66, Table J2.2 / D1.1, 2.3.1.4 / D1.1, Table 2.1 / UBC, 2-689, J2.1

Table 2.1. When required by the engineer, test sections shall be used to verify that the effective weld size is consistently obtained. For a given set of procedural conditions, if the contractor has demonstrated consistent production of larger effective weld sizes than those shown in Table 2.1, the contractor may establish such larger effective weld sizes by qualification.

**Table 2.1**
**Effective Weld Sizes of Flare Groove Welds**
**(see 2.3.1.4)**

| Flare-Bevel-Groove Welds | Flare-V-Groove Welds |
|---|---|
| 5/16 R | 1/2 R* |

NOTE:  R = radius of outside surface

*Use 3/8 R for GMAW (except short circuiting transfer) process when R is 1/2 in. (13 mm) or greater.

End Returns

AISC, 5-67, J2.2.b

D1.1, . . . Side or end fillet welds terminating at ends or sides of header angles, brackets, beam seats, and similar connections shall be returned continuously around the corner for a distance at least twice the nominal size of the weld, except as provided in 8.8.5.

UBC, 2-692, J2.2.b
D1.1, 8.8.6.1

*Boxing* shall be indicated on the drawings.

Gaps in (Fit Up)

D1.1, 3.3.1

D1.1, . . . The parts to be joined by fillet welds shall be brought into as close contact as practicable. The root opening shall not exceed 3/16 in. except in cases involving either shapes or plates 3 in. or greater in thickness if, after straightening and in assembly, the root opening cannot be closed sufficiently to meet this tolerance. In such cases, a maximum root opening of 5/16 in. is acceptable provided suitable backing is used. If the

*(Continued)*

Fillet Welds, End Returns *(Continued)*

separation is greater than $1/16$ in., the leg of the fillet weld shall be increased by the amount of the root opening, or the contractor shall demonstrate that the required effective throat has been obtained.

In Holes and Slots                                            AISC, 5-67, J2.2.b

D1.1, . . . Fillet welds in holes or slots may be            D1.1, 2.7.1.3
used to transfer shear or to prevent buckling or            UBC, 2-692, J2.2.b
separation of lapped parts. . . .

*Fillet* welds in holes or slots are not to be considered as plug or slot welds.

Intermittent                                                 AISC, 5-67, J2.2.b

D1.1, . . . The minimum length of an intermit-              UBC, 2-692, J2.2.b
tent fillet weld shall be 1 $1/2$ in.                        D1.1, 2.7.1.4

*The minimum* effective length of a fillet weld             D1.1, 2.3.2.3
shall be at least four times the nominal size, or
the size of the weld shall be considered not to exceed 25% of its effective length.

Length

Intermittent                                                 AISC, 5-67, J2.2.b

D1.1, . . . The minimum length of an inter-                UBC, 2-692, J2.2.b
mittent fillet weld shall be 1 $1/2$ in.                     D1.1, 2.7.1.4

Minimum                                                      AISC, 5-67, J2.2.b

D1.1, . . . The minimum effective length of a              D1.1, 2.3.2.1
fillet weld shall be at least four times the                UBC, 2-692, J2.2.b
nominal size, or the size of the weld shall be
considered not to exceed 25% of its effective
length.

Longitudinal Spacing of                                      D1.1, 8.8.1

D1.1, . . . The transverse spacing of longitudi-            AISC, 5-67, J2.2.b
nal fillet welds used in end connections shall not
exceed 8 in. unless end traverse welds or intermediate plug or slot welds are used.

Maximum Single Pass

S.M.A.W.                                                    D1.1, 4.6.6

D1.1, . . . The maximum size of single pass
fillet welds and root passes of multi-pass fillet
welds shall be:

1. $^3/_8$ in. for the flat position.
2. $^5/_{16}$ in. for the horizontal or overhead positions.
3. $^1/_2$ in. for the vertical position.

G.M.A.W.                                                    D1.1, 4.14.1.3

D1.1, . . . The maximum size of a fillet weld
made in one pass shall be:

1. $^1/_2$ in. for the flat and vertical positions.
2. $^3/_8$ in. for the horizontal position.
3. $^5/_{16}$ in. for the overhead position.

F.C.A.W. (see G.M.A.W.)                                     D1.1, 4.14.1.3

Maximum Single Pass and Root Pass                           D1.1, 4.6.6,

Maximum Size Along Edge                                     D1.1, 2.7.1.2

D1.1, . . . The maximum fillet weld size de-     D1.1, Fig. 2.3
tailed along edges of material shall be:         UBC, 2-691, J2.2.b

1. The thickness of the base metal, for metal less     AISC, 5-67, J2.2.b
   than $^1/_4$ in. thick (see Figure 2.3, Detail A).

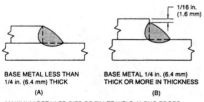

1/16 in.
(1.6 mm)

BASE METAL LESS THAN          BASE METAL 1/4 in. (6.4 mm)
1/4 in. (6.4 mm) THICK        THICK OR MORE IN THICKNESS
        (A)                              (B)
MAXIMUM DETAILED SIZE OF FILLET WELD ALONG EDGES

Figure 2.3 — Details for Prequalified
Fillet Welds (see 2.7.1.2)

*(Continued)*

Fillet Welds, Maximum Size Along Edge
*(Continued)*

2. $1/16$ in. less than the thickness of the base
metal, for metal 1/4 in. or more in thickness
(see Figure 2.3, Detail B), unless the weld is
designated on the drawing to be built out to
obtain full throat thickness. . . .

Minimum Length                                    D1.1, 2.3.2.3

D1.1, . . . The minimum effective length of a fillet weld shall be at least four times the nominal
size, or the size of the weld shall be considered
not to exceed 25% of its effective length.

Minimum Size                                      AISC, 5-67, J2.2.b

D1.1, . . . The minimum fillet weld size, except    AISC, 5-67, Table J2.4
for fillet welds used to reinforce groove welds,    UBC, 2-691, Table J2.4
shall be as shown in Table 2.2. In both cases, the  D1.1, 2.7.1.1
minimum size applies if it is sufficient to satisfy D1.1, Table 2.2
design requirements,
*except* that the weld size need not exceed the
thickness of the thinner part joined.

### Table 2.2
### Minimum Fillet Weld Size for
### Prequalified Joints (see 2.7.1.1)

| Base Metal Thickness (T)* | | Minimum Size of Fillet Weld** | | |
|---|---|---|---|---|
| in. | mm | in. | mm | |
| T≤1/4 | T≤ 6.4 | 1/8*** | 3 | ⎫ |
| 1/4<T≤1/2 | 6.4<T≤12.7 | 3/16 | 5 | Single-pass |
| 1/2<T≤3/4 | 12.7<T≤19.0 | 1/4 | 6 | welds must |
| 3/4<T | 19.0<T | 5/16 | 8 | be used |

*For non-low hydrogen processes without preheat calculated in
accordance with 4.2.2, T equals thickness of the thicker part joined.
For non-low hydrogen processes using procedures established to
prevent cracking in accordance with 4.2.2, and for low hydrogen
processes, T equals thickness of the thinner part joined; single pass
requirement does not apply.

**Except that the weld size need not exceed the thickness of the
thinner part joined.

***Minimum size for dynamically loaded structures is 3/16 in. (5 mm).

Root Opening (Maximum)                          D1.1, 3.3.1

D1.1, . . . The parts to be joined by fillet welds
shall be brought into as close contact as practicable. The root opening shall not exceed ³/₁₆ in. except in cases involving either shapes or plates
3 in. or greater in thickness if, after straightening and in assembly, the root opening cannot be
closed sufficiently to meet this tolerance. In such
cases, a maximum root opening of ⁵/₁₆ in. is acceptable provided suitable backing is used. If the
separation is greater than ¹/₁₆ in., the leg of the
fillet weld shall be increased by the amount of
the root opening, or the contractor shall demonstrate that the required effective throat has been
obtained.

Sizes for Small Studs                           D1.1, 7.5.5.4, 7.5.5.6

D1.1, . . . When fillet welds are used, the mini-    D1.1, Table 7.2
mum size shall be the larger of those required in
Table 2.2 or Table 7.2. . . .

S.M.A.W. welding shall be performed using low
hydrogen electrodes ⁵/₃₂ or ³/₁₆ in. in diameter, except that a smaller diameter electrode may be
used on studs ⁷/₁₆ in. or less in diameter for out-
of-position welds.

### Table 7.2
### Minimum Fillet Weld Size for Small Diameter
### Studs (see 7.5.5.1)

| Stud Diameter | | Min. Size Fillet | |
|---|---|---|---|
| in. | mm | in. | mm |
| 1/4 thru 7/16 | 6.4 thru 11.1 | 3/16 | 5 |
| 1/2 | 12.7 | 1/4 | 6 |
| 5/8, 3/4, 7/8 | 15.9, 19, 22.2 | 5/16 | 8 |
| 1 | 25.4 | 3/8 | 10 |

*(Continued)*

Fillet Welds, Sizes for Small Studs *(Continued)*

Underrun In (Undersize)                                    D1.1, 8.15.1.7

> D1.1, . . . A fillet weld in any continuous weld shall be permitted to underrun the nominal fillet size by $1/16$ in. without correction, provided that the undersize portion of the weld does not exceed 10% of the length of the weld. On web-to-flange welds on girders, no underrun is permitted at the ends for a length equal to twice the width of the flange.

## Fit of Column Compression Joints                       AISC, 5-90, M4.4

A gap of $1/16$ in. is allowed in column compression joints. If the gap is greater than $1/16$ in. but is less than $1/4$ in., and with engineering approval, the gap can be packed with non-tapered steel shims. These shims may be mild steel, regardless of the grade of the main member.

## Fit of Intermediate Stiffeners                          D1.1, 3.5.3.1

> D1.1, . . . Where tight fit of intermediate stiffeners is specified, it shall be defined as allowing a gap of up to $1/16$ in. between stiffener and flange. . . .

> *The* out-of-straightness variation of intermediate stiffeners shall not exceed $1/2$ in. for girders up to 6 ft. deep, and $3/4$ in. for girders over 6 ft. deep, with due regard for members that frame into them.

Bearing Stiffeners                                         D1.1, 3.5.2 - 3.5.1.12

> D1.1, . . . The bearing ends of bearing stiffeners shall be square with the web and shall have at least 75% of the stiffener bearing cross sectional area in contact with the inner surface of the flanges. . . .

AISC, 5-82, K1.8
UBC, 2-704, K1.8

> *The* out-of-straightness variation of bearing stiffeners shall not exceed 1/4 in. up to 6 ft. deep or 1/2 in. over 6 ft. . . .

## Flanges

UBC, 2-669, B10

UBC, . . . The total cross-sectional area of cover plates of bolted or riveted girders shall not exceed 70% of the total flange area.

## Flange Development

AISC, 5-37, B10
UBC, 2-669, B10

## Flare Groove Welds (Effective Throats)

UBC, 2-690, J2.2a

D1.1, . . . The effective weld size for flare groove welds when filled flush to the surface of a bar, a 90 deg. bend in a formed section, or a rectangular tube, shall be as shown in Table 2.1. When required by the engineer, test sections shall be used to verify that the effective weld size is consistently obtained. For a given set of procedural conditions, if the contractor has demonstrated consistent production of larger effective weld sizes than those shown in Table 2.1, the contractor may establish such larger effective weld sizes by qualification.

UBC, 2-690, Table J2.2
AISC, 5-66, J2.1.a
AISC, 5-66, Table J2.2
D1.1, 2.3.1.4

### TABLE J2.2
### Effective Throat Thickness of Flare Groove Welds

| Type of Weld | Radius ($R$) of Bar or Bend | Effective Throat Thickness |
|---|---|---|
| | | × 25.4 for mm |
| Flare bevel groove | All | $5/16 R$ |
| Flare V-groove | All | $1/2 R$ [a] |

[a] Use $3/8 R$ for Gas Metal Arc Welding (except short circuiting transfer process) when $R \geq 1/2$-in. (13 mm).

*(Continued)*

Flare Groove Welds *(Continued)*

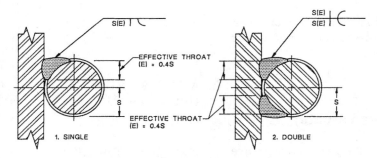

**A - FLARE-BEVEL-GROOVE WELD**

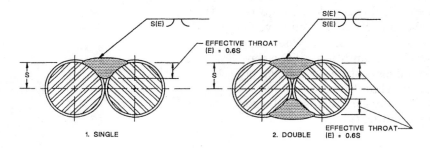

**B - FLARE-V-GROOVE WELD**

NOTE:  1.  RADIUS OF REINFORCING BAR = S

2.  THESE ARE SECTIONAL VIEWS.  BAR DEFORMATIONS ARE SHOWN ONLY FOR ILLUSTRATIVE PURPOSES.

**Figure 2.1 — Effective Weld Sizes for Flare-Groove Welds (See 2.3.2.3)**

## Flux (Studs, with or without)

D1.1, 7.2.3

D1.1, . . . A suitable deoxidizing and arc stabilizing flux for welding shall be furnished with each stud of $5/16$ in. diameter or larger. Studs less than $5/16$ in. in diameter may be furnished with or without flux.

## Formed Steel Decking *(See Studs)*

AISC, 2-246 and 2-255

The maximum diameter of a welded shear connector shall be $2\frac{1}{2}$ times the thickness of the flange to which it is welded. Larger studs may be used if welded directly over the web of the member.

AISC, 5-60, I5 - I5.3
UBC, 2-686, I5, (f) and (g)
AWS, D1.3

*Composite* construction shall, where applicable, be limited to decking having a 3 in. maximum rib height and an average rib height of not less than 2 in.

*Welded shear* connectors shall be $3/4$ in. or less in diameter.

*After installation* by welding, shear connectors shall extend a minimum of $1^{1}/_{2}$ in. above the top of the steel deck.

*The top* of the composite slab shall be a minimum of 2 in. above the top of the decking.

*The maximum* center-to-center spacing of shear connectors on a beam or girder shall not be greater than 36 in.

*Decking shall* be anchored, by welds or shear connectors, at a maximum center-to-center spacing of 16 in. Welds and studs are used as deck anchors to help resist uplifting of the deck.

# Friction Connections (Classes of)

AISC, 5-271, Table 3

**Allowable Loads for Different Classes of Slip-Critical Connections**

| | Direction of Load for Different Hole Types | | | | | | | |
| --- | --- | --- | --- | --- | --- | --- | --- | --- |
| | All Directions | | | | Transverse Direction | | Parallel Direction | |
| | Standard Holes | | Oversize and Short Slots | | Long Slotted Holes | | Long Slotted Holes | |
| Faying Surfaces | A325 | A490 | A325 | A490 | A325 | A490 | A325 | A490 |
| Class A | 17 | 21 | 15 | 18 | 12 | 15 | 10 | 13 |
| Class B | 28 | 34 | 24 | 29 | 20 | 24 | 17 | 20 |
| Class C | 22 | 27 | 19 | 23 | 16 | 19 | 14 | 16 |

Class A:   1. Clean mill scale is allowed.
2. Blast cleaned surfaces with Class A protective coatings.

Class B:   1. Blast cleaned surfaces.
2. Blast cleaned surfaces with Class B protective coatings.

Class C:   1. Hot dipped galvanized and roughened surfaces.

## Gaps

In Bearing Joints                                           UBC, 2-687, J1.4

UBC, . . . Lack of contact bearing not exceeding      UBC, 2-708, M4.4
a gap of $1/16$ in., regardless of the type of splice
used (partial-penetration, groove-welded, or
bolted) shall be acceptable. If the gap exceeds
$1/16$ in., but is less than $1/4$ in., and if an engineering investigation shows sufficient contact area
does not exist, the gap shall be packed out with
non-tapered steel shims. Shims may be of mild
steel, regardless of the grade of the main
material.

In Plug and Slots                                          D1.1, 3.3.1

D1.1, . . . The separation between faying surfaces of plug and slot welds, and of butt joints
landing on a backing, shall not exceed $1/16$ in. The
use of fillers is prohibited except as specified on
the drawings or as specially approved by the engineer and made in accordance with 2.4.

## Gas Shielding (Dew Point)                               D1.1, 4.13

D1.1, . . . A gas or gas mixture used for shielding
in gas metal arc welding or flux cored arc welding
shall be of a welding grade having a dew point of
−40 deg. F or lower. . . .

## Girders (Riveted)                                       UBC, 2-669, B10

UBC, . . . The total cross sectional area of cover
plates of bolted or riveted girders shall not exceed
70% of the total flange area. . . .

## Gouging (Oxygen)                                        D1.1, 3.2.5

D1.1, . . . Machining, thermal cutting, gouging,
chipping, or grinding may be used for joint preparation, or the removal of unacceptable work or metal,
except that oxygen gouging shall not be used on
steels that are ordered as quenched and tempered
or normalized.

# Groove Welds

Backing                                                    D1.1, 3.13.3

D1.1, . . . Groove welds made with the use of             D1.1 Comm., 3.13
steel backing shall have the weld metal thor-
oughly fused with the backing. . . .

*Steel backing* shall be made continuous for the
full length of the weld. . . .

*The suggested* minimum nominal thickness of
backing bars, provided that the backing shall be
of sufficient thickness to prevent melt-through, is
shown in the following table:

|              | Thickness, min | |
|--------------|------|------|
| Process      | in.  | mm   |
| SMAW         | 3/16 | 4.8  |
| GMAW         | 1/4  | 6.4  |
| FCAW-SS      | 1/4  | 6.4  |
| FCAW-G       | 3/8  | 9.5  |
| SAW          | 3/8  | 9.5  |

Note: Commercially available steel backing for pipe and tubing is
acceptable, provided there is no evidence of melting on exposed interior surfaces.

D1.1 Comm, . . . All prequalified complete joint
penetration groove welds made from one side
only, except as permitted for tubular structures,
are required to have complete fusion of the weld
metal with a steel backing. . . .

Butt Joints

Alignment of                                               D1.1, 3.3.3

D1.1, . . . Parts to be joined at butt joints
shall be carefully aligned. Where the parts are
effectively restrained against bending due to
eccentricity in alignment, an offset not exceeding 10% of the thickness of the thinner part
joined, but in no case more than $1/8$ in., shall be
permitted as a departure from the theoretical
alignment. In correcting misalignment in such
cases, the parts shall not be drawn into a
greater slope than $1/2$ in. in 12 in. . . .

*(Continued)*

Groove Welds, Butt Joints *(Continued)*

Variance in                                                      D1.1, 3.3.4

D1.1, . . . With the exclusion of electroslag            D1.1, Fig. 3.3
and electrogas welding, and with the excep-
tion of 3.3.4.1 for root openings in excess of
those permitted in Figure 3.3, the dimensions
of the cross section of the groove welded joints
that vary from those on the detail drawings by
more than the following tolerances shall be referred to the engineer for approval or correction.

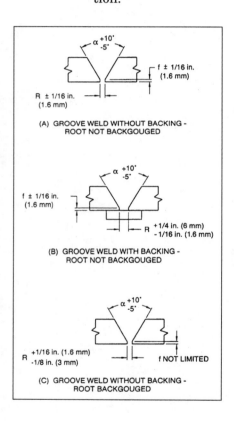

(A) GROOVE WELD WITHOUT BACKING -
ROOT NOT BACKGOUGED

(B) GROOVE WELD WITH BACKING -
ROOT NOT BACKGOUGED

(C) GROOVE WELD WITHOUT BACKING -
ROOT BACKGOUGED

|  | Root not back gouged* | | Root back gouged | |
|---|---|---|---|---|
|  | in. | mm | in. | mm |
| (1) Root face of joint | ±1/16 | 1.6 | Not limited | |
| (2) Root opening of joints without backing | ±1/16 | 1.6 | +1/16 −1/8 | 1.6 3 |
| Root opening of joints with backing | +1/4 −1/16 | 6 1.6 | Not applicable | |
| (3) Groove angle of joint | +10° −5° | | +10° −5° | |

*see 10.13.1 for tolerances for complete joint penetration tubular groove welds made from one side without backing.

**Figure 3.3 — Workmanship Tolerances in Assembly of Groove Welded Joints (see 3.3.4)**

Effective Areas of                                      AISC, 5-65, J2.1

D1.1, . . . The effective area shall be the effec-       D1.1, 2.3
tive weld length multiplied by the weld size.            UBC, 2-690, J2.1

*The weld* size of a complete joint penetration
weld shall be the thickness of the thinner part
joined. No increase is permitted for weld reinforcement.

Flare Grooves                                           UBC, 2-690, Table J2.2

D1.1, . . . The effective weld size for flare            D1.1, 2.3.1.4
groove welds when filled flush to the surface of a
bar, 90 deg. bend in a formed section, or a rectangular tube shall be as shown in Table 2.1.

### TABLE J2.2
### Effective Throat Thickness of Flare Groove Welds

| Type of Weld | Radius (*R*) of Bar or Bend | Effective Throat Thickness |
| --- | --- | --- |
| | | × 25.4 for mm |
| Flare bevel groove | All | $5/_{16}R$ |
| Flare V-groove | All | $1/_2R$[a] |

[a] Use $3/_8R$ for Gas Metal Arc Welding (except short circuiting transfer process) when $R \geq 1/_2$-in. (13 mm).

Reinforcement of                                        D1.1, 3.6.2 - 3.6.3

D1.1, . . . Groove welds shall preferably be
made with slight or minimum reinforcement except as may be otherwise provided. In the case of
butt joints and corner joints, the face reinforcement shall not exceed $1/_8$ in. in height and shall
have gradual transition to the plane of the base
metal surface. . . .

*Surfaces* of butt joints required to be flush shall
be finished so as not to reduce the thickness of
the thinner base metal or weld metal by more
than $1/_{32}$ in. or 5% of the thickness, whichever is
smaller, nor leave reinforcement that exceeds
$1/_{32}$ in. . . .

*(Continued)*

## Groove Welds, Reinforcement of *(Continued)*

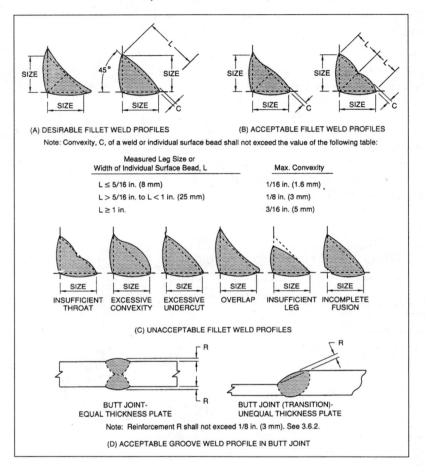

(A) DESIRABLE FILLET WELD PROFILES   (B) ACCEPTABLE FILLET WELD PROFILES

Note: Convexity, C, of a weld or individual surface bead shall not exceed the value of the following table:

| Measured Leg Size or Width of Individual Surface Bead, L | Max. Convexity |
|---|---|
| L ≤ 5/16 in. (8 mm) | 1/16 in. (1.6 mm) |
| L > 5/16 in. to L < 1 in. (25 mm) | 1/8 in. (3 mm) |
| L ≥ 1 in. | 3/16 in. (5 mm) |

| INSUFFICIENT THROAT | EXCESSIVE CONVEXITY | EXCESSIVE UNDERCUT | OVERLAP | INSUFFICIENT LEG | INCOMPLETE FUSION |
|---|---|---|---|---|---|

(C) UNACCEPTABLE FILLET WELD PROFILES

BUTT JOINT-
EQUAL THICKNESS PLATE

BUTT JOINT (TRANSITION)-
UNEQUAL THICKNESS PLATE

Note:  Reinforcement R shall not exceed 1/8 in. (3 mm). See 3.6.2.

(D) ACCEPTABLE GROOVE WELD PROFILE IN BUTT JOINT

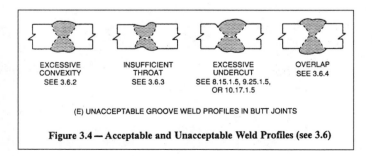

| EXCESSIVE CONVEXITY SEE 3.6.2 | INSUFFICIENT THROAT SEE 3.6.3 | EXCESSIVE UNDERCUT SEE 8.15.1.5, 9.25.1.5, OR 10.17.1.5 | OVERLAP SEE 3.6.4 |
|---|---|---|---|

(E) UNACCEPTABLE GROOVE WELD PROFILES IN BUTT JOINTS

**Figure 3.4 — Acceptable and Unacceptable Weld Profiles (see 3.6)**

## Heated Areas (Temperatures of)

D1.1,3.7.3

D1.1, . . . Members distorted by welding shall be straightened by mechanical means or by the application of a limited amount of localized heat. The temperature of heated areas as measured by approved methods shall not exceed 1100 deg. F for quenched and tempered steel nor 1200 deg. F for other steels. . . .

D1.1 Comm., 3.7.3

*Quenched and tempered* steels should not be heated above 1100 deg. F because deterioration of mechanical properties may possibly result from the formation of an undesirable microstructure when cooled to room temperature. Other steels should not be heated above 1200 deg. F to avoid the possibility of undesirable transformation products or grain coarsening, or both.

## Heat Input (Quenched and Tempered Steel)

D1.1 Comm., 4.3

D1.1, . . . The strength and toughness of the heat-affected zone of welds in quenched and tempered steels are related to the cooling rate. Contrary to principles applicable to other steels, the fairly rapid dissipation of welding heat is needed to retain adequate strength and toughness. . . .

## High-Strength Bolts

UBC, 2-605, M2.2.5

UBC, . . . Bolts shall conform to the requirements of U.B.C. Standard 22-1, except as provided in Section 2221.4. . . .

UBC, 2-445, Division IV
UBC, 2-446, 2222
SBCCI, 1508

Bolted Construction

AISC, 5-88, M2.5

During erection, the use of drift pins to align bolt holes in members is allowed, provided their use does not distort, oblong, or otherwise enlarge the holes.

*Improper* or poor matching of bolt holes shall be cause for rejection of the member.

*(Continued)*

### High-Strength Bolts *(Continued)*

*Holes may* be punched when the material is not more than $1/8$ in. thicker than the diameter of the bolt. For thicker materials, the holes can be drilled or sub-punched and reamed.

*Holes in* A514 steel plates over $1/2$ in. thick shall be drilled. . . .

*If the slope* of the faying surfaces in the connection exceeds 1:20, with regard to the bolt or nut face, a single hardened beveled washer shall be installed to compensate for the slope.

*Foreign material,* loose scale, and rust are not permitted in the faying surface of bolted connections. Burrs, fins, or any other material that prevents proper seating of the connected parts shall be removed.

### Bolts (Slip-Critical Connection)                    AISC, 5-64, J1.11

D1.1, . . . Rivets or bolts used in bearing type connections shall not be considered as sharing the stress in combination with welds. Welds, if used, shall be provided to carry the entire stress in the connection. However, connections that are welded to one member and riveted or bolted to the other member are permitted. High-strength bolts properly installed as a friction type connection prior to welding may be considered as sharing the stress with the welds.                    AISC, 5-71, J3
D1.1, 8.7
D1.1, 9.14

*When ASTM A*449 bolts are used in bearing-type and tension-type connections, and are to be tightened to a condition exceeding 50% of their minimum specified tensile strength, an ASTM F436 hardened washer shall be installed under the bolt head and the nuts shall be ASTM A563.

### Galvanized                    AISC, 5-291, para. 4

Galvanized high-strength bolts must be manufactured and shipped as a matched assembly.

*Some of the areas* of concern that must be considered when deciding whether or not to use galvanized bolts are:

1. The effect the galvanizing process might have on the high-strength bolt.

2. Whether the nut stripping strength will be reduced due to the galvanized coatings.

3. How will the galvanizing effect the torque induced in tensioning.

4. Shipping requirements.

In Combination with Rivets and Welds — AISC, 5-302, para.2

UBC, . . . In both new work and alterations, high-strength bolts in slip-critical connections may be considered as sharing the load with rivets. . . . — UBC, 2-689, J1.11

*Because rivets* are rarely used in modern steel construction, high-strength bolts used in conjunction with rivets are seldom, if ever, encountered. (see pp. 238 to 240 of the specification)

*(See Bolt Index)*

# Holes
AISC, 5-88, M2.5

Enlargement of — UBC, 2-448, 2226

During erection, the use of drift pins to align bolt holes in members is allowed, provided their use does not distort, oblong, or otherwise enlarge the holes.

*Improper* or poor matching of bolt holes shall be cause for rejection of the member.

Fillet Welds in — UBC, 2-692, J2.2.b

UBC, . . . Fillet welds in holes and slots are permitted to transmit shear in lap joints or to prevent the buckling or separation of lapped parts and to join components of built-up members. Such fillet welds may overlap, subject to the provisions of Section J2. Fillet welds in holes or slots are not to be considered plug or slot welds. — AISC, 5-67, J2.2.b / D1.1, 2.7.1.3

*(Continued)*

Holes *(Continued)*

Misalignment of                                    AISC, 5-88, M2.5

> *Improper* or poor matching of bolt holes shall be
> cause for rejection of the member.

Nominal Dimensions of                              AISC, 5-268, Table 1

> Maximum bolt hole sizes are found in the follow-
> ing table:

### Bolt Hole Dimensions in Inches

| Bolt Diameter | 1/2 in. | 5/8 in. | 3/4 in. | 7/8 in | 1 in |
|---|---|---|---|---|---|
| Standard | 9/16 | 11/16 | 13/16 | 15/16 | 1-1/16 |
| Oversize | 5/8 | 13/16 | 15/16 | 1-1/16 | 1-1/4 |
| Short Slot | 9/16 x 1 1/4 | 11/16 x 7/8 | 13/16 x 1 | 15/16 x 1-1/8 | 1-1/16 x 2-1/2 |
| Long Slot | 9/16 x 1-1/4 | 11/16 x 1-9/16 | 13/16 x 1-7/8 | 15/16 x 2-3/16 | 1-1/16 x 2-1/2 |

| Bolt Diameter | 1-1/8 | 1-1/4 | 1-3/8 | 1-1/2 |
|---|---|---|---|---|
| Standard | 1-3/16 | 1-5/16 | 1-7/16 | 1-9/16 |
| Oversize | 1-7/16 | 1-9/16 | 1-11/16 | 1-13/16 |
| Short Slot | 1-3/16 x 1-1/2 | 1-5/16 x 1-5/8 | 1-7/16 x 1-3/4 | 1-9/16 x 1-7/8 |
| Long Slot | 1-3/16 x 2-13/16 | 1-5/16 x 3-1/8 | 1-7/17 x 3-7/16 | 1-9-16 x 3-3/4 |

Oversized and Slotted                              UBC, 2-448, 2226

> UBC, . . . In the absence of approval by the en-      UBC, 2-594, J3.7 - J3.9
> gineer for use of other hole types, standard holes    UBC, 2-594, Table J3.5
> shall be used in high-strength bolted connec-          UBC, 2-694, J3.2
> tions. . . .

> *Hole sizes* for different diameter bolts can be       UBC, 2-698, J3.8.B
> found in this specification. For convenience, these    UBC, 2-699, Table J3.6
> sizes are provided in the previous table.             AISC, 5-271, para. 7 (b)
>                                                        AISC, 5-294, C3, para. 2

*Bear in* mind that base plate holes for anchor bolts may be oversized as follows (also, see Anchor Bolts in Section A):

AISC, 5-71, J3.3a - 2.e
AISC, 5-268, Table 1
AISC, 5-71, J3.1

| Bolt Size, inches (mm) | Hole Size, inches (mm) |
|---|---|
| $3/4$ (19.1) | $5/16$ (7.9) oversized |
| $7/8$ (22.2) | $5/16$ (7.9) oversized |
| 1 < 2 (25.4 < 50.8) | $1/2$ (12.7) oversized |
| >2 (> 50.8) | 1 (25.4) > bolt diameter |

*Finger* shims up to $1/4$ in. may be introduced into slip-critical connections designed on the basis of standard holes without reducing the nominal shear strength of the fastener to that specified for slotted holes. . . .

*Short-slotted* holes may be used in any or all plies of slip-critical connections, but they shall not be used in bearing-type connections. Hardened washers shall be installed over oversized holes in an outer ply. . . .

*Long-slotted* holes may be used in only one of the connected parts of either a slip-critical or bearing-type connection at an individual faying surface. . . .

*If a slotted* or oversized hole is used in an outer ply of the connection, and ASTM A490 bolts over 1 in. in diameter are used, a $5/16$ in. minimum thickness F436 washer must be used.

*When A490* bolts over 1 in. in diameter are used in slotted or oversized holes in external plies, a single hardened washer conforming to F436, except with $5/26$ in. minimum thickness, shall be used in lieu of the standard washer. . . .

*The minimum* distance from the center of a standard size bolt hole to the edge of a connected part shall be as indicated in the following table:

*(Continued)*

Holes, Oversized and Slotted *(Continued)*

### Minimum Edge Distance

| Bolt Diameter in inches | From a sheared Edge | From a Rolled Edge |
|:---:|:---:|:---:|
| 1/2 | 7/8 | 3/4 |
| 5/8 | 1-1/8 | 7/8 |
| 3/4 | 1-1/4 | 1 |
| 7/8 | 1-1/2 | 1-1/8 |
| 1 | 1-3/4 | 1-1/4 |
| 1-1/8 | 2 | 1-1/2 |
| 1-1/4 | 2-1/4 | 1-5/8 |
| 1-1/2 | 2-5/8 | 1-7/8 |

Edge distances are measured from the edge of the connected material to the centerline of the hole.

*Because* an increase in hole size generally reduces the net area of a connected part, the use of oversize holes is subject to approval by the engineer. . . .

*Otherwise* the clear distance between holes shall not be less than one bolt diameter.

Repair of

D1.1, . . . The technique for making plug welds set forth in 4.21 of this code is not satisfactory for restoring the entire cross section of the base metals at mislocated holes. Plug welds are intended to transmit shear from one plane surface to another and not to develop the full cross section of the hole. One method of restoring unacceptable holes is to fill $\frac{1}{2}$ the depth or less with steel backing of the same material specification as the base metal, gouge an elongated boat-shaped cavity down to the backing, then fill the cavity by welding using the stringer bead technique. After the first side is welded, gouge another elongated boat-shaped cavity completely removing the temporary backing on the second side, and complete by welding using the stringer bead technique.

D1.1, 3.7.7

Size for Rivets and Bolts

> UBC, 2-605, M2.2.5

> The diameter of a hole for a rivet or bolt shall be ¹⁄₁₆ in. greater than the nominal diameter of the rivet or bolt to be installed .

> AISC, 5-33, B2
> AISC, 5-71, J3.2A

## Identification of Steel

UBC, 2-356, 2202.2.2

> UBC, . . . Steel that is not readily identifiable as to grade from markings and test records shall be tested to determine conformity to such standards. . . .

> UBC, 2-555, A3.1, (b)
> AISC, 5-92, M5.5

> *When* structural steel is furnished to a specified minimum yield point greater than 36,000 psi, the ASTM or other specification designation shall be included near the erection mark on each shipping assembly or important construction component over any shop coat of paint prior to shipment from the fabricator's plant. Pieces of such steel that are to be cut to smaller sizes, shall, before cutting, be legibly marked with the fabricator's identification mark on each of the smaller size pieces to provide continuity of identification. . . .

> *Any identification* method used shall at least verify and document the following items:

> 1. Material specification.
>
> 2. Heat number.
>
> 3. Material test reports for special requirements.

## Improperly Fitted or Welded Members

D1.1, 3.7.5

> D1.1, . . . The engineer shall be notified before improperly fitted and welded members are cut apart.

## Inspection

UBC, 2-45, 1701.1, .2, .3, .5

> UBC, . . . In addition to the inspections required by Section 108, the owner or the engineer or architect of record acting as the owner's agent shall em-

> UBC, 1701.6.1, .2, .7;
> UBC, 1702
> D1.1, Chap. 6

*(Continued)*

Inspection *(Continued)*

ploy one or more special inspectors who shall provide inspections during construction on the types of
work listed under Section 1701.5. . . .

*Except* as provided in Section 1701.1, the types of
work listed below shall be inspected by a special inspector.

Structural Welding

SBCCI, A103.8 - A103.8.8
SBCCI, 101.2.3
BOCA, 109.3 and 109.4

BOCA, 115.2 and 115.2.3.2
BOCA, 110.1 - 110.8

BOCA, 1306.3.2 - 1306.4

During the welding of any member or connection
that is designed to resist loads and forces required by this code work shall be inspected by a
special inspector. Exceptions include welding
done in an approved fabricator's shop in accordance with Section 1701.7. 2. The special inspector need not be continuously present during
welding of the following items, provided the materials, qualifications of welding procedures, and
welders are verified prior to the start of work;
periodic inspections are made of work in
progress; and a visual inspection of all welds is
made prior to completion or prior to shipment of
shop welding:

2.1  Single-pass fillet welds not exceeding $5/16$ in.
     in size.

2.2  Floor and roof deck welding.

2.3  Welded studs when used for structural di-
     aphragm or composite systems.

2.4  Welded sheet steel for cold-formed framing
     members such as studs and joists.

2.5  Welding of stairs and railing systems.

5.2  Special moment-resisting steel frames. . . .

5.3  Welding of reinforcing steel. . . .

High-Strength Bolting.

As required by Chapter 22, Division IV. Such inspections may be performed on a periodic basis
in accordance with the requirements of Section
1701.6.

Current, Polarity, Position

D1.1, . . . The inspector shall make certain that electrodes are used only in the positions and with the type of welding current and polarity for which they are classified.

General Requirements

Intervals

D1.1, . . . The inspector shall, at suitable intervals, observe joint preparation, assembly practice, the welding technique, and performance of each welder, welding operator, and tack welder to make certain that the applicable requirements of this code are met.

Of Materials

D1.1, . . . The inspector shall make certain that only materials conforming to the requirements of this code are used. . . .

*This* code is all-encompassing. It requires inspection of materials certification and mill test reports. It is important that this work be done in a timely manner so that the unacceptable materials are not incorporated into the work.

Of Welder and Tacker

D1.1, . . . The inspector shall permit welding to be performed only by welders, welding operators, and tack welders who are qualified in accordance with the requirements of 5.3 or 5.4, or shall make certain that each welder, welding operator, or tack welder has previously demonstrated such qualification under other suitable supervision.

*When* the quality of a qualified welder's, welding operator's, or tack welder's work appears to be below the standards of this code, the inspector may require that the welder, welding operator, or tack welder demonstrate an ability to produce sound welds by means of a simple test, such as a fillet-weld break test, or by requiring complete requalification in accordance with 5.3.

D1.1, 6.5.3

D1.1 Comm., 6.

D1.1, 6.5.4

D1.1, 6.2
D1.1 Comm., 6.2
BOCA, Table 1307.3.2

D1.1, 6.4
D1.1 Comm., 6.4 - 6.4.3

*(Continued)*

Inspection, Of Welds *(Continued)*

| Of Welds | D1.1, 8.15.1.9 |
|---|---|

D1.1, . . . Visual inspection of welds in all steels may begin immediately after the completed welds have cooled to ambient temperature. Acceptance criteria for ASTM A514 and A517 steels shall be based on visual inspection performed not less than 48 hours after completion of the weld.

BOCA, 1307.3.3.2

| Special | D1.1, 2.1.5 |
|---|---|

D1.1, . . . Any special inspection requirements shall be noted on the drawings or in the specifications. . . .

UBC, 2-43, 1701
BOCA, 1307 - 1307.3.3.3
BOCA, 108.2 - 108.2.1
BOCA, 201.0

## Inspector

Prints for

D1.1 Comm., 6.1.5

D1.1, . . . Inspectors need a complete set of approved drawings to enable them to properly do their work. They need be furnished only the portion of the contract documents describing the requirements of products that they will inspect. Much of the contract documents deal with matters that are not the responsibility of the inspector; these portions need not be furnished.

Responsibilities of

D1.1, 6.5 - 6.5.7

D1.1, . . . The inspector shall make certain that the size, length, and location of all welds conform to the requirements of this code and to the detail drawings and that no unspecified welds have been added without approval. . . .

UBC, 2-45, 1701.1 / .2 / .3 / .5
UBC, 2-45, 1701.6.1 / .2 / .7
UBC, 2-45, 1702

*The inspector* shall make certain that electrodes are used only in the positions and with the type of welding current and polarity for which they are classified. . . .

*The Inspector* shall make certain that only welding procedures are employed that meet the provisions of 5.1 or are qualified in accordance with 5.2 and 5.5. . . .

*The inspector* shall, at suitable intervals, observe joint preparation, assembly practice, the welding techniques, and performance of each welder, welding operator, and tack welder to make certain that the applicable requirements of this code are met. . . .

*Inspectors* shall identify with a distinguishing mark or other recording methods all parts of joints that they have inspected and accepted. Any recording method that is mutually agreeable may be used. Die stamping of dynamically loaded members is not permitted without the approval of the engineer. . . .

*The inspector* shall keep a record of qualifications of all welders, welding operators, and tack welders; all procedure qualifications or other tests that are made; and such other information as may be required. . . .

*The special* inspector shall observe the work assigned for conformance with the approved designed drawings and specifications. . . .

*The special* inspector shall furnish inspection reports to the building official, the engineer or architect of record, and other designated persons. All discrepancies shall be brought to the immediate attention of the contractor for correction, then, if uncorrected, to the proper design authority and to the building official. . . .

*Continuous* special inspection means that the special inspector is on the site at all times observing the work requiring special inspection. . . .

*Some* inspections may be made on a periodic basis and satisfy the requirements of continuous inspection, provided this periodic scheduled inspection is performed as outlined in the project plans and specifications and approved by the building official. . . .

*(Continued)*

Inspection, Responsibilities of *(Continued)*

Verification                                                                                 D1.1, 6.1.2

D1.1, . . . The verification inspector is the duly          D1.1 Comm., 6.1.4
designated person who acts on behalf of the
owner or engineer on all inspections and quality
matters within the scope of the contract documents. . . .

## Interpass Temperatures                                          D1.4, 5.2.1

Reinforcement                                                                         D1.4, Table 5.2

D1.4, . . . Minimum preheat and interpass temperatures shall be in accordance with Table 5.2
using the highest carbon equivalent number of
the base metal as determined in accordance
with 1.3.4.

## Table 5.2
## Minimum Preheat and Interpass Temperature[1,2]
## (see 5.2.1)

| Carbon Equivalent[3,4] (C.E.) Range, % | Size of Reinforcing Bar | Shielded Metal Arc Welding with Low Hydrogen Electrodes, Gas Metal Arc Welding, or Flux Cored Arc Welding | |
|---|---|---|---|
| | | Minimum Temperature | |
| | | °F | °C |
| Up to 0.40 | up to 11 inclusive | none[5] | none[5] |
| | 14 and 18 | 50 | 10 |
| Over 0.40 to 0.45 inclusive | up to 11 inclusive | none[5] | none[5] |
| | 14 and 18 | 100 | 40 |
| Over 0.45 to 0.55 inclusive | up to 6 inclusive | none[5] | none[5] |
| | 7 to 11 inclusive | 50 | 10 |
| | 14 to 18 | 200 | 90 |
| Over 0.55 to 0.65 inclusive | up to 6 inclusive | 100 | 40 |
| | 7 to 11 inclusive | 200 | 90 |
| | 14 to 18 | 300 | 150 |
| Over 0.65 to 0.75 | up to 6 inclusive | 300 | 150 |
| | 7 to 18 inclusive | 400 | 200 |
| Over 0.75 | 7 to 18 inclusive | 500 | 260 |

1. When reinforcing steel is to be welded to main structural steel, the preheat requirements of the structural steel shall also be considered (see ANSI/AWS D1.1, table titled "Minimum Preheat and Interpass Temperature." The minimum preheat requirement to apply in this situation shall be the higher requirement of the two tables. However, extreme caution shall be exercised in the case of welding reinforcing steel to quenched and tempered steels, and such measures shall be taken as to satisfy the preheat requirements for both. If not possible, welding shall not be used to join the two base metals.

2. Welding shall not be done when the ambient temperature is lower than 0°F (-18°C). When the base metal is below the temperature listed for the welding process being used and the size and carbon equivalent range of the bar being welded, it shall be preheated (except as otherwise provided) in such a manner that the cross section of the bar for not less than 6 in. (150 mm) on each side of the joint shall be at or above the specified minimum temperature. Preheat and interpass temperatures shall be sufficient to prevent crack formation.

3. After welding is complete, bars shall be allowed to cool naturally to ambient temperature. Accelerated cooling is prohibited.

4. Where it is impractical to obtain chemical analysis, the carbon equivalent shall be assumed to be above 0.75%. See also 1.3.4.

5. When the base metal is below 32°F (0°C), the base metal shall be preheated to at least 70°F (20°C), or above, and maintained at this minimum temperature during welding.

*(Continued)*

Interpass Temperatures *(Continued)*

Structural                                                      D1.1, 4.2

D1.1, . . . The preheat and interpass must be          D1.1, Table 4.3
sufficient to prevent cracking. Experience has
shown that the minimum temperatures specified
in Table 4.3 are adequate to prevent cracking in
most cases. However, increased preheat temperatures may be necessary in situations involving
higher restraint, higher hydrogen, lower welding
heat input, or steel composition at the top end of
the specification. Conversely, lower preheat temperatures may be adequate to prevent cracking
depending on restraint, hydrogen level, and actual steel composition or higher welding heat
input. . . .

*With* the exclusion of stud welding and electroslag and electrogas welding, the minimum
preheat and interpass temperatures shall be either in accordance with Table 4.3 for the welding
process being used and higher strength steel being welded or in accordance with 4.2.2.

## Table 4.3
### Minimum Preheat and Interpass Temperature[1,2] (see 4.2)

| Category | Steel Specification | | Welding Process | Thickness of Thickest Part at Point of Welding, in. | mm | Minimum Temperature, °F | °C |
|---|---|---|---|---|---|---|---|
| A | ASTM A36[3] | | Shielded metal arc welding with other than low hydrogen electrodes | Up to 3/4 | 19 incl. | None[4] | |
| | ASTM A53 | Grade B | | Over 3/4 thru 1-1/2 | 19 / 38.1 incl. | 150 | 66 |
| | ASTM A106 | Grade B | | Over 1-1/2 thru 2-1/2 | 38.1 / 63.5 | 225 | 107 |
| | ASTM A131 | Grades A, B, CS, D, DS, E | | Over 2-1/2 | 63.5 | 300 | 150 |
| | ASTM A139 | Grade B | | | | | |
| | ASTM A381 | Grade Y35 | | | | | |
| | ASTM A500 | Grade A | | | | | |
| | ASTM A501 | Grade B | | | | | |
| | ASTM A516 | Grades I & II | | | | | |
| | ASTM A524 | | | | | | |
| | ASTM A529 | | | | | | |
| | ASTM A570 | Grade 65 | | | | | |
| | ASTM A573 | Grade 36[3] | | | | | |
| | ASTM A709 | Grade B | | | | | |
| | API 5L | Grades X42 | | | | | |
| | ABS | Grades A, B, D, CS, DS | | | | | |
| | ABS | Grade E | | | | | |
| B | ASTM A36[3] | | Shielded metal arc welding with low hydrogen electrodes, submerged arc welding,[5] gas metal arc welding, gas tungsten arc welding, flux cored arc welding | Up to 3/4 | 19 incl. | None[4] | |
| | ASTM A53 | Grade B | | Over 3/4 thru 1-1/2 | 19 / 38.1 incl. | 50 | 10 |
| | ASTM A106 | Grade B | | Over 1-1/2 thru 2-1/2 | 38.1 / 63.5 incl. | 150 | 66 |
| | ASTM A131 | Grades A, B, CS, D, DS, E / AH 32 & 36 / DH 32 & 36 / EH 32 & 36 | | Over 2-1/2 | 63.5 | 225 | 107 |
| | ASTM A139 | Grade B | | | | | |
| | ASTM A242 | | | | | | |
| | ASTM A381 | Grade Y35 | | | | | |
| | ASTM A441 | | | | | | |
| | ASTM A500 | Grade A | | | | | |
| | | Grade B | | | | | |
| | ASTM A501 | | | | | | |
| | ASTM A516 | Grades 55 & 60, 65 & 70 | | | | | |
| | ASTM A524 | Grades I & II | | | | | |
| | ASTM A529 | | | | | | |
| | ASTM A537 | Classes 1 & 2 | | | | | |
| | ASTM A570 | All grades | | | | | |
| | ASTM A572 | Grades 42, 50 | | | | | |
| | ASTM A573 | Grade 65 | | | | | |
| | ASTM A588 | | | | | | |
| | ASTM A595 | Grades A, B, C | | | | | |
| | ASTM A606 | | | | | | |
| | ASTM A607 | Grades 45, 50, 55 | | | | | |
| | ASTM A618 | Grades A, B | | | | | |
| | | Grades C, D | | | | | |
| | ASTM A633 | Grades 36, 50, 50W | | | | | |
| | ASTM A709 | Grade B | | | | | |
| | ASTM A808 | Grade X42 | | | | | |
| | API 5L | Grades 42, 50 | | | | | |
| | API Spec. 2H | Grades AH 32 & 36 / DH 32 & 36 / EH 32 & 36 | | | | | |
| | ABS | Grades A, B, D, CS, DS | | | | | |
| | ABS | Grade E | | | | | |

(continued)

## Table 4.3 (continued)

| Category | Steel Specification | Welding Process | Thickness of Thickest Part at Point of Welding, in. | mm | Minimum Temperature, °F | °C |
|---|---|---|---|---|---|---|
| C | ASTM A572 Grades 60 & 65<br>ASTM A633 Grade E<br>API 5L Grade X52 | Shielded metal arc welding with low hydrogen electrodes, submerged arc welding,5 gas metal arc welding, gas tungsten arc welding, flux cored arc welding | Up to 3/4 | 19 incl. | 50 | 10 |
| | | | Over 3/4 thru 1-1/2 | 19 38.1 incl. | 150 | 66 |
| | | | Over 1-1/2 thru 2-1/2 | 38.1 63.5 incl. | 225 | 107 |
| | | | Over 2-1/2 | 63.5 | 300 | 150 |
| D | ASTM A514<br>ASTM A517<br>ASTM A709 Grades 100 & 100W | Shielded metal arc welding with low hydrogen electrodes, submerged arc welding5 with carbon or alloy steel wire, neutral flux, gas metal arc welding, gas tungsten arc welding, or flux cored arc welding | Up to 3/4 | 19 incl. | 50 | 10 |
| | | | Over 3/4 thru 1-1/2 | 19 38.1 incl. | 125 | 50 |
| | | | Over 1-1/2 thru 2-1/2 | 38.1 63.5 incl. | 175 | 80 |
| | | | Over 2-1/2 | 63.5 | 225 | 107 |
| E | ASTM A710 Grade A (All classes) | no preheat is required6 | | | | |

Notes:

1. Welding shall not be done when the ambient temperature is lower than 0° F (–18° C). Zero °F (–18° C) does not mean the ambient environmental temperature but the temperature in the immediate vicinity of the weld. The ambient environmental temperature may be below 0° F, but a heated structure or shelter around the area being welded could maintain the temperature adjacent to the weldment at 0° F or higher. When the base metal is below the temperature listed for the welding process being used and the thickness of material being welded, it shall be preheated (except as otherwise provided) in such manner that the surfaces of the parts on which weld metal is being deposited are at or above the specified minimum temperature for a distance equal to the thickness of the part being welded, but not less than 3 in. (75 mm) in all directions from the point of welding. Preheat and interpass temperatures must be sufficient to prevent crack formation. Temperature above the minimum shown may be required for highly restrained welds. For A514, A517, and A709 Grades 100 and 100W steel, the maximum preheat and interpass temperature shall not exceed 400° F (205° C) for thickness up to 1-1/2 in. (38.1 mm) inclusive, and 450° F (230° C) for greater thickness. Heat input when welding A514, A517, and A709 Grades 100 and 100W steel shall not exceed the steel producer's recommendations.

2. In joints involving combinations of base metals, preheat shall be as specified for the higher strength steel being welded.

3. Only low hydrogen electrodes shall be used when welding A36 or A709 Grade 36 steel more than 1 in. (25.4 mm) thick for dynamically loaded structures.

4. When the base metal temperature is below 32° F (0° C), the base metal shall be preheated to at least 70° F (21° C) and this minimum temperature maintained during welding.

5. For modification of preheat requirements for submerged arc welding with parallel or multiple electrodes, see 4.10.6 or 4.11.6.

6. Preheat is not required for the base metal. Preheat for E80XX-X filler metal shall be as for Grade 65; for higher strength filler metal treat as Group D.

# Joints

### Butt

D1.1, 3.6.3

D1.1, . . . Surfaces of butt joints required to be flush shall be finished so as not to reduce the thickness of the thinner base metal by more than $1/32$ in. or 5% of the thickness, whichever is smaller, nor leave reinforcement that exceeds $1/32$ in. However, all reinforcement must be removed where the weld forms part of a faying or contact surface.

### Alignment of

D1.1, 3.3.3

D1.1, . . . Parts to be joined at butt joints shall be carefully aligned. Where the parts are effectively restrained against bending due to eccentricity in alignment, an offset not exceeding 10% of the thickness of the thinner part joined, but in no case more than $1/8$ in. shall be permitted as a departure from the theoretical alignment. In correcting misalignment in such cases, the parts shall not be drawn in to a greater slope than $1/2$ in. in 12 in.

### Variance in

D1.1, 3.3.4

D1.1, . . . With the exclusion of electroslag and electrogas welding, and with the exception of 3.3.4.1, for root openings in excess of those permitted in Figure 3.3, the dimensions of the cross section of the groove welded joints that vary from those shown on the detail drawings by more than the following tolerances shall be referred to the engineer for approval or correction.

*(Continued)*

Joints *(Continued)*

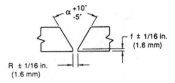

(A)  GROOVE WELD WITHOUT BACKING -
ROOT NOT BACKGOUGED

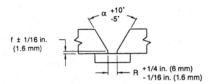

(B)  GROOVE WELD WITH BACKING -
ROOT NOT BACKGOUGED

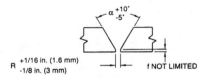

(C)  GROOVE WELD WITHOUT BACKING -
ROOT BACKGOUGED

|  | Root not back gouged* | | Root back gouged | |
|---|---|---|---|---|
|  | in. | mm | in. | mm |
| (1) Root face of joint | ±1/16 | 1.6 | Not limited | |
| (2) Root opening of joints without backing | ±1/16 | 1.6 | +1/16 −1/8 | 1.6 3 |
| Root opening of joints with backing | +1/4 −1/16 | 6 1.6 | Not applicable | |
| (3) Groove angle of joint | +10° −5° | | +10° −5° | |

*see 10.13.1 for tolerances for complete joint penetration
tubular groove welds made from one side without backing.

**Figure 3.3 — Workmanship Tolerances in
Assembly of Groove Welded Joints (see 3.3.4)**

Lap

D1.1, . . . For lap joints, the minimum amount
of lap shall be five times the thickness of the
thinner part joined but not less than 1 in. . . .

*Lap* joints in parts carrying axial stress shall be
double fillet welded except where deflection of
the joint is sufficiently restrained to prevent it
from opening under load.

D1.1, 8.8.3

D1.1, 9.10
AISC, 5-67, J2.2.b

Surfaces of

UBC, . . . When assembled, all joint surfaces, including surfaces adjacent to the bolt head and nut, shall be free of scale, except tight mill scale, and shall be free of dirt or other foreign material. Burrs that would prevent solid seating of the connected parts in the snug tight condition shall be removed. . . .

*Paint* is permitted on the faying surfaces unconditionally in connections, except in slip-critical connections as defined in Section 2224.1. . . .

UBC, 2-446, 2222.2

UBC, 2-446, 2221.7

# Joists (Open-Web Steel)

UBC, 2-377, Division III
UBC, 2-384, Section 5.4

Bridging

UBC, 2-407, Section 104.5
UBC, 2-385, Bridging Table

### Number Of Rows Of Bridging***
### Distances are Span Lengths
### (See "Definition of Span" in front of load table)

| *Section Number | 1 Row | 2 Rows | 3 Rows | 4 Rows** | 5 Rows** |
|---|---|---|---|---|---|
| | Multiply table span lengths × 304.8 for mm | | | | |
| #1 | Up thru 16' | Over 16' thru 24' | Over 24' thru 28' | | |
| #2 | Up thru 17' | Over 17' thru 25' | Over 25' thru 32' | | |
| #3 | Up thru 18' | Over 18' thru 28' | Over 28' thru 38' | Over 38' thru 40' | |
| #4 | Up thru 19' | Over 19' thru 28' | Over 28' thru 38' | Over 38' thru 48' | |
| #5 | Up thru 19' | Over 19' thru 29' | Over 29' thru 39' | Over 39' thru 50' | Over 50' thru 52' |
| #6 | Up thru 19' | Over 19' thru 29' | Over 29' thru 39' | Over 39' thru 51' | Over 51' thru 56' |
| #7 | Up thru 20' | Over 20' thru 33' | Over 33' thru 45' | Over 45' thru 58' | Over 58' thru 60' |
| #8 | Up thru 20' | Over 20' thru 33' | Over 33' thru 45' | Over 45' thru 58' | Over 58' thru 60' |
| #9 | Up thru 20' | Over 20' thru 33' | Over 33' thru 46' | Over 46' thru 59' | Over 59' thru 60' |
| #10 | Up thru 20' | Over 20' thru 37' | Over 37' thru 51' | Over 51' thru 60' | |
| #11 | Up thru 20' | Over 20' thru 38' | Over 38' thru 53' | Over 53' thru 60' | |
| #12 | Up thru 20' | Over 20' thru 39' | Over 39' thru 53' | Over 53' thru 60' | |

*Last digit(s) of joist designation shown in Load Tables.
**Where 4 or 5 rows of bridging are required, a row nearest the mid-span of the joist shall be diagonal bridging with bolted connections at chords and intersection.
***See Section 5.11 for additional bridging required for uplift design.

*(Continued)*

Joints *(Continued)*

Camber of                                                                            UBC, 2-383, 4.7

**4.7  CAMBER**

Camber is optional with the manufacturer but, when provided, recommended approximate camber is as follows:

| Top Chord Length | Approximate Camber |
|---|---|
| 20 feet (6096 mm) | $1/4$ inch (7 mm) |
| 30 feet (9144 mm) | $3/8$ inch (10 mm) |
| 40 feet (12 192 mm) | $5/8$ inch (16 mm) |
| 50 feet (15 240 mm) | 1 inch (26 mm) |
| 60 feet (18 288 mm) | $1^1/2$ inches (38 mm) |

In no case will joists be manufactured with negative camber.

End Anchorages                                                            UBC, 2-385, 5.6

UBC. . . Ends of K-Series joists resting on steel bearing plates on masonry or structural concrete shall be attached thereto with a minimum of two $1/8$-in. fillet welds 1 in. long, or with a $1/2$ in. bolt. . . .

*Ends of K-Series* joists resting on steel supports shall be attached thereto with a minimum of two $1/8$ in. fillet welds 1 in. long, or with a $1/2$ in. bolt. In steel frames, where columns are not framed in at least two directions with structural steel members, joists at column lines shall be field bolted at the columns to provide lateral stability during construction.

Handling and Erection                                                  UBC, 2-387, Section 6

UBC . . . Care should be exercised at all times to avoid damage to the joists and accessories through careless handling during unloading, storing, and erecting.

** **IMPORTANT**** UNDER NO CIRCUM-STANCES SHALL ANY PERSONNEL AT-TEMPT TO WALK ON UNBRIDGED JOISTS. AS SOON AS THE JOISTS ARE ERECTED, ALL BRIDGING SHALL BE COMPLETELY IN-STALLED AND ANCHORED, THEN THE JOISTS PERMANENTLY FASTENED INTO PLACE. UNTIL THIS IS DONE, NO CON-STRUCTION LOADS SHALL BE APPLIED TO THE JOISTS. . . .

Inspection of Welding

UBC, 2-383, 4.5
UBC, 2-387, 5.12
UBC, 2-386, 5.8(b)

Slab Thickness

UBC . . . Cast-in-place slabs shall not be less than 2 in. thick.

Span

UBC, 2-384, 5.2

UBC . . . The span of a joist shall not exceed 24 times its depth.

Welding of

UBC, 2-382, 4.5
UBC, 2-385, 5.6
UBC, 2-404, 103.5
UBC, 2-426, 1003.5

## Kips (Minimum Tension in)

AISC, 5-77, Table J3.7
UBC, 2-454, Table 22-IV-D
UBC, 2-591, Table J3.1

### Minimum Bolt Tension in Kips

| Bolt Diameters in inches | A325 | A490 |
|---|---|---|
| 1/2 | 12 | 15 |
| 5/8 | 19 | 24 |
| 3/4 | 28 | 35 |
| 7/8 | 39 | 49 |
| 1 | 51 | 64 |
| 1-1/8 | 56 | 80 |
| 1-1/4 | 71 | 102 |
| 1-3/8 | 85 | 121 |
| 1-1/2 | 103 | 148 |

## Lap Joints

AISC, 5-67, J2.2.b

D1.1, . . . For lap joints, the minimum amount of lap shall be five times the thickness of the thinner part joined but not less than 1 in. . . .

D1.1, 8.8.3
D1.1, 9.10

## Limitations (Thickness)

D1.1, 1.2.3

D1.1, . . . The provisions of this code are not intended to apply to welding base metals less than $\frac{1}{8}$ in. thick.

## Link Beam Column Connections

UBC, 2-368, 2211.10.12

UBC, . . . Where a link beam is connected to a column flange, the following requirements shall be met:

1. The beam flanges shall have full-penetration welds to the column.

2. Where the link beam strength is controlled by shear in conformance with Section 2211.10.8, the web connection shall be welded to develop the full link beam web shear strength.

3. Where the link beam is connected to the column web, the beam flanges shall have full-penetration welds to the connection plates and the web connection shall be welded to develop the link beam shear strength. . . .

## Longitudinal Fillet Weld Spacing

D1.1, 8.8.1

D1.1, . . . If longitudinal fillet welds are used alone in end connections of flat bar tension members, the length of each fillet weld shall be no less than the perpendicular distance between them. The transverse spacing of longitudinal fillet welds used in end connections shall not exceed 8 in. unless end transverse welds or intermediate plug or slot welds are used.

D1.1, 8.12.1
D1.1, 9.17
AISC, 5-43, E4

D1.1, . . . In built-up tension members, the longitudinal spacing of stitch welds connecting a plate component to other components, or connecting two plate components to each other, shall not exceed 12 in. or 24 times the thickness of the thinner plate.

*The maximum* longitudinal spacing of stitch welds connecting two or more rolled shapes in contact with one another shall not exceed 24 in.

*Connections* or splices of tension members made by groove welds shall have complete joint penetration welds. . . .

## Low Carbon Steel Bolts

UBC Std., 22-1, 3-659

## Low Hydrogen Electrode Storage

D1.1, 4.5.2

D1.1, . . . All electrodes having low hydrogen coverings conforming to ANSI/AWS A5.1 shall be purchased in hermetically sealed containers or shall be dried for at least 2 hours between 500 and 800 deg. F before they are used. Electrodes having low hydrogen coverings conforming to ANSI/AWS A5.5 shall be purchased in hermetically sealed containers or shall be dried at least 1 hour between 700 and 800 deg. F before being used. Electrodes shall be dried prior to use if the hermetically sealed container shows evidence of damage. . . .

D1.1, Table 4.6
D1.1 Comm., 4.5

*Immediately* after opening of the hermetically sealed container or removal of the electrodes from the drying ovens, electrodes shall be stored in ovens held at temperatures of at least 250 deg. F. . . .

*The* ability of low hydrogen electrodes to prevent underbead cracking is dependent on the moisture content in the coating. During welding, the moisture dissociates into hydrogen and oxygen; hydrogen is absorbed into the molten metal and porosity and cracks may appear in the weld after the weld metal solidifies. The provisions of the code for handling, storage, drying, and use of low hydrogen electrodes should be strictly adhered to in order to prevent moisture absorption by the coating material.

## Table 4.6
### Permissible Atmospheric Exposure of Low Hydrogen Electrodes (see 4.5.2.1)

| Electrode | Column A (hours) | Column B (hours) |
|---|---|---|
| **A5.1** | | |
| E70XX | 4 max | Over 4 to 10 max |
| E70XXR | 9 max | |
| E70XXHZR | 9 max | |
| E7018M | 9 max | |
| **A5.5** | | |
| E70XX-X | 4 max | Over 4 to 10 max |
| E80XX-X | 2 max | Over 2 to 10 max |
| E90XX-X | 1 max | Over 1 to 5 max |
| E100XX-X | 1/2 max | Over 1/2 to 4 max |
| E110XX-X | 1/2 max | Over 1/2 to 4 max |

Notes:
1. Column A: Electrodes exposed to atmosphere for longer periods than shown shall be redried before use.
2. Column B: Electrodes exposed to atmosphere for longer periods than those established by testing shall be redried before use.
3. Entire table: Electrodes shall be issued and held in quivers, or other small open containers. Heated containers are not mandatory.
4. The optional supplemental designator, R, designates a low hydrogen electrode which has been tested for covering moisture content after exposure to a moist environment for 9 hours and has met the maximum level permitted in ANSI/AWS A5.1-91, *Specification for Carbon Steel Electrodes for Shielded Metal Arc Welding.*

## Matching Filler Metal Requirements

AISC 5-28, A3.6
SBCCI 1507

Reinforcing Steel (see Tables)

D1.4, 5.1

D1.4, . . . For direct butt joints, indirect butt joints, lap joints, and anchorage, base plates, and insert connections, the filler metal, filler metal-gas combinations, or classification of filler metal shall be in accordance with Table 5.1. . . .

D1.4, Table 5.1

*When* joining different grades of steels, the filler metal shall be selected for the lower strength base metal.

See turned table on pages 175 and 176.

Restrictions for A242 and A588

D1.1, 4.1.4

D1.1, . . . For exposed, bare, unpainted applications of ASTM A242 and A588 steel requiring

D1.1, Table 4.2

**Table 5.1**
**Matching Filler Metals Requirements**
**(See 5.1)**

| Group | Steel Specification | | Steel Specification Requirements | | | | Filler Metal Requirements | | | | |
|---|---|---|---|---|---|---|---|---|---|---|---|
| | | | Minimum Yield Point/Strength | | Minimum Tensile Strength | | Electrode Specification[4] | Yield Point/Strength[1] | | Tensile Strength[1] | |
| | | | ksi | MPa | ksi | MPa | | ksi | MPa | ksi | MPa |
| I | ASTM A615 | Grade 40 | 40 | — | 70 | — | SMAW AWS A5.1 and A5.5 E7015, E7016, E7018, E7028, E7015-X, E7016-X, E7018-X | 60 / 57 | 415 / 390 | 72 / 70 | 495 / 480 |
| | ASTM A615M | Grade 300 | — | 300 | — | 500 | GMAW AWS A5.18 ER70S-X | 60 | 415 | 72 | 495 |
| | ASTM A617 | Grade 40 | 40 | — | 70 | — | FCAW AWS A5.20 E7XT-X (Except -2, -3, -10, -GS) | 60 | 415 | 72 | 495 |
| | ASTM A617M | Grade 300 | — | 300 | — | 500 | | | | | |
| II | ASTM A616 | Grade 50 | 50 | — | 80 | — | SMAW AWS A5.5 E8015-X, E8016-X, E8018-X | 67 | 460 | 80 | 550 |
| | ASTM A616M | Grade 350 | — | 350 | — | 550 | GMAW AWS A5.28 ER80S-X | 68 | 470 | 80 | 550 |
| | ASTM A706 | Grade 60 | 60 | — | 80 | — | FCAW AWS A5.29 E8XTX-X | 68 | 470 | 80-100 | 550-690 |
| | ASTM A706M | Grade 400 | — | 400 | — | 550 | | | | | |
| III | ASTM A615 | Grade 60 | 60 | — | 90 | — | SMAW AWS A5.5 E9015-X, E9016-X, E9018-X | 77 | 530 | 90 | 620 |
| | ASTM A615M | Grade 400 | — | 400 | — | 600 | GMAW AWS A5.28 ER90S-X | 78 | 540 | 90 | 620 |
| | ASTM A616 | Grade 60 | 60 | — | 90 | — | FCAW AWS A5.29 E9XTX-X | 78 | 540 | 90-110 | 620-760 |
| | ASTM A616M | Grade 400 | — | 400 | — | 600 | | | | | |
| | ASTM A617 | Grade 60 | 60 | — | 90 | — | | | | | |
| | ASTM A617M | Grade 400 | — | 400 | — | 600 | | | | | |

**(Continued)**

## Table 5.1
### (Continued)

| | Steel Specification Requirements | | | | | Filler Metal Requirements | | | | |
|---|---|---|---|---|---|---|---|---|---|---|
| Group | Steel Specification | Minimum Yield Point/Strength ksi | MPa | Minimum Tensile Strength ksi | MPa | Electrode Specification[4] | Yield Point/Strength[1] ksi | MPa | Tensile Strength[1] ksi | MPa |
| | | | | | | **SMAW AWS A5.5** | | | | |
| | | | | | | E10015-X, E10016-X, E10018-X | 87 | 600 | 100 | 690 |
| | | | | | | E10018-M | 88-100 | 610-690 | 100 | 690 |
| | | | | | | **GMAW AWS A5.28** | | | | |
| IV | ASTM A615 Grade 75[2] | 75 | — | 100 | — | ER100S-X | 88-102 | 610-700 | 100 | 690 |
| | | | | | | **FCAW AWS A5.29** | | | | |
| | ASTM A615M Grade 500[3] | — | 500 | — | 700 | E10XTX-X | 88 | 610 | 100-120 | 690-830 |

Notes:

1. This table is based on filler metal as-welded properties. Single values are minimums. Hyphenated values indicated minimum and maximum.
2. Applicable to bar sizes Nos. 11, 14, and 18.
3. Applicable to bar sizes Nos. 35, 45, and 55.
4. Filler metals classified in the postweld heat treated (PWHT) condition by the AWS filler metal specification may be used when given prior approval by the Engineer. Consideration shall be made of the differences in tensile strength, ductility and hardness between the PWHT versus as-welded condition.

weld metal with atmospheric corrosion resistance
and coloring characteristics similar to that of the
base metal, the electrode or electrode-flux combination shall be in accordance with Table 4.2. . . .

**Table 4.2**
**Filler Metal Requirements for Exposed Bare Applications of**
**ASTM A242 and A588 Steel (see 4.1.4)**

| Welding Processes | | | |
|---|---|---|---|
| Shielded Metal Arc | Submerged Arc | Gas Metal Arc or Gas Tungsten Arc | Flux Cored Arc |
| A5.5 | A5.23 [1,4] | A5.28 [4] | A5.29 |
| E7018-W | F7AX-EXXX-W | | |
| E8018-W | | | E8XT1-W |
| E8016-C3 or E8018-C3 | F7AX-EXXX-Ni1 [2] | ER80S-Ni1 | E8XTX-Ni1 |
| E8016-C1 or E8018-C1 | F7AX-EXXX-Ni4 [2] | | |
| E8016-C2 or E8018-C2 | | | |
| E7016-C1L or E7018-C1L | F7AX-EXXX-Ni2 [2] | ER80S-Ni2 | E8XTX-Ni2 |
| E7016-C2L or E7018-C2L | F7AX-EXXX-Ni3 [2] | ER80S-Ni3 | E80T5-Ni3 |
| E8018-B2L [1] | | ER80S-B2L [1] | E80T5-B2L [1] |
| | | ER80S-G [1,3] | |
| | | | E71T8-Ni1 |
| | | | E71T8-Ni2 |
| | | | E7XTX-K2 |

Restrictions for A514 and A517            D1.1, 4.5.3

D1.1, . . . When used for welding ASTM A514
or A517 steels, electrodes of any classification
lower than E100XX-X, except for E7018M and
E70XXH4R, shall be dried at least 1 hour at temperatures between 700 and 800 deg. F before being used, whether furnished in hermetically
sealed containers or otherwise.

Structural Steel (see Tables)             D1.1, Table 4.1

D1.1, . . . The electrode or electrode-flux combi-        AISC 5-28, A3.6
nation for complete joint penetration or partial
joint penetration groove welds, and for fillet
welds shall be as specified in Tables 8.1, 9.1, and
10.1, as applicable. . . .

*When* matching weld metal as required in
Tables 8.1, 9.1, or 10.1, the electrode or electrode-
flux combination shall be in accordance with
Table 4.1.

*(Continued)*

# Materials

Approved for Use

AISC, 5-25, A3.1

D1.1, . . . The base metals to be welded under this code are carbon and low alloy steels commonly used in the fabrication of steel structures. Steels complying with the specifications listed in 8.2, 9.2, and 10.2, together with special requirements applicable individually to each type of structure, are approved for use with this code.

D1.1, 1.2.2
D1.1, 8.2
D1.1, 9.2
D1.1, 10.2
UBC, 2-555, A3
UBC, 3-659, 22-1

1. ASTM A36, Specification for Structural Steel

2. Pipe, ASTM A53, Grade B, Specification for Pipe, Steel, Black and Hot-dipped, Zinc-Coated, Welded and Seamless

3. ASTM A242, Specification for High-Strength Low-Alloy Structural Steel (if the properties are suitable for welding)

4. ASTM A441, Specification for High-Strength Low-Alloy Structural Manganese-Vanadium Steel

5. ASTM A500, Specification for Cold-Formed Welded and Seamless Carbon Steel Structural Tubing in Rounds and Shapes

6. ASTM A501, Specification for Hot-Formed Welded and Seamless Carbon Steel Structural Tubing

7. ASTM A514, Specification for High-Yield Strength, Quenched and Tempered Alloy-Steel Plate, Suitable for Welding

8. ASTM A516, Specification for Pressure Vessel Plates, Carbon Steel, for Moderate- and Lower-Temperature Service

9. ASTM A517, Specification for Pressure Vessel Plates, Alloy Steel, High-Strength, Quenched and Tempered

10. ASTM A529, Specification for Standard Steel with 42 ksi Minimum Yield Point ($\frac{1}{2}$ in. [13mm] Maximum Thickness)

11. ASTM A570, Specification for Hot-Rolled Carbon Steel Sheet and Strip

12. ASTM A572, Specification for High-Strength, Low-Alloy Columbium-Vanadium Steels of Structural Quality

13. ASTM A588, Specification for High-Strength, Low-Alloy Structural Steel with 50 ksi Minimum Yield Point to 4-in. Thick

14. ASTM A606, Type 2 (Type 4 if the properties are suitable for welding), Specification for Steel Sheet and Strip, Hot-Rolled and Cold-Rolled High-Strength, Low-Alloy, with Improved Corrosion Resistance

15. ASTM A607, Grades 45, 50, and 55, Specification for Steel Sheet and Strip, Hot-Rolled or Cold-Rolled, High-Strength, Low-Alloy, Columbium and/or Vanadium

16. ASTM A618, Grades II and III (Grade I if the properties are suitable for welding), Specification for Hot-Formed Welded and Seamless High-Strength Low-Alloy Structural Tubing

17. ASTM A633, Specification for Normalized High-Strength Low-Alloy Structural Steel

18. ASTM A709 Specification for Structural Steel for Bridges

19. ASTM A710, Grade A, Specification for Low Carbon Age-Hardening Nickle-Copper-Chromium-Molybdenum-Columbium and Nickle-Copper Columbium Alloy Steels

20. ASTM A808, Specification for High-Strength Low-Alloy Carbon, Manganese, Columbium, Vanadium Steel, of Structural Quality with Improved Notch Toughness.

*Certified mill* test reports or certified reports of tests made by the fabricator or a testing laboratory in accordance with ASTM A6 or A568, as applicable, and the governing specification shall constitute sufficient evidence of conformity with one of the above ASTM standards. Additionally, the fabricator shall, if requested, provide an affidavit stating the structural steel furnished meets the requirements of the grade specified.

*(Continued)*

Materials *(Continued)*

Inspection of                                                                    D1.1 Comm., 6.2

D1.1, . . . This code provision is all-encompassing. It requires inspection of materials and re-
view of materials certification and mill test reports. It is important that this work be done in a
timely manner so that the unacceptable materials are not incorporated in the work.

Standards for                                                                    SBCCI 1509

# Maximum

Diameter of Root Pass (S.M.A.W.)                                                 D1.1, 4.6.5 - 4.6.6

D1.1, . . . The maximum thickness of root
passes in groove welds shall be $\frac{1}{4}$ in. . . .

Edge Distance                                                                    AISC, 5-76, J3.10

When fasteners are used to connect a plate with                                  UBC, 2-699, J3.10
a shape, or two plates together, the center-to-
center spacing shall not exceed a maximum of 14
times the thickness of the thinner part, nor 7 in.

*For unpainted* built-up members made of
weather resistant steel, the maximum edge distance shall not be greater than 8 times the thickness of the thinner part, or 5 in.

*The maximum* edge distance as measured from
the center of a bolt hole to the nearest edge of
the connection shall not be greater than 12 times
the thickness of the connected part and shall not
be greater than 6 in.

UBC, . . . The maximum distance from the center of any rivet or bolt to the nearest edge of
parts in contact shall be 12 times the thickness
of the connected part under consideration, but
shall not exceed 6 in. . . .

Effective Weld Sizes                                                             UBC, 2-689, J2

UBC, . . . The maximum size of fillet welds that                                 UBC, 2-691, Table J2.4
is permitted along edges of connected parts shall
be:

- Material less than $\frac{1}{4}$ in. thick, not greater than
  the thickness of the material.

- Material $1/4$ in. or more in thickness, not greater than the thickness of the material minus $1/16$ in., unless the weld is especially designated on the drawings to be built out to obtain full-throat thickness.

### TABLE J2.4
### Minimum Size of Fillet Welds

| Material Thickness of Thicker Part Joined (in.) | Minimum Size of Fillet Weld[a] (in.) |
|---|---|
| × 25.4 for mm | |
| To $1/4$ inclusive | $1/8$ |
| Over $1/4$ to $1/2$ | $3/16$ |
| Over $1/2$ to $3/4$ | $1/4$ |
| Over $3/4$ | $5/16$ |

[a]Leg dimension of fillet welds. Single-pass welds must be used.

Electrode Diameter

S.M.A.W.                                                    D1.1, 4.6.3

D1.1, . . . The maximum diameter of electrodes shall be:

1. $5/16$ in. for all welds made in the flat position, except root passes.

2. $1/4$ in. for horizontal fillet welds.

3. $1/4$ in. for root passes of fillet welds made in the flat position and groove welds made in the flat position with backing and with a root opening of $1/4$ in. or more.

4. $5/32$ in. for welds made with EXX14 and low hydrogen electrodes in the vertical and overhead positions.

5. $3/16$ in. for root passes of groove welds and for all other welds not included under 4.6.3(1) - (4).

G.M.A.W.                                                    D1.1, 4.14.1.2

D1.1, . . . The maximum diameter of welding electrodes shall be:

1. $5/32$ in. for the flat and horizontal positions.

2. $3/32$ in. for the vertical position.

3. $5/64$ in. for the overhead position.

*(Continued)*

Maximum, Electrode Diameter *(Continued)*

F.C.A.W. (Same as G.M.A.W.)                          D1.1, 4.14.1.2

S.A.W.                                               D1.1, 4.7.3

D1.1, . . . The diameter of electrodes shall
not exceed $1/4$ in.

Root Opening of a Fillet Weld                        D1.1, 3.3.1

D1.1, . . . The parts to be joined by fillet welds
shall be brought into as close contact as practicable. The root opening shall not exceed $3/16$ in. except in cases involving either shapes or plates 3
in. or greater in thickness if, after straightening
and in assembly, the root opening cannot be
closed sufficiently to meet this tolerance. In such
cases, a maximum root opening of $5/16$ in. is acceptable provided suitable backing is used. If the
separation is greater than $1/16$ in., the leg of the
fillet weld shall be increased by the amount of
the root opening, or the contractor shall demonstrate that the required effective throat has been
obtained.

Single Pass and Root Pass (S.M.A.W.)                 D1.1, 4.6.6 - 4.6.6.7.2

D1.1, . . . The maximum size of single pass fillet welds and root passes of multi-pass fillet
welds shall be:

1. $3/8$ in. for the flat position.

2. $5/16$ in. for the horizontal or overhead positions.

3. $1/2$ in. for the vertical position.

Size Fillet Weld                                     AISC, 5-67, J2.2B

D1.1, . . . The maximum fillet weld size de-        UBC, 2-691, J2.2.b
tailed along edges of material shall be:            D1.1, 2.7.1.2

1. The thickness of the base metal, for metal less
   than $1/4$ in. thick. . . .

2. $1/16$ in. less than the thickness of the base
   metal, for metal $1/4$ in. or more in thickness,
   unless the weld is designated on the drawings
   to be built out to obtain full throat thickness.
   In the as-welded condition, the distance be-

tween the edge of the base metal and the toe of the weld may be less than $1/16$ in., provided the weld size is clearly verifiable.

Thickness of Root Pass (S.M.A.W.)                    D1.1, 4.6.5

D1.1, . . . The maximum thickness of root passes in groove welds shall be $1/4$ in.

Wind Velocity                                        D1.1, 4.14.3

D1.1, . . . Gas metal arc or flux cored arc weld-    D1.4, 5.2.2
ing with external gas shielding shall not be done in a draft or wind unless the weld is protected by a shelter. Such shelter shall be of a material and shape appropriate to reduce wind velocity in the vicinity of the weld to a maximum of 5 miles per hour.

D1.4, . . . Welding shall not be done when the ambient temperature is lower than 0 deg. F, when surfaces to be welded are exposed to rain, snow, or wind velocities greater than 5 miles per hour, or when welders are exposed to inclement conditions.

## Metals (Base)

Reinforcing                                          D1.4, 1.3
                                                     SBCCI 1602.5
Sheet Steel                                          D1.3, 1.2 - 1.2.3

Structural                                           D1.1, 1.2
                                                     D1.1, 8.2.1
                                                     D1.1 Comm., 1.2
                                                     D1.1 Comm., 3.2.1

Preparation of

D1.1 Comm., . . . Girder web-to-flange welds are usually minimum size fillet welds deposited at relatively high speeds; these welds may ex-hibit piping porosity when welded over heavy mill scale often found on thick flange plates. It is only for these flange-to-web welds in girders that the mandatory requirement to completely re-move mill scale applies.                            D1.1, 10.2

Tubular

## Metal Reinforcement

## Mill Scale

D1.1, . . . Surfaces on which weld metal is to be deposited shall be smooth, uniform, and free from fins, tears, cracks, and other discontinuities that would adversely affect the quality or strength of the weld. Surfaces to be welded, and surfaces adjacent to a weld, shall also be free from loose or thick scale, slag, rust, moisture, grease, and other foreign material that would prevent proper welding or produce objectionable fumes. Mill scale that can withstand vigorous wire brushing, a thin rust-inhibitive coating, or antispatter compound may remain with the following exception: for girders in dynamically loaded structures, all mill scale shall be removed from the surfaces on which flange-to-web welds are to be made by submerged arc welding or by shielded metal arc welding with low hydrogen electrodes.

D1.1 Comm. . . It is only for these flange-to-web welds in girders that the mandatory requirement to completely remove mill scale applies.

## Minimum

Ambient Temperature

D1.1, . . . Welding shall not be done when the ambient temperature is lower than 0 deg. F, when surfaces are wet or exposed to rain, snow, or high wind velocities, or when welding personnel are exposed to inclement conditions.

*D1.1 Comm., . . .* Experience has shown that welding personnel cannot produce optimum results when working in an environment where the temperature is lower than 0 deg. F. Reference is made in 3.1.4 to 4.2 relative to the use of a heated structure or shelter to protect the welder, and the area being welded, from inclement weather conditions.

Bolt Tension in kips

AISC 5-77, Table J3.7

UBC, . . . In addition, all connections specified to be slip-critical or subject to axial tension, the special inspector shall assure that the specified procedure was followed to achieve the pretension specified in Table 22-IV-D. . . .

UBC, 2-454, Table 22-IV-D,<br>2-591, Table J3.1

**TABLE 22-IV-D—FASTENER TENSION REQUIRED FOR SLIP-CRITICAL CONNECTIONS AND CONNECTIONS SUBJECT TO DIRECT TENSION**

| NOMINAL BOLT SIZE, INCHES | MINIMUM TENSION[1] IN 1000's OF POUNDS (kips) | |
|---|---|---|
| | × 4448 for N | |
| × 25.4 for mm | A 325 Bolts | A 490 Bolts |
| 1/2 | 12 | 15 |
| 5/8 | 19 | 24 |
| 3/4 | 28 | 35 |
| 7/8 | 39 | 49 |
| 1 | 51 | 64 |
| 1 1/8 | 56 | 80 |
| 1 1/4 | 71 | 102 |
| 1 3/8 | 85 | 121 |
| 1 1/2 | 103 | 148 |

[1]Equal to 70 percent of specified minimum tensile strength of bolts (as specified for tests of full-size A 325 and A 490 bolts with UNC threads loaded in axial tension) rounded to the nearest kip.

Edge Distance

UBC, 2-698, J3.9

The minimum distance from the center of a standard size bolt hole to the edge of a connected part shall be as indicated in the following table:

UBC, 2-699, Table J3.5<br>AISC, 5-75, J3.9<br>AISC, 5-76, Table J3.5

**Minimum Edge Distance**

| Bolt Diameter in inches | From a sheared Edge | From a Rolled Edge |
|---|---|---|
| 1/2 | 7/8 | 3/4 |
| 5/8 | 1-1/8 | 7/8 |
| 3/4 | 1-1/4 | 1 |
| 7/8 | 1-1/2 | 1-1/8 |
| 1 | 1-3/4 | 1-1/4 |
| 1-1/8 | 2 | 1-1/2 |
| 1-1/4 | 2-1/4 | 1-5/8 |
| 1-1/2 | 2-5/8 | 1-7/8 |

Edge distances are measured from the edge of the connected material to the centerline of the hole.

Effective Throat of Partial Penetration Weld

UBC, 2-691, Table J2.3

UBC, . . . The minimum effective throat thickness of a partial-penetration groove weld shall be as shown in Table J2.3. Minimum effective throat thickness is determined by the thicker of

D1.1, Table 2.3

*(Continued)*

Minimum, Effective Throat of Partial Penetration
Weld *(Continued)*

the two parts joined, except that the weld size
need not exceed the thickness of the thinner part
joined.

### TABLE J2.3
### Minimum Effective Throat Thickness of
### Partial-penetration Groove Welds

| Material Thickness of Thicker Part Joined (in.) | Minimum Effective Throat Thickness[a] (in.) |
|---|---|
| × 25.4 for mm | |
| To 1/4 inclusive | 1/8 |
| Over 1/4 to 1/2 | 3/16 |
| Over 1/2 to 3/4 | 1/4 |
| Over 3/4 to 1 1/2 | 5/16 |
| Over 1/2 to 2 1/4 | 3/8 |
| Over 2 1/4 to 6 | 1/2 |
| Over 6 | 5/8 |

[a]See Sect. J2.

### Table 2.3
### Minimum Weld Size for Partial Joint
### Penetration Groove Welds (see 2.10.3)

| Base Metal Thickness of Thicker Part Joined | | Minimum Weld Size* | |
|---|---|---|---|
| in. | mm | in. | mm |
| 1/8 (3.2) to 3/16 (4.8) incl. | | 1/16 | 2 |
| Over 3/16 (4.8) to 1/4 (6.4) incl. | | 1/8 | 3 |
| Over 1/4 (6.4) to 1/2 (12.7) incl. | | 3/16 | 5 |
| Over 1/2 (12.7) to 3/4 (19.0) incl. | | 1/4 | 6 |
| Over 3/4 (19.0) to 1-1/2 (38.1) incl. | | 5/16 | 8 |
| Over 1-1/2 (38.1) to 2-1/4 (57.1) incl. | | 3/8 | 10 |
| Over 2-1/4 (57.1) to 6 (152) incl. | | 1/2 | 13 |
| Over 6 (152) | | 5/8 | 16 |

*Except the weld size need not exceed the thickness of the thinner part.

Fillet Weld Length                                          D1.1, 2.3.2.3

D1.1, . . . The minimum effective length of a fillet weld shall be at least four times the nominal
size, or the size of the weld shall be considered
not to exceed 25% of its effective length.

Intermittent Fillet Weld Length                             D1.1, 2.7.1.4

D1.1, . . . The minimum length of an intermittent fillet weld shall be 1 1/2 in.

Preheat and Interpass Temperature

| | |
|---|---|
| Reinforcing (see Tables) | D1.4, Table 5.2 |
| Structural (see Tables) | D1.1, Table 4.3 |

Root Pass Size                                           D1.1, 4.6.4

D1.1, . . . The minimum size of a root pass
shall be sufficient to prevent cracking.

Size of Fastener Hole                                    AISC, 5-71, Table J3.1

Hole sizes for different diameter bolts can be
found in this specification. For convenience, these
sizes are provided in the following table.

**Bolt Hole Dimensions in Inches**

| Bolt Diameter | 1/2 in. | 5/8 in. | 3/4 in. | 7/8 in | 1 in |
|---|---|---|---|---|---|
| Standard | 9/16 | 11/16 | 13/16 | 15/16 | 1-1/16 |
| Oversize | 5/8 | 13/16 | 15/16 | 1-1/16 | 1-1/4 |
| Short Slot | 9/16 x 1 1/4 | 11/16 x 7/8 | 13/16 x 1 | 15/16 x 1-1/8 | 1-1/16 x 2-1/2 |
| Long Slot | 9/16 x 1-1/4 | 11/16 x 1-9/16 | 13/16 x 1-7/8 | 15/16 x 2-3/16 | 1-1/16 x 2-1/2 |

| Bolt Diameter | 1-1/8 | 1-1/4 | 1-3/8 | 1-1/2 |
|---|---|---|---|---|
| Standard | 1-3/16 | 1-5/16 | 1-7/16 | 1-9/16 |
| Oversize | 1-7/16 | 1-9/16 | 1-11/16 | 1-13/16 |
| Short Slot | 1-3/16 x 1-1/2 | 1-5/16 x 1-5/8 | 1-7/16 x 1-3/4 | 1-9/16 x 1-7/8 |
| Long Slot | 1-3/16 x 2-13/16 | 1-5/16 x 3-1/8 | 1-7/17 x 3-7/16 | 1-9-16 x 3-3/4 |

*Bear in* mind that base plate holes for anchor
bolts may be oversized as follows (also, see Anchor Bolts in Section A):

Spacing for Rivets and Bolts                             UBC, 2-698, J3.8

The minimum distance from the center of a stan-     AISC, 5-75, J3.8
dard size bolt hole to the edge of a connected part
shall be as indicated in the following table:

*(Continued)*

Minimum, Spacing for Rivets and Bolts *(Continued)*

**Minimum Edge Distance**

| Bolt Diameter in inches | From a sheared Edge | From a Rolled Edge |
|:---:|:---:|:---:|
| 1/2 | 7/8 | 3/4 |
| 5/8 | 1-1/8 | 7/8 |
| 3/4 | 1-1/4 | 1 |
| 7/8 | 1-1/2 | 1-1/8 |
| 1 | 1-3/4 | 1-1/4 |
| 1-1/8 | 2 | 1-1/2 |
| 1-1/4 | 2-1/4 | 1-5/8 |
| 1-1/2 | 2-5/8 | 1-7/8 |

Edge distances are measured from the edge of the connected material to the centerline of the hole.

*But* the clear distance between holes shall not be less than one bolt diameter.

Temperature for Stud Welding                      D1.1, 7.5.4

D1.1, . . . Welding shall not be done when the base metal temperature is below 0 deg. F or when the surface is wet or exposed to rain or snow. When the temperature of the base metal is below 32 deg. F, one additional stud in each 100 studs welded shall be tested by methods specified in 7.7.1.3 and 7.7.1.4, except that the angle of testing shall be approximately 15 deg. This is in addition to the first two studs tested for each start of a new production period or change in set-up.

Weld Size for Small Studs                         D1.1, 7.5.5.4

D1.1, . . . When fillet welds are used, the mini-    D1.1, Table 7.2
mum size shall be the larger of those required in
Table 2.2 or Table 7.2.

**Table 7.2**
**Minimum Fillet Weld Size for Small Diameter**
**Studs (see 7.5.5.1)**

| Stud Diameter | | Min. Size Fillet | |
|:---:|:---:|:---:|:---:|
| in. | mm | in. | mm |
| 1/4 thru 7/16 | 6.4 thru 11.1 | 3/16 | 5 |
| 1/2 | 12.7 | 1/4 | 6 |
| 5/8, 3/4, 7/8 | 15.9, 19, 22.2 | 5/16 | 8 |
| 1 | 25.4 | 3/8 | 10 |

## Moisture Content (Electrodes)

D1.1, 4.5.2.2

D1.1, . . . Additionally, E70XX or E70XX-X (ANSI/AWS A5.1 or A.5.5) low hydrogen electrode coverings shall be limited to a maximum moisture content not exceeding 0.4% by weight.

## Moment Resisting Space Frames

UBC, 2-360, 2211.6, and 2211.7

## Multiple Electrodes (S.A.W.)

D1.1, 4.11.1 - 4.11.6

D1.1, . . . Multiple electrodes are defined as the combination of two or more single or parallel electrode systems. Each of the component systems has its own independent power source and its own electrode feeder. . . .

*Submerged* arc welds with multiple electrodes, except fillet welds, shall be made in the flat position. Fillet welds may be made in either the flat or horizontal position, except that single-pass multiple electrode fillet welds made in the horizontal position shall not exceed ½ in. . . .

## Nondestructive Testing (NDT)

UBC, 2-46, 1703

UBC, . . . In Seismic Zones 3 and 4, welded connections between the primary members of special moment-resisting frames shall be tested by nondestructive methods for compliance with approved standards and job specifications. This testing shall be part of the special inspection requirements of Section 1701.5. A program for this testing shall be established by the person responsible for the structural design and as shown on the plans and specifications. As a minimum, this program shall include the following:

D1.1, 6.7
D1.1, Fig. 8.5, 9.7, 9.8
D1.1, Table 8.2, 9.3

1. All complete penetration groove welds contained in joints and splices shall be tested 100% either by ultrasonic testing or by radiography.

Nondestructive Testing (NDT) *(Continued)*

EXCEPTION: When approved, the nondestructive testing rate for an individual welder or welding operator may be reduced to 25%, provided the rejection rate is demonstrated to be 5% or less of the welds tested for the welder or welding operator. A sampling of at least 40 completed welds for a job shall be made for such reduction evaluation. Reject rate is defined as the number of welds containing rejectable defects divided by the number of welds completed. For evaluating the reject rate of continuous welds over 3 ft. in length where the effective throat thickness is 1 in. or less, each 12 in. increment or fraction thereof shall be considered as one weld. For evaluating the reject rate on continuous welds over 3 ft. in length where the effective throat thickness is greater than 1 in., each 6 in. of length or fraction thereof shall be considered one weld. . . .

*For* complete penetration groove welds on materials less than 5/16 in, thick, nondestructive testing is not required; for this welding, continuous inspection is required. . . .

*When* approved by  the building official and outlined in the project plans and specifications, this nondestructive ultrasonic testing may be performed in the shop of an approved fabricator utilizing qualified test techniques in the employment of the fabricator.

2. Partial penetration groove welds when used in column splices shall be tested either by ultrasonic testing or radiography when required by the plans and specifications. For partial penetration groove welds when used in column splices, with an effective throat less than 3/4 in. thick, nondestructive testing is not required.

3. Base metal thicker than 1 1/2 in., when subjected to through-thickness weld shrinkage strains, shall be ultrasonically inspected for discontinuities directly behind such welds after joint completion. . . .

*Any* material discontinuities shall be accepted or rejected on the basis of the defect rating in accordance with the (larger reflector) criteria of approved national standard.

| Radiographic Testing (X-Ray) | D1.1, 6.9 |
| Ultrasonic Testing (UT) | D1.1, 6.13<br>D1.1, Table 8.2, 9.3<br>D1.1, Fig.8.5, 9.7, 9.8 |

## Table 8.2
## Ultrasonic Acceptance-Rejection Criteria (see 8.15.4)

| Discontinuity Severity Class | Weld Thickness* in in. (mm) and Search Unit Angle | | | | | | | | | | |
|---|---|---|---|---|---|---|---|---|---|---|---|
| | 5/16(8) thru 3/4(19) | > 3/4 thru 1-1/2(38) | > 1-1/2 thru 2-1/2(64) | | | > 2-1/2 thru 4(100) | | | > 4 thru 8(200) | | |
| | 70° | 70° | 70° | 60° | 45° | 70° | 60° | 45° | 70° | 60° | 45° |
| Class A | +5 & lower | +2 & lower | −2 & lower | +1 & lower | +3 & lower | −5 & lower | −2 & lower | 0 & lower | −7 & lower | −4 & lower | −1 & lower |
| Class B | +6 | +3 | −1 0 | +2 +3 | +4 +5 | −4 −3 | −1 0 | +1 +2 | −6 −5 | −3 −2 | 0 +1 |
| Class C | +7 | +4 | +1 +2 | +4 +5 | +6 +7 | −2 to +2 | +1 +2 | +3 +4 | −4 to +2 | −1 to +2 | +2 +3 |
| Class D | +8 & up | +5 & up | +3 & up | +6 & up | +8 & up | +3 & up | +3 & up | +5 & up | +3 & up | +3 & up | +4 & up |

Notes:
1. Class B and C discontinuities shall be separated by at least 2L, L being the length of the longer discontinuity, except that when two or more such discontinuities are not separated by at least 2L, but the combined length of discontinuities and their separation distance is equal to or less than the maximum allowable length under the provisions of Class B or C, the discontinuity shall be considered a single acceptable discontinuity.
2. Class B and C discontinuities shall not begin at a distance less than 2L from weld ends carrying primary tensile stress, L being the discontinuity length.
3. Discontinuities detected at "scanning level" in the root face area of complete joint penetration double groove weld joints shall be evaluated using an indication rating 4 dB more sensitive than described in 6.19.6.5 when such welds are designated as "tension welds" on the drawing (subtract 4 dB from the indication rating "d").
4. Electroslag or electrogas welds: discontinuities detected at "scanning level" which exceed 2 in. (51 mm) in length shall be suspected as being piping porosity and shall be further evaluated with radiography.
5. For indications that remain on the display as the search unit is moved, refer to 8.15.4.

*Weld thickness shall be defined as the nominal thickness of the thinner of the two parts being joined.

**Class A (large discontinuities)**
Any indication in this category shall be rejected (regardless of length).

**Class B (medium discontinuities)**
Any indication in this category having a length greater than 3/4 inch (19 mm) shall be rejected.

**Class C (small discontinuities)**
Any indication in this category having a length greater than 2 inches (51 mm) shall be rejected.

**Class D (minor discontinuities)**
Any indication in this category shall be accepted regardless of length or location in the weld.

| Scanning Levels | |
|---|---|
| Sound path** in in. (mm) | Above Zero Reference, dB |
| through 2-1/2 (64 mm) | 14 |
| > 2-1/2 through 5 (64-127 mm) | 19 |
| > 5 through 10 (127-254 mm) | 29 |
| > 10 through 15 (254-381 mm) | 39 |

**This column refers to sound path distance; NOT material thickness.

Nondestructive Testing (NDT) *(Continued)*

## Table 9.3
## Ultrasonic Acceptance-Rejection Criteria (see 9.25.3.1)

| Discontinuity Severity Class | 5/16 (8) thru 3/4 (19) | >3/4 thru 1-1/2 (38) | >1-1/2 thru 2-1/2 (64) | | | >2-1/2 thru 4 (100) | | | >4 thru 8 (200) | | |
|---|---|---|---|---|---|---|---|---|---|---|---|
| | 70° | 70° | 70° | 60° | 45° | 70° | 60° | 45° | 70° | 60° | 45° |
| Class A | +10 & lower | +8 & lower | +4 & lower | +7 & lower | +9 & lower | +1 & lower | +4 & lower | +6 & lower | -2 & lower | +1 & lower | +3 & lower |
| Class B | +11 | +9 | +5 +6 | +8 +9 | +10 +11 | +2 +3 | +5 +6 | +7 +8 | -1 0 | +2 +3 | +4 +5 |
| Class C | +12 | +10 | +7 +8 | +10 +11 | +12 +13 | +4 +5 | +7 +8 | +9 +10 | +1 +2 | +4 +5 | +6 +7 |
| Class D | +13 & up | +11 & up | +9 & up | +12 & up | +14 & up | +6 & up | +9 & up | +11 & up | +3 & up | +6 & up | +8 & up |

Notes:

1. Class B and C discontinuities shall be separated by at least 2L, L being the length of the longer discontinuity, except that when two or more such discontinuities are not separated by at least 2L, but the combined length of discontinuities and their separation distance is equal to or less than the maximum allowable length under the provisions of Class B or C, the discontinuity shall be considered a single acceptable discontinuity.

2. Class B and C discontinuities shall not begin at a distance less than 2L from the end of the weld, L being the discontinuity length.

3. Discontinuities detected at "scanning level" in the root face area of complete joint penetration double groove weld joints shall be evaluated using an indication rating 4 dB more sensitive than that described in 6.19.6.5 when such welds are designated as "tension welds" on the drawing (subtract 4 dB from the indication rating "d").

4. For indications that remain on the display as the search unit is moved, refer to 9.25.3.2.

*Weld thickness shall be defined as the nominal thickness of the thinner of the two parts being joined.

**Class A (large discontinuities)**
Any indication in this category shall be rejected (regardless of length).

**Class B (medium discontinuities)**
Any indication in this category having a length greater than 3/4 inch (19 mm) shall be rejected.

**Class C (small discontinuities)**
Any indication in this category having a length greater than 2 in. (51 mm) in the middle half or 3/4 inch (19 mm) length in the top or bottom quarter of weld thickness shall be rejected.

**Class D (minor discontinuities)**
Any indication in this category shall be accepted regardless of length or location in the weld.

| Scanning Levels | |
|---|---|
| Sound Path** in in. (mm) | Above Zero Reference, dB |
| through 2-1/2 (64 mm) | 20 |
| >2-1/2 through 5 (64-127 mm) | 25 |
| >5 through 10 (127-254 mm) | 35 |
| >10 through 15 (254-381 mm) | 45 |

**This column refers to sound path distance; NOT material thickness.

Tubular                                                      D1.1, 10.17.6

D1.1, . . . For welds subject to nondestructive testing in accordance with 10.17.2, 10.17.3, 10.17.4, and 10.17.5, the testing may begin immediately after the completed welds have cooled to ambient temperature. Acceptance criteria for

ASTM A514 and A517 steels shall be based on nondestructive testing performed not less than 48 hours after completion of the welds.

## Nuts *(See Bolt Section)*

## Obligations of Contractor

D1.1 Comm., . . . If the inspector(s) find deficiencies in the materials and workmanship, regardless of whether the inspector(s) is a representative of the owner or an employee of the contractor, the contractor shall be responsible for all necessary corrections.

D1.1 Comm., 6.6 - 6.6.5

## Open Web Steel Joists *(See Joists)*

UBC, 2-377, Division III
SBCCI 1505

## Overlap

D1.1, Fig. 3.4

D1.1, . . . Welds shall be free from overlap. . . .

*Excessive* weld metal shall be removed.

D1.1, 3.6.4 and 3.7.2

## Owners Responsibility (Line and Grade)

AISC, 5-235, 7.6

It is the responsibility of the owner to set all leveling nuts, leveling plates, and loose bearing plates that can be set, without the use of heavy equipment, to proper line and grade. All other bearing members are set by the erector to lines and grades established by the owner.

*Acceptance of* the final location and the proper grouting of base and bearing plates is the responsibility of the owner.

*The elevation* tolerance on bearing devices is plus or minus $1/8$ in.

## Oxygen Cut Surfaces (Allowable Roughness)

D1.1, 3.2.2

UBC, . . . Thermally cut free edges that will be subject to substantial tensile stress shall be free of gouges greater than $3/16$ in. deep and sharp notches; gouges greater than $3/16$ in. deep and sharp notches shall be removed by grinding or repaired by welding. Thermally cut edges that are to have weld deposited upon them shall be reasonably free from notches and gouges. . . .

D1.1 Comm., 3.2.2<br>UBC, 2-604, M2.2

*All* reentrant corners shall be shaped to a smooth transition. If specific contour is required it must be shown on the contract documents. . . .

### Table 3.1
### Limits on Acceptability and Repair of Mill Induced Laminar Discontinuities in Cut Surfaces (see 3.2.3)

| Description of Discontinuity | Repair Required |
|---|---|
| Any discontinuity 1 in. (25 mm) in length or less | None, need not be explored. |
| Any discontinuity over 1 in. (25 mm) in length and 1/8 in. (3 mm) maximum depth | None, but the depth should be explored.* |
| Any discontinuity over 1 in. (25 mm) in length with depth over 1/8 in. (3 mm) but not greater then 1/4 in. (6 mm) | Remove, need not weld. |
| Any discontinuity over 1 in. (25 mm) in length with depth over 1/4 in. (6 mm) but not greater than 1 in. | Completely remove and weld. |
| Any discontinuity over 1 in. (25 mm) in length with depth greater than 1 in. | See 3.2.3.2. |

*A spot check of 10% of the discontinuities on the cut surface in question should be explored by grinding to determine depth. If the depth of any one of the discontinuities explored exceeds 1/8 in. (3 mm), then all of the discontinuities over 1 in. (25 mm) in length remaining on that cut surface shall be explored by grinding to determine depth. If none of the discontinuities explored in the 10% spot check have a depth exceeding 1/8 in. (3 mm), then the remainder of the discontinuities on that cut surface need not be explored.

## Oxygen Gouging on Quenched and Tempered Steel

D1.1, 3.2.6

D1.1, . . . Machining, thermal cutting, gouging, chipping, or grinding may be used for joint preparation, or the removal of unacceptable work or metal, except that oxygen gouging shall not be used on steels that are ordered as quenched and tempered or normalized.

D1.1 Comm., 3.2.5

*D1.1 Comm., . . .* Oxygen gouging on quenched and tempered or normalized steel is prohibited because of the high heat input of the process (see C4.3).

## Paint in Contact Surfaces

AISC, 5-89, M3.3

UBC, . . . Paint is permitted on the faying surfaces unconditionally in connections, except in slip-critical connections as defined in Section 2224.1. The faying surfaces of slip-critical connections shall meet the requirements of the following paragraphs, as applicable:

AISC, 5-295, para. 2
UBC, 2-446, 2222.2

1.  In noncoated joints, paint, including any inadvertent overspray, shall be excluded from areas closer than one bolt diameter but not less than 1 in. from the edge of any hole and all areas within the bolt pattern.

2.  Joints specified to have painted faying surfaces shall be blast cleaned and coated with a paint that has been qualified as Class A or B in accordance with the requirements of Section 2221.7, except as provided in Section 2222.2, Item 3.

3.  Subject to the approval of the building official, coatings providing a slip coefficient less than 0.33 may be used, provided. . . .

4.  Coated joints shall not be assembled before the coatings have cured for the minimum time used in the qualifying test.

5.  Galvanized faying surfaces shall be hot dip galvanized in accordance with Table 22-IV-O and shall be roughened by means of hand wire brushing. Power wire brushing is not permitted.

## Parallel Deck Ribs

AISC, 5-60, I5.3

The maximum diameter of a welded shear connector shall be 2 1/2 times the thickness of the flange to which it is welded. Larger studs may be used if welded directly over the web of the member.

AISC, 2-246 and 2-255
UBC, 2-687, I5.3

## Partial Penetration Welds (Minimum Effective Throat)

UBC, 2-690, J21; J2.1b

UBC, . . . The effective throat thickness of a partial-penetration groove weld shall be as shown in Table J2.1. . . .

UBC, 2-691, Table J2.3
D1.1, Table 2.3

*The minimum* effective throat thickness of a partial-penetration groove weld shall be as shown in Table J2.3. Minimum effective throat thickness is determined by the thicker of the two parts joined, except that the weld size need not exceed the thickness of the thinnest part joined. For this exception, particular care shall be taken to provide sufficient preheat for soundness of the weld.

### TABLE J2.3
### Minimum Effective Throat Thickness of
### Partial-penetration Groove Welds

| Material Thickness of Thicker Part Joined (in.) | Minimum Effective Throat Thickness[a] (in.) |
|---|---|
| × 25.4 for mm | |
| To $1/4$ inclusive | $1/8$ |
| Over $1/4$ to $1/2$ | $3/16$ |
| Over $1/2$ to $3/4$ | $1/4$ |
| Over $3/4$ to $1^1/2$ | $5/16$ |
| Over $1/2$ to $2^1/4$ | $3/8$ |
| Over $2^1/4$ to 6 | $1/2$ |
| Over 6 | $5/8$ |

[a]See Sect. J2.

## Peening

D1.1, 3.8

D1.1, . . . Peening may be used on intermediate weld layers for control of shrinkage stresses in thick welds to prevent cracking or distortion, or both. No peening shall be done on the root or surface layer of the weld or the base metal at the edges of the weld as provided in 10.7.5(3). Care should be taken to prevent overlapping or cracking of the weld or base metal. . . .

*The* use of manual slag hammers, chisels, and lightweight vibrating tools for the removal of slag and spatter is permitted and is not considered peening.

## Perpendicular Deck Ribs

AISC, 5-60, I5.2

The maximum center-to-center spacing of shear connectors on a beam or girder shall not be greater than 36 in.

AISC, 2-246 and 2-255
UBC, 2-686, I5.2

*Decking shall* be anchored, by welds or shear connectors, at a maximum center-to-center spacing of 16 in. Welds and studs are used as deck anchors to help resist uplifting of the deck.

## Pin Connected Members

UBC, 2-671, D3

## Placement of Rivets, Bolts, and Welds

UBC, 2-688, J9

## Plans and Specifications

BOCA, 1101.0

## Plate Flatness

AISC, 1-157 - 1-158

If satisfactory contact is obtained, rolled base plates 2 in. or under in thickness need not be milled.

AISC, 5-89, M2.8
AISC, 3-111

*Rolled base* plates over 2 in. and under 4 in. in thickness may be straightened by the use of presses or by milling all bearing surfaces.

UBC, 2-707, M2.8
UBC, 2-605, M2.2.6

*Rolled base* plates greater in thickness than 4 in. shall have all bearing surfaces milled except when the following provisions apply:

1. When the bearing and/or base plates are to be grouted, milling is not required.
2. When full penetration welds are used to connect columns to base plates, milling is not required on the top surface.

*If column* base plates are of material other than rolled steel, then these base plates shall be milled for all bearing surfaces.

UBC, . . . Compression joints that depend on contact bearing as part of the splice capacity shall have the bearing surfaces of individual fabricated pieces prepared by milling, sawing, or other suitable means.

*(Continued)*

## Plate Flatness *(Continued)*

### Variations From Flatness in
### CARBON STEEL PLATE

| Material Thickness in Inches | Flatness Variations for Different Width Materials in Inches | | | | | | | |
|---|---|---|---|---|---|---|---|---|
| | < 36 | 36 to < 48 | 48 to < 60 | 60 to < 72 | 72 to < 84 | 84 to < 96 | 96 to < 108 | 108 to < 120 |
| < 1/4 | 9/16 | 3/4 | 15/16 | 1-1/4 | 1-3/8 | 1-1/2 | 1-5/8 | 1-3/4 |
| 1/4 to < 3/8 | 1/2 | 5/8 | 3/4 | 15/16 | 1-1/8 | 1-1/4 | 1-3/8 | 1-1/2 |
| 3/8 to < 1/2 | 1/2 | 9/16 | 5/8 | 5/8 | 3/4 | 7/8 | 1 | 1-1/8 |
| 1/2 to < 3/4 | 7/16 | 1/2 | 9/16 | 5/8 | 5/8 | 3/4 | 1 | 1 |
| 3/4 to < 1 | 7/16 | 1/2 | 9/16 | 5/8 | 5/8 | 5/8 | 3/4 | 7/8 |
| 1 to < 2 | 3/8 | 1/2 | 1/2 | 9/16 | 9/16 | 5/8 | 5/8 | 5/8 |
| 2 to < 4 | 5/16 | 3/8 | 7/16 | 1/2 | 1/2 | 1/2 | 1/2 | 9/16 |
| 4 to < 6 | 3/8 | 7/16 | 1/2 | 1/2 | 9/16 | 9/16 | 5/8 | 3/4 |
| 6 to < 8 | 7/16 | 1/2 | 1/2 | 5/8 | 11/16 | 3/4 | 7/8 | 7/8 |

### Variations From Flatness in
### High-Strength Low Alloy Steel,
### Hot Rolled or Heat Treated

| Material Thickness in Inches | Flatness Variations for Different Width Materials in Inches | | | | | | | |
|---|---|---|---|---|---|---|---|---|
| | < 36 | 36 to < 48 | 48 to < 60 | 60 to < 72 | 72 to < 84 | 84 to < 96 | 96 to < 108 | 108 to < 120 |
| < 1/4 | 13/16 | 1-1/8 | 1-3/8 | 1-7/8 | 2 | 2-1/4 | 2-3/8 | 2-5/8 |
| 1/4 to < 3/8 | 3/4 | 15/16 | 1-1/8 | 1-3/3 | 1-3/4 | 1-7/8 | 2 | 2-1/4 |
| 3/8 to < 1/2 | 3/4 | 7/8 | 15/16 | 15/16 | 1-1/8 | 1-5/16 | 1-1/2 | 1-5/8 |
| 1/2 to < 3/4 | 5/8 | 3/4 | 13/16 | 7/8 | 1 | 1-1/8 | 1-1/4 | 1-3/8 |
| 3/4 to < 1 | 5/8 | 3/4 | 7/8 | 7/8 | 15/16 | 1 | 1-1/8 | 1-5/16 |
| 1 to < 2 | 9/16 | 5/8 | 3/4 | 13/16 | 7/8 | 15/16 | 1 | 1 |
| 2 to < 4 | 1/2 | 9/16 | 11/16 | 3/4 | 3/4 | 3/4 | 3/4 | 7/8 |
| 4 to < 6 | 9/16 | 11/16 | 3/4 | 3/4 | 7/8 | 7/8 | 15/16 | 1-1/8 |
| 6 to < 8 | 5/8 | 3/4 | 3/4 | 15/16 | 1 | 1-1/8 | 1-1/4 | 1-5/16 |

## Plate Girders and Rolled Beams

AISC, 5-37, B10

## Plates (Cover Plates on Flanges)

D1.1, 9.21.6 - 9.21.5.3

D1.1, . . . Cover plates shall preferably be limited to one on any flange. The maximum thickness of cover plates on a flange (total thickness of all cover plates if more than one is used) shall not be greater than 1 $\frac{1}{2}$ times the thickness of the flange to which the cover plate is attached. . . .

AISC, 5-37, B10
UBC, 2-669, B10

*Fillet* welds connecting a cover plate to the flange in the region between terminal developments shall be continuous welds of sufficient size to transmit the incremental longitudinal shear between the cover plate and the flange . . . *and in* no case shall the welds be smaller than the minimum size permitted by 2.7.1.1.

UBC, . . . The total cross sectional area of cover plates of bolted or riveted girders shall not exceed 70% of the total flange area.

## Plug and Slot Welds

Effective Areas of

UBC, 2-690, J2.3a

UBC, . . . The effective shearing area of plug and slot welds shall be considered as the nominal cross sectional area of the hole or slot in the plane of the faying surface.

D1.1, 2.3.3
AISC, 5-68, J2.3B

D1.1, . . . The effective area shall be the nominal area of the hole or slot in the plane of the faying surface.

Faying Surface Gaps

D1.1, 3.3.1

D1.1, . . . The separation between faying surfaces of plug and slot welds, and of butt joints landing on a backing, shall not exceed $\frac{1}{16}$ in. The use of fillers is prohibited except as specified on the drawings or as specially approved by the engineer and made in accordance with 2.4.

*(Continued)*

Plug and Slot Welds *(Continued)*

Fillet Welds in                                                      AISC, 5-67, J2.2.b

D1.1, . . . Fillet welds in holes or slots in lap          D1.1, 2.7.1.3
joints may be used to transfer shear or to pre-           UBC, 2-692, J2.2.b
vent buckling or separation of lapped parts.
These fillet welds may overlap, subject to the
provisions of 2.3.2.2. Fillet welds in holes or slots
are not to be considered plug or slot welds.

Specifications for                                                  D1.1, 2.8 - 2.8.8

D1.1, . . . The minimum diameter of the hole
for a plug weld shall be no less than the thickness of the part containing it plus $5/16$ in., preferably rounded off to the next greater odd $1/16$ in.
The maximum diameter shall equal the mini-
mum diameter plus $1/8$ in. or $2\ 1/4$ times the thickness of the member, whichever is greater. . . .
*The* minimum center-to-center spacing of plug
welds shall be four times the diameter of the
hole. . . .

*The* length of the slot for a slot weld shall not exceed ten times the thickness of the part containing it. The width of the slot shall be no less
than the thickness of the part containing it plus
$5/16$ in., preferably rounded off to the next
greater odd $1/16$ in. The maximum width shall
equal the minimum width plus $1/8$ in. or $2\ 1/4$
times the thickness of the member, whichever is
greater. . . .

*Plug* and slot welds are not permitted in
quenched and tempered steels. . . .

*The* minimum spacing of lines of slot welds in a
direction transverse to their length shall be four
times the width of the slot. . . .

*The* depth of filling of plug or slot welds in metal
$5/8$ in. thick or less shall be equal to the thickness
of the material. In metal over $5/8$ in. thick, it shall
be at least $1/2$ the thickness of the material, but
no less than $5/8$ in.

Technique                                                           D1.1, 4.21 - 4.22

## Plumb (Determining in Columns)

AISC, 5-238, 7.11.3.1

Subject to the limitations set forth in the specification, individual columns are considered plumb if the working line does not deviate from the plumb line by more than 1 in. in 500 in. or about 1 in. in 41 ft.

## Porosity

D1.1, 3.7.2.3; 8.15.1.6; 9.25.1.6; 10.17.1.6
B1.10, 2.3 - 2.3.4

D1.1, . . . The sum of diameters of visible piping porosity $\frac{1}{32}$ in. or greater in fillet welds shall not exceed $\frac{3}{8}$ in. in any linear inch of weld and shall not exceed $\frac{3}{4}$ in. in any 12 in. length of weld.

B1.10, . . . Unless porosity is excessive, it is not as critical a discontinuity as sharp discontinuities that cause stress concentrations. Excessive porosity is a sign that the welding parameters, welding consumables, or joint fit-up are not being properly controlled for the welding process selected or that the base metal is contaminated or of a composition incompatible with the weld filler metal being used. . . .

*Porosity* is not caused exclusively by hydrogen, but the presence of porosity does indicate that there is a possibility of hydrogen in the weld and heat-affected zones that may lead to cracking in ferrous materials.

Cluster Porosity

B1.10, 2.3.2

B1.10, . . . Cluster porosity is a localized grouping of pores. It often results from improper initiation or termination of the welding arc.

Linear Porosity

B1.10, 2.3.3

B1.10, . . . Linear porosity is a series of pores that are aligned. It often occurs along a weld interface, the interface of weld beads, or near the weld root, and is caused by contamination that leads to gas evolution at those locations.

*(Continued)*

Porosity *(Continued)*

| | |
|---|---|
| Piping Porosity | D1.1, 8.15.1.6 |

B1.10, . . . Piping porosity is an elongated gas pore. Piping porosity in fillet welds extends from the weld root toward the weld surface. When one or two pores are seen in the weld surface, careful excavation may also reveal subsurface porosity. Much of the piping porosity found in welds does not extend all the way to the surface. Piping porosity in electroslag welds can become very long; for example, 20 in.        B1.10, 2.3.4

Uniformly Scattered Porosity                B1.10, 2.3.1

B1.10, . . . Uniformly scattered porosity is porosity uniformly distributed throughout the weld metal. When excessive uniformly scattered porosity is encountered, the cause is generally faulty welding technique or materials. The joint preparation technique or materials used may result in conditions that may cause porosity. . . .

*If a weld* cools slowly enough to allow most of the gas to pass to the surface before weld solidification, there will be few pores in the weld.

## Preheat and Interpass Temperatures

Reinforcement                          D1.4, 5.2

D1.4, . . . Minimum preheat and interpass temperatures shall be in accordance with Table 5.2 using the highest carbon equivalent number of the base metal, as determined in accordance with 1.3.4.        D1.4, Table 5.2

### Table 5.2
### Minimum Preheat and Interpass Temperature[1,2]
### (see 5.2.1)

| Carbon Equivalent[3,4] (C.E.) Range, % | Size of Reinforcing Bar | Shielded Metal Arc Welding with Low Hydrogen Electrodes, Gas Metal Arc Welding, or Flux Cored Arc Welding | |
|---|---|---|---|
| | | Minimum Temperature | |
| | | °F | °C |
| Up to 0.40 | up to 11 inclusive | none[5] | none[5] |
| | 14 and 18 | 50 | 10 |
| Over 0.40 to 0.45 inclusive | up to 11 inclusive | none[5] | none[5] |
| | 14 and 18 | 100 | 40 |
| Over 0.45 to 0.55 inclusive | up to 6 inclusive | none[5] | none[5] |
| | 7 to 11 inclusive | 50 | 10 |
| | 14 to 18 | 200 | 90 |
| Over 0.55 to 0.65 inclusive | up to 6 inclusive | 100 | 40 |
| | 7 to 11 inclusive | 200 | 90 |
| | 14 to 18 | 300 | 150 |
| Over 0.65 to 0.75 | up to 6 inclusive | 300 | 150 |
| | 7 to 18 inclusive | 400 | 200 |
| Over 0.75 | 7 to 18 inclusive | 500 | 260 |

1. When reinforcing steel is to be welded to main structural steel, the preheat requirements of the structural steel shall also be considered (see ANSI/AWS D1.1, table titled "Minimum Preheat and Interpass Temperature." The minimum preheat requirement to apply in this situation shall be the higher requirement of the two tables. However, extreme caution shall be exercised in the case of welding reinforcing steel to quenched and tempered steels, and such measures shall be taken as to satisfy the preheat requirements for both. If not possible, welding shall not be used to join the two base metals.

2. Welding shall not be done when the ambient temperature is lower than 0°F (-18°C). When the base metal is below the temperature listed for the welding process being used and the size and carbon equivalent range of the bar being welded, it shall be preheated (except as otherwise provided) in such a manner that the cross section of the bar for not less than 6 in. (150 mm) on each side of the joint shall be at or above the specified minimum temperature. Preheat and interpass temperatures shall be sufficient to prevent crack formation.

3. After welding is complete, bars shall be allowed to cool naturally to ambient temperature. Accelerated cooling is prohibited.

4. Where it is impractical to obtain chemical analysis, the carbon equivalent shall be assumed to be above 0.75%. See also 1.3.4.

5. When the base metal is below 32°F (0°C), the base metal shall be preheated to at least 70°F (20°C), or above, and maintained at this minimum temperature during welding.

*(Continued)*

Preheat and Interpass Temperatures *(Continued)*

Structural                                                    D1.1, 4.2

The preheat and interpass temperature must be                D1.1, Table 4.2
sufficient to prevent cracking. Experience has               D1.1 Comm., 4.2
shown that the minimum temperatures specified
in Table 4.3 are adequate to prevent cracking in
most cases. However, increased preheat temperatures may be necessary in situations involving
higher restraint, higher hydrogen, lower welding
heat input, or steel composition at the top end of
the specification. Conversely, lower preheat temperatures may be adequate to prevent cracking
depending on restraint, hydrogen level, and actual steel composition or higher welding input
heat.

**Table 4.2**
**Filler Metal Requirements for Exposed Bare Applications of**
**ASTM A242 and A588 Steel (see 4.1.4)**

| Welding Processes | | | |
|---|---|---|---|
| Shielded Metal Arc | Submerged Arc | Gas Metal Arc or Gas Tungsten Arc | Flux Cored Arc |
| A5.5 | A5.23[1,4] | A5.28[4] | A5.29 |
| E7018-W | F7AX-EXXX-W | | |
| E8018-W | | | E8XT1-W |
| E8016-C3 or E8018-C3 | F7AX-EXXX-Ni1[2] | ER80S-Ni1 | E8XTX-Ni1 |
| E8016-C1 or E8018-C1 | F7AX-EXXX-Ni4[2] | | |
| E8016-C2 or E8018-C2 | | | |
| E7016-C1L or E7018-C1L | F7AX-EXXX-Ni2[2] | ER80S-Ni2 | E8XTX-Ni2 |
| E7016-C2L or E7018-C2L | F7AX-EXXX-Ni3[2] | ER80S-Ni3 | E80T5-Ni3 |
| E8018-B2L[1] | | ER80S-B2L[1] | E80T5-B2L[1] |
| | | ER80S-G[1,3] | |
| | | | E71T8-Ni1 |
| | | | E71T8-Ni2 |
| | | | E7XTX-K2 |

## Table 4.3
### Minimum Preheat and Interpass Temperature[1,2] (see 4.2)

| Category | Steel Specification | Welding Process | Thickness of Thickest Part at Point of Welding, in. | mm | Minimum Temperature, °F | °C |
|---|---|---|---|---|---|---|
| A | ASTM A36[3]; ASTM A53 — Grade B; ASTM A106 — Grade B; ASTM A131 — Grades A, B, CS, D, DS, E; ASTM A139 — Grade B; ASTM A381 — Grade Y35; ASTM A500 — Grade A, Grade B; ASTM A501; ASTM A516 — Grades I & II; ASTM A524; ASTM A529 — All grades; ASTM A570 — Grade 65; ASTM A573 — Grade 36[3]; ASTM A709 — Grade B; API 5L — Grades X42; ABS — Grades A, B, D, CS, DS, Grade E | Shielded metal arc welding with other than low hydrogen electrodes | Up to 3/4 | 19 incl. | None[4] | |
| | | | Over 3/4 thru 1-1/2 | 19 / 38.1 incl. | 150 | 66 |
| | | | Over 1-1/2 thru 2-1/2 | 38.1 / 63.5 | 225 | 107 |
| | | | Over 2-1/2 | 63.5 | 300 | 150 |
| B | ASTM A36[3] — All grades; ASTM A53 — Grade B; ASTM A106 — Grade B; ASTM A131 — Grades A, B, C, CS, D, DS, E, AH 32 & 36, DH 32 & 36, EH 32 & 36; ASTM A139 — Grade B; ASTM A242; ASTM A381 — Grade Y35; ASTM A441 — Grade A; ASTM A500 — Grade B; ASTM A501; ASTM A516 — Grades 55 & 60, 65 & 70; ASTM A524 — Grades I & II; ASTM A529; ASTM A537 — Classes 1 & 2; ASTM A570 — Grades 42, 50, Grade 65; ASTM A572 — Grades 42, 50; ASTM A573 — Grades 45, 50, 55; ASTM A588; ASTM A595 — Grades A, B, C; ASTM A606; ASTM A607 — Grades 45, 50, 55; ASTM A618; ASTM A633 — Grades A, B, Grades C, D; ASTM A709 — Grades 36, 50, 50W; ASTM A808; API 5L — Grade X42; API Spec. 2H — Grades 42, 50; ABS — Grades AH 32 & 36, DH 32 & 36, EH 32 & 36; ABS — Grades A, B, D, CS, DS, Grade E | Shielded metal arc welding with low hydrogen electrodes,[5] submerged arc welding, gas metal arc welding, gas tungsten arc welding, flux cored arc welding | Up to 3/4 | 19 incl. | None[4] | |
| | | | Over 3/4 thru 1-1/2 | 19 / 38.1 incl. | 50 | 10 |
| | | | Over 1-1/2 thru 2-1/2 | 38.1 / 63.5 incl. | 150 | 66 |
| | | | Over 2-1/2 | 63.5 | 225 | 107 |

(continued)

## Table 4.3 (continued)

| Category | Steel Specification | Welding Process | Thickness of Thickest Part at Point of Welding, in. | mm | Minimum Temperature, °F | °C |
|---|---|---|---|---|---|---|
| C | ASTM A572 Grades 60 & 65<br>ASTM A633 Grade E<br>API 5L Grade X52 | Shielded metal arc welding with low hydrogen electrodes, submerged arc welding,[5] gas metal arc welding, gas tungsten arc welding, flux cored arc welding | Up to 3/4 | 19 incl. | 50 | 10 |
| | | | Over 3/4 thru 1-1/2 | 19–38.1 incl. | 150 | 66 |
| | | | Over 1-1/2 thru 2-1/2 | 38.1–63.5 incl. | 225 | 107 |
| | | | Over 2-1/2 | 63.5 | 300 | 150 |
| D | ASTM A514<br>ASTM A517<br>ASTM A709 Grades 100 & 100W | Shielded metal arc welding with low hydrogen electrodes, submerged arc welding[5] with carbon or alloy steel wire, neutral flux, gas metal arc welding, gas tungsten arc welding, or flux cored arc welding | Up to 3/4 | 19 incl. | 50 | 10 |
| | | | Over 3/4 thru 1-1/2 | 19–38.1 incl. | 125 | 50 |
| | | | Over 1-1/2 thru 2-1/2 | 38.1–63.5 incl. | 175 | 80 |
| | | | Over 2-1/2 | 63.5 | 225 | 107 |
| E | ASTM A710 Grade A (All classes) | no preheat is required[6] | | | | |

Notes:
1. Welding shall not be done when the ambient temperature is lower than 0° F (–18° C). Zero °F (–18° C) does not mean the ambient environmental temperature but the temperature in the immediate vicinity of the weld. The ambient environmental temperature may be below 0° F, but a heated structure or shelter around the area being welded could maintain the temperature adjacent to the weldment at 0° F or higher. When the base metal is below the temperature listed for the welding process being used and the thickness of material being welded, it shall be preheated (except as otherwise provided) in such manner that the surfaces of the parts on which weld metal is being deposited are at or above the specified minimum temperature for a distance equal to the thickness of the part being welded, but not less than 3 in. (75 mm) in all directions from the point of welding. Preheat and interpass temperatures must be sufficient to prevent crack formation. Temperature above the minimum shown may be required for highly restrained welds. For A514, A517, and A709 Grades 100 and 100W steel, the maximum preheat and interpass temperature shall not exceed 400° F (205° C) for thickness up to 1-1/2 in. (38.1 mm) inclusive, and 450° F (230° C) for greater thickness. Heat input when welding A514, A517, and A709 Grades 100 and 100W steel shall not exceed the steel producer's recommendations.
2. In joints involving combinations of base metals, preheat shall be as specified for the higher strength steel being welded.
3. Only low hydrogen electrodes shall be used when welding A36 or A709 Grade 36 steel more than 1 in. (25.4 mm) thick for dynamically loaded structures.
4. When the base metal temperature is below 32° F (0° C), the base metal shall be preheated to at least 70° F (21° C) and this minimum temperature maintained during welding.
5. For modification of preheat requirements for submerged arc welding with parallel or multiple electrodes, see 4.10.6 or 4.11.6.
6. Pre..... t is not required for the base metal. Preheat for E80XX-X filler metal shall be as for Gr....; for higher strength filler metal treat as Group D.

# Preparation of Surfaces

Base Metals

D1.1, 3.2

D1.1, . . . Surfaces on which weld metal is to be deposited shall be smooth, uniform, and free from fins, tears, cracks, and other discontinuities that would adversely affect the quality or strength of the weld. Surfaces to be welded, and surfaces adjacent to a weld, shall also be free from loose or thick scale, slag, rust, moisture, grease, and other foreign material that would prevent proper welding or produce objectionable fumes. Mill scale that can withstand vigorous wire brushing, a thin rust-inhibitive coating, or antispatter compound may remain with the following exception: for girders in dynamically loaded structures, all mill scale shall be removed from the surfaces on which flange-to-web welds are to be made by submerged arc welding or shielded metal arc welding with low hydrogen electrodes. . . .

D1.1 Comm., 3.2.1

Stud Welding

D1.1, 7.4.3; 7.5.4; 7.5.5.1; 7.5.5.5

D1.1, . . . The areas to which studs are to be welded shall be free of scale, rust, moisture, or other injurious material to the extent necessary to obtain satisfactory welds. . . .

*For* fillet welds, the stud base shall be prepared so that the base of the stud fits against the base metal. . . .

*The* stud base shall not be painted, galvanized, or cadmium-plated prior to welding. . . .

*For* fillet welds, the stud base shall be prepared so that the base of the stud fits against the base metal.

## Prequalification Tables

Full Penetration Groove Welds                          D1.1, pp. 8 - 33

Partial Penetration Groove Welded Joints               D1.1, pp. 35 - 54

Weld Joints                                            AISC 4-155–4-173

## Progression in Welding                              D1.1, 4.6.8; 4.14.1.7

D1.1, . . . The progression for all passes in vertical   D1.1, 3.4.4
position welding shall be upward, except that un-
dercut may be repaired vertically downwards when
preheat is in accordance with Table 4.3, but not
lower than 70 deg. F.

However, when tubular products are welded, the
progression of vertical welding may be upwards,
but only in the direction(s) for which the welder is
qualified.

Tubular                                                D1.1, 4.6.8
                                                       D1.1, 4.14.1.7

## Quality of Welds

Statically Loaded Structures                           D1.1, 8.15

D1.1, . . . All welds shall be visually inspected
and shall be acceptable if the following conditions are met: . . .

*The* weld shall have no cracks. . . .

*Thorough* fusion shall exist between adjacent
layers of weld metal and between weld metal
and base metal. . . .
*All* craters shall be filled to the full cross section
of the weld, except for the ends of intermittent
fillet welds outside their effective length. . . .

*Weld* profiles shall be in accordance with
3.6. . . .

*For* material less than 1 in. thick, undercut shall
not exceed $1/32$ in., except that a maximum $1/16$ in.
is permitted for an accumulated length of 2 in. in
any 12 in. For material equal to or greater than

1 in. thick, undercut shall not exceed $\frac{1}{16}$ in. for any length of weld. . . .

A *fillet* weld in any single continuous weld shall be permitted to underrun the nominal fillet size specified by $\frac{1}{16}$ in. without correction, provided that the undersize portion of the weld does not exceed 10% of the length of the weld. On web-to-flange welds on girders, no underrun is permitted at the ends for a length equal to twice the width of the flange. . . .

*Complete* joint penetration groove welds in butt joints transverse to the direction of computed tensile stress shall have no visible piping porosity. For all other groove welds, the sum of the visible piping porosity $\frac{1}{32}$ in. or greater in diameter shall not exceed $\frac{3}{8}$ in. in any linear inch of weld and shall not exceed $\frac{3}{4}$ in. in any 12 in. length of weld. . . .

Dynamically Loaded Structures                   D1.1, 9.25

D1.1, . . . In primary members, undercut shall be no more than 0.01 in. deep when the weld is transverse to tensile stress under any design loading condition. Undercut shall be no more than $\frac{1}{32}$ in. deep for all other cases. . . .

*The* frequency of piping porosity in fillet welds shall not exceed one in each 4 in. of weld length and the maximum diameter shall not exceed $\frac{3}{32}$ in. Exception: for fillet welds connecting stiffeners to web, the sum of diameters of piping porosity shall not exceed $\frac{3}{8}$ in. in any linear inch of weld and shall not exceed $\frac{3}{4}$ in. in any 12 in. length of weld. . . .

# Quenched and Tempered Steel                   D1.1, 8.2

Oxygen Gouging on                   D1.1, 4.3

D1.1, . . . When quenched and tempered steels are welded, the heat input shall be restricted in conjunction with the maximum preheat and interpass temperatures required (because of base metal thickness). The above limitations shall be

*(Continued)*

Quenched and Tempered Steel, Oxygen Gouging on
*(Continued)*

in strict accordance with the steel producer's recommendations. The use of stringer beads to
avoid overheating is strongly recommended. Oxygen gouging of quenched and tempered steels is
not permitted.

## Redrying Electrodes

D1.1, 4.5.4

D1.1, . . . Electrodes that conform to the provisions of 4.5.2 shall subsequently be redried no more
than one time. Electrodes that have been wet shall
not be used.

## Reentrant Corners

D1.1, 3.2.4

D1.1, . . . Reentrant corners of cut material shall
be formed to provide a gradual transition with a radius of not less than 1 in. Adjacent surfaces shall
meet without offset or cutting past the point of tangency.

D1.1 Comm., 3.2.4
AISC, 5-87, M2.2

## Rejections

AISC, 5-91, M5.2

Material and/or workmanship that does not reasonably conform to the requirements of this specification may, at any time, be rejected.

UBC, 2-607, M5.2

## Removal of Mill Scale

D1.1, 3.2.1

D1.1, . . . Mill scale that can withstand vigorous
wire brushing, a thin rust-inhibitive coating, or antispatter compound may remain with the following
exception: for girders in dynamically loaded structures, all mill scale shall be removed from the surfaces on which flange-to-web welds are to be made
by submerged arc welding or by shielded arc welding with low hydrogen electrodes.

D1.1 Comm., 3.2.1

## Repair of Welds

D1.1, 3.7.1 - 3.7.7

D1.1, . . . Weld metal shall be deposited to compensate for any deficiency in size. . . .

*The* contractor has the option of either repairing an unacceptable weld or removing and replacing the entire weld, except as modified by 3.7.4. The repaired or replaced weld shall be retested by the method originally used, and the same technique and quality acceptance criteria shall be applied. . . .

## Repairs of

### Plate Cut Edges

D1.1, 3.2.3.1

D1.1, . . . The limits of acceptability and the repair of visually observed cut surface discontinuities shall be in accordance with Table 3.1, in which the length of discontinuity is the visible long dimension on the cut surface of material and the depth is the distance that the discontinuity extends into the material from the cut surface. . . .

D1.1, Table 3.1
D1.1 Comm., 3.2.3

**Table 3.1**
**Limits on Acceptability and Repair of Mill Induced**
**Laminar Discontinuities in Cut Surfaces (see 3.2.3)**

| Description of Discontinuity | Repair Required |
|---|---|
| Any discontinuity 1 in. (25 mm) in length or less | None, need not be explored. |
| Any discontinuity over 1 in. (25 mm) in length and 1/8 in. (3 mm) maximum depth | None, but the depth should be explored.* |
| Any discontinuity over 1 in. (25 mm) in length with depth over 1/8 in. (3 mm) but not greater then 1/4 in. (6 mm) | Remove, need not weld. |
| Any discontinuity over 1 in. (25 mm) in length with depth over 1/4 in. (6 mm) but not greater than 1 in. | Completely remove and weld. |
| Any discontinuity over 1 in. (25 mm) in length with depth greater than 1 in. | See 3.2.3.2. |

*A spot check of 10% of the discontinuities on the cut surface in question should be explored by grinding to determine depth. If the depth of any one of the discontinuities explored exceeds 1/8 in. (3 mm), then all of the discontinuities over 1 in. (25 mm) in length remaining on that cut surface shall be explored by grinding to determine depth. If none of the discontinuities explored in the 10% spot check have a depth exceeding 1/8 in. (3 mm), then the remainder of the discontinuities on that cut surface need not be explored.

Repairs of *(Continued)*

Studs                                                                D1.1, 7.7.1.5

> D1.1, . . . If on visual examination the test studs do not exhibit a 360 deg. flash, or if on testing failure occurs in the weld zone of either stud, the procedure shall be corrected, and two more studs shall be welded to separate material or on the production member and tested in accordance with the provisions of 7.7.1.3 and 7.7.1.4. If either of the second two studs fails, additional welding shall be continued on separate plates until two consecutive studs are tested and found to be satisfactory before any more production studs are welded to the member.

Welds                                                                D1.1, 3.7.2

> D1.1, . . . The contractor has the option of either repairing an unacceptable weld or removing and replacing the entire weld, except as modified by 3.7.4. The repaired or replaced weld shall be retested by the method originally used, and the same technique and quality acceptance criteria shall be applied. . . .

                                                                     D1.1 Comm., 3.7.2

## Reinforcing Steel *(See Reinforcing Section)*

D1.4
BOCA 1804.0

## Reinforcement of Groove Welds

D1.1, 3.6.2 - 3.6.3

> D1.1, . . . Groove welds shall preferably be made with slight or minimum face reinforcement except as may be otherwise provided. In the case of butt and corner joints, the face reinforcement shall not exceed $1/8$ in. in height and shall have gradual transition to the plane of the base metal surface. . . .

D1.1, Fig. 3.4

> *Surfaces* of butt joint welds required to be flush shall be finished so as not to reduce the thickness of the thinner base metal by more than $1/32$ in. However, all reinforcement must be removed where the weld forms part of a faying or contact surface. Any reinforcement must blend smoothly into the plate surfaces with transition areas free from undercut. . . .

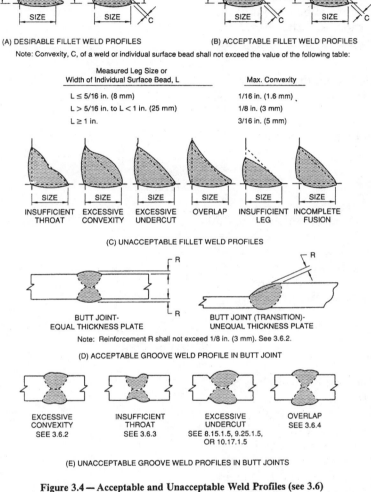

(A) DESIRABLE FILLET WELD PROFILES          (B) ACCEPTABLE FILLET WELD PROFILES

Note: Convexity, C, of a weld or individual surface bead shall not exceed the value of the following table:

| Measured Leg Size or Width of Individual Surface Bead, L | Max. Convexity |
|---|---|
| L ≤ 5/16 in. (8 mm) | 1/16 in. (1.6 mm) |
| L > 5/16 in. to L < 1 in. (25 mm) | 1/8 in. (3 mm) |
| L ≥ 1 in. | 3/16 in. (5 mm) |

INSUFFICIENT THROAT    EXCESSIVE CONVEXITY    EXCESSIVE UNDERCUT    OVERLAP    INSUFFICIENT LEG    INCOMPLETE FUSION

(C) UNACCEPTABLE FILLET WELD PROFILES

BUTT JOINT- EQUAL THICKNESS PLATE          BUTT JOINT (TRANSITION)- UNEQUAL THICKNESS PLATE

Note: Reinforcement R shall not exceed 1/8 in. (3 mm). See 3.6.2.

(D) ACCEPTABLE GROOVE WELD PROFILE IN BUTT JOINT

EXCESSIVE CONVEXITY SEE 3.6.2    INSUFFICIENT THROAT SEE 3.6.3    EXCESSIVE UNDERCUT SEE 8.15.1.5, 9.25.1.5, OR 10.17.1.5    OVERLAP SEE 3.6.4

(E) UNACCEPTABLE GROOVE WELD PROFILES IN BUTT JOINTS

**Figure 3.4 — Acceptable and Unacceptable Weld Profiles (see 3.6)**

# Requalification of Welder

D1.1, 6.4.3

D1.1, . . . The inspector shall require requalification of any qualified welder, welding operator, or tack welder who has for a period exceeding 6 months not used the process for which the welder, welding operator, or tack welder was qualified. (See 5.30 and 5.41.)

## Responsibility of Owner (Line and Grade)                    AISC, 5-235, 7.6

It is the responsibility of the owner to set all leveling nuts, leveling plates, and loose bearing plates that can be set, without the use of heavy equipment, to proper line and grade. All other bearing members are set by the erector to lines and grades established by the owner.

*Acceptance of* the final location and the proper grouting of base and bearing plates is the responsibility of the owner.

*The elevation* tolerance on bearing devices is plus or minus $1/8$ in.

## Retest (Studs)                                              D1.1, 7.7.3

D1.1, . . . In production, studs on which a full 360 deg. flash is not obtained may, at the option of the contractor, be repaired by adding the minimum fillet weld as required by 7.5.5 in place of the missing flash. The repair weld shall extend at least $3/8$ in. beyond each end of the discontinuity being repaired.

## Revisions (Design)                                          D1.1, 3.7.6

D1.1, . . . If, after an unacceptable weld has been made, work is performed that has rendered that weld inaccessible or has created new conditions that make correction of the unacceptable weld dangerous or ineffectual, then the original conditions shall be restored by removing welds or members, or both, before the corrections are made. If this is not done, the deficiency shall be compensated for by additional work performed according to an approved revised design.

## Rivets and

| | |
|---|---|
| Bolts (Shear and Tension) | UBC, 2-695, J3.4 |
| Bolts and Threaded Parts | UBC, 2-695, J3.4 |

Bolts and Welds (Weld First or Bolt First)      AISC, 5-163, J1.10

The specification advises against making the      AISC 5-64, J1.10
welds before accomplishing the final tightening      D1.1, 8.7
of the bolts. The reason being that the welds
might not allow the faying surfaces of the connection to develop the high contact pressure required in a high-strength bolted connection. In
extreme cases, the welds could prevent the faying surfaces from making contact altogether.

Conformance to      AISC, 5-27, A3.3

Hole Enlargement, Misalignment, Rivet Tempera-      AISC, 5-88, M2.5
ture, and Slope Ratios

During erection, the use of drift pins to align bolt
holes in members is allowed, provided their use
does not distort, oblong, or otherwise enlarge the
holes.

*Improper* or poor matching of bolt holes shall be
cause for rejection of the member.

*If the slope* of the faying surfaces in the connection exceeds 1:20, with regard to the bolt or nut
face, a single hardened beveled washer shall be
installed to compensate for the slope.

Minimum Spacing of      UBC, 2-698, J3.8

UBC, . . . The distance between centers of standard, oversized, or slotted holes shall not be less
than $2 \frac{2}{3}$ times the nominal diameter of the fastener or less than that required by the following
paragraph, if applicable. . . .

Placement of      UBC, 2-688, J9

Size of Holes for      UBC, 2-605, M2.2.5

Hole sizes for different diameter bolts can be      UBC, Table 22-IV-A
found in this specification. For convenience, these      AISC, 5-33, B2
sizes are provided in the following table;      AISC, 5-71, J3.2a, Table J3.1

*(Continued)*

Rivets and, Size of Holes for *(Continued)*

### Bolt Hole Dimensions in Inches

| Bolt Diameter | 1/2 in. | 5/8 in. | 3/4 in. | 7/8 in | 1 in |
|---|---|---|---|---|---|
| Standard | 9/16 | 11/16 | 13/16 | 15/16 | 1-1/16 |
| Oversize | 5/8 | 13/16 | 15/16 | 1-1/16 | 1-1/4 |
| Short Slot | 9/16 x 1 1/4 | 11/16 x 7/8 | 13/16 x 1 | 15/16 x 1-1/8 | 1-1/16 x 2-1/2 |
| Long Slot | 9/16 x 1-1/4 | 11/16 x 1-9/16 | 13/16 x 1-7/8 | 15/16 x 2-3/16 | 1-1/16 x 2-1/2 |

| Bolt Diameter | 1-1/8 | 1-1/4 | 1-3/8 | 1-1/2 |
|---|---|---|---|---|
| Standard | 1-3/16 | 1-5/16 | 1-7/16 | 1-9/16 |
| Oversize | 1-7/16 | 1-9/16 | 1-11/16 | 1-13/16 |
| Short Slot | 1-3/16 x 1-1/2 | 1-5/16 x 1-5/8 | 1-7/16 x 1-3/4 | 1-9/16 x 1-7/8 |
| Long Slot | 1-3/16 x 2-13/16 | 1-5/16 x 3-1/8 | 1-7/17 x 3-7/16 | 1-9-16 x 3-3/4 |

*Bear in* mind that base plate holes for anchor bolts may be oversized as follows (also, see Anchor Bolts in Section A):

**Bolt Size, inches (mm)**
$3/4$ (19.1)
$7/8$ (22.2)
1 < 2 (25.4 < 50.8)
>2 (> 50.8)

**Hole Size, inches (mm)**
$5/16$ (7.9) oversized
$5/16$ (7.9) oversized
$1/2$ (12.7) oversized
1 (25.4) > bolt diameter

Steel (Conformance to)

UBC, 2-555, A3
UBC, 3-659, STD. 22-1
D1.1, 8.7

Welds and Bolts in Combination with

D1.1, . . . Rivets or bolts used in bearing-type connections shall not be considered as sharing the stress in combination with welds. Welds, if used, shall be provided to carry the entire stress in the connection. However, connections that are welded to one member and riveted or welded to

UBC, 2-688, J1.10; 2-689, J1.12
AISC 5-64, J1.10

the other are permitted. High-strength bolts properly installed as a friction type connection prior to welding may be considered as sharing the stress with the welds.

UBC, . . . In new work, A307 bolts or high-strength bolts used in bearing-type connections shall not be considered as sharing the stress in combination with welds. Welds, if used, shall be provided to carry the entire stress in the connection. High-strength bolts proportioned for slip-critical connections may be considered as sharing the stress with welds.

## Root Opening (Buttering)                              D1.1, 3.3.4.1

D1.1, . . . Root openings greater than those per-     D1.1 Comm., 3.3.4.1
mitted in 3.3.4, but not greater than twice the
thickness of the thinner part or $^3/_4$ in., whichever is
less, may be corrected by welding to acceptable dimensions prior to joining the parts by welding. . . .

*Root* openings greater than those permitted by
3.3.4.1 may be corrected by welding only with the
approval of the engineer.

## Root Pass

Maximum Thickness (S.M.A.W.)                           D1.1, 4.6.5

   D1.1, . . . The maximum thickness of root
   passes in groove welds shall be $^1/_4$ in.

Minimum Thickness (S.M.A.W.)                           D1.1, 4.6.4

   D1.1, . . . The minimum size of a root pass
   shall be sufficient to prevent cracking.

## Runoff Tabs                                           D1.1, 3.12.2

D1.1, . . . Welds shall be terminated at the end of     D1.1, 6.10.3.1
a joint in a manner that will ensure sound welds.       D1.1, 8.2.4
Whenever necessary, this shall be done by the use       D1.1, 9.2.5
of weld tabs aligned in such a manner to provide an     D1.1, 10.2.4
extension of the joint preparation. . . .

*(Continued)*

Runoff Tabs *(Continued)*

*For statically* loaded structures, weld tabs need not be removed unless required by the engineer. . . .

D1.1 Comm., 3.12
AISC, 5-231, 6.3.2

*For dynamically* loaded structures, weld tabs shall be removed upon completion and cooling of the weld, and the ends of the weld shall be made smooth and flush with the edges of the abutting parts.

*Weld tabs* shall be removed prior to radiographic inspection unless other approved by the engineer.

## Safety Precautions

D1.1, 1.8

## Structural Shape Size Groupings

AISC, 1-8, Table 2

### Structural Shape Sizes Per Tensile Group Classifications

| W Shapes | | | | |
|---|---|---|---|---|
| Group 1 | Group 2 | Group 3 | Group 4 | Group5 |
| W24 x 55 | W44 x 198 | W44 x 248 | W40 x 362 to W40 x 655 | W36 x 848 |
| W24 x 62 | W44 x 224 | W44 x 285 | W36 x 328 to W36 x 798 | W14 x 605 to W14 x 730 |
| W21 x 44 to W21 x 57 | W40 x 149 to W40 x 268 | W40 x 277 to W40 x 328 | W33 x 318 to W33 x 619 | |
| W18 x 35 to W 18 x 71 | W36 x 135 to W36 x 210 | W36 x 230 to W36 x 300 | W30 x 292 to W30 x 581 | |
| W16 x 26 to W16 x 57 | W33 x 118 to W33 x 152 | W33 x 201 to W33 x 291 | W27 x 281 to W27 x 539 | |
| W14 x 22 to W14 x 53 | W30 x 90 to W30 x 211 | W30 x 235 to W30 x 261 | W24 x 250 to W24 x 492 | |
| W12 x 14 to W12 x 58 | W27 x 84 to W27 x 178 | W27 x 194 to W27 x 258 | W21 x 248 to W21 x 402 | |
| W10 x 12 to W10 x 45 | W24 x 68 to W24 x 162 | W24 x 176 to W24 x 229 | W18 x 211 to W18 x 311 | |
| W8 x 10 to W8 x 48 | W21 x 62 to W21 x 147 | W21 x 166 to W21 x 223 | W14 x 233 to W14 x 550 | |
| W6 x 9 to W6 x 25 | W18 x 76 to W18 x 143 | W18 x 158 to W18 x 192 | W12 x 21- to W12 x 336 | |
| W5 x 16 W5 x 19 | W16 x 67 to W16 x 100 | W14 x 145 to W14 x 211 | | |
| W4 x 13 | W14 x 61 to W14 x 132 | W12 x 120 to W12 x 190 | | |
| | W12 x 65 to W12 x 106 | | | |
| | W10 x 49 to W10 x 112 | | | |
| | W8 x 58 | | | |
| | W8 x 67 | | | |

**M Shapes**

| to 37.7 lb./ft. | | | | |
|---|---|---|---|---|

**S Shapes**

| to 35 lb./ft. | | | | |
|---|---|---|---|---|

**HP Shapes**

| | to 102 lb./ft | > 102 lb./ft. | | |
|---|---|---|---|---|

**Standard Channel**

| to 20.7 lb./ft. | > 20.7 lb./ft. | | | |
|---|---|---|---|---|

**Miscellaneous Channel**

| to 28.5 lb./ft. | >28.5 lb./ft. | | | |
|---|---|---|---|---|

**Angle Iron**

| to 1/2 in. | >1/2 to 3/4 in. | >3/4 in. | | |
|---|---|---|---|---|

## Submerged Arc Welding

Maximum Electrode Size                                    D1.1, 4.7.3

D1.1, . . . The diameter of electrode shall not
exceed $1/4$ in.

Multiple Electrodes                                        D1.1, 4.11.2

D1.1, . . . Submerged arc welds with multiple
electrodes, except fillet welds, shall be made in
the flat position. Fillet welds may be made in either the flat or horizontal position, except that
single-pass multiple electrode fillet welds made
in the horizontal position shall not exceed $1/2$ in.

## Shear and Tension (Rivets and Bolts)          UBC, 2-695, J3.4

## Shear Connectors                              UBC, 2-684, I4

UBC, . . . Shear connectors shall have at least            AISC, 5-29, A3.7
1 in. of lateral concrete cover, except for connectors     AISC, 5-60, I5.1
installed in the ribs of formed steel decks. Unless        AISC, 2-246 and 2-255
located directly over the web, the diameter of studs       AISC, 5-156, I4
shall not be greater than 2 $1/2$ times the thickness of
the flange to which they are welded. The minimum
center-to-center spacing of stud connectors shall be
6 diameters along the longitudinal axis of the supporting composite beam.

*(Continued)*

Shear Connectors *(Continued)*

*The maximum* center-to-center spacing of shear connectors on a beam or girder shall not be greater than 36 in.

*Decking shall* be anchored, by welds or shear connectors, at a maximum center-to-center spacing of 16 in. Welds and studs are used as deck anchors to help resist uplifting of the deck.

*Composite construction* shall, where applicable, be limited to decking having a 3 in. maximum rib height and an average rib height of not less than 2 in.

*Welded shear* connectors shall be $3/4$ in. or less in diameter.

*After installation* by welding, shear connectors shall extend a minimum of 1 $1/2$ in. above the top of the steel deck.

*The top* of the composite slab shall be a minimum of 2 in. above the top of the decking.

*Composite* construction shall, where applicable, be limited to decking having a 3 in. maximum rib height and an average rib height of not less than 2 in.

*Welded shear* connectors shall be $3/4$ in. or less in diameter.

*After installation* by welding, shear connectors shall extend a minimum of 1 $1/2$ in. above the top of the steel deck.

*The top* of the composite slab shall be a minimum of 2 in. above the top of the decking.

*Decking shall* be anchored, by welds or shear connectors, at a maximum center-to-center spacing of 16 in. Welds and studs are used as deck anchors to help resist uplifting of the deck.

# Shielding Gas                                        D1.1, 4.13

D1.1, . . . A gas or gas mixture used for shielding in gas metal arc welding or flux cored arc welding

shall be of a welding grade having a dew point of
−40 deg. F or lower.

## Shields (Arc)

D1.1, 7.2.2

D1.1, . . . An arc shield (ferrule) of heat resistant
ceramic or other suitable material shall be furnished with each stud.

## Shop Drawings

BOCA 1801.3

D1.1, . . . The inspector shall be furnished complete detailed drawings showing size, length, type,
and location of all welds to be made. The inspector
shall be furnished the portion of the contract documents that describe material and quality requirements for the products to be fabricated or erected,
or both.

D1.1, 6.1.5

## Shop Paint

AISC, 5-232, 6.5

UBC, . . . Shop painting and surface preparation
shall be in accordance with the provisions of the
*Code of Standard Practice* of the American Institute of Steel Construction, Inc., 1986 edition.

UBC, 2-605, M3.
BOCA 1801.4

*Unless otherwise* specified, steelwork that will be
concealed by interior building finish or will be in
contact with concrete need not be painted. Unless
specifically excluded, all other steelwork shall be
given one coat of shop paint.

*Except for* contact surfaces, surfaces inaccessible after shop assembly shall be cleaned and painted
prior to assembly, if required by the design documents.

*Paint is* permitted unconditionally in bearing-type
connections. . . .

*Machine-finished* surfaces shall be protected
against corrosion by a rust-inhibiting coating that
can be removed prior to erection, or that has characteristics that make removal prior to erection un-
necessary.

*(Continued)*

Shop Paint *(Continued)*

*Unless otherwise* specified in the design documents, surfaces within 2 in. of any field weld location shall be free of materials that would prevent proper welding or produce toxic fumes during welding.

## Shrinkage and Distortion

D1.1, 3.4.1

D1.1, . . . In assembling and joining parts of a structure or of built-up members and in welding re-inforcing parts to members, the procedure and sequence shall be such as will minimize distortion and shrinkage. Insofar as practicable, all welds shall be made in a sequence that will balance the applied heat of welding while the welding progresses. . . .

D1.1 Comm., 3.3.4.1

## Single Pass Fillet Welds

S.M.A.W.

D1.1, 4.6.6 - 4.6.6.3

D1.1, . . . The maximum size of single-pass fillet welds and root passes of multiple-pass fillet welds shall be:

1. $3/8$ in. for the flat position.
2. $5/16$ in. for the horizontal or overhead positions.
3. $1/2$ in. for the vertical position.

F.C.A.W.

D1.1, 4.14.1.3

D1.1, . . . The maximum size of a fillet weld made in one pass shall be:

1. $1/2$ in. for the flat and vertical positions.
2. $3/8$ in. for the horizontal position.
3. $5/16$ in. for the overhead position.

G.M.A.W.

D1.1, 4.14.1.3

D1.1, . . . The maximum size of a fillet weld made in one pass shall be:

1. $1/2$ in. for the flat and vertical positions.
2. $3/8$ in. for the horizontal position.
3. $5/16$ in. for the overhead position.

# Slag

D1.1, . . . Unacceptable portions shall be removed and rewelded. . . .

*Before* welding over previously deposited metal, all slag shall be removed and the weld and adjacent base metal shall be brushed clean. . . .

*Slag* shall be removed from all completed welds, and the weld and adjacent base metal shall be cleaned by brushing or other suitable means. Tightly adherent spatter remaining after the cleaning operation is acceptable, unless its removal is required for the purpose of nondestructive testing. . . .

D1.1, 3.7.2.3

D1.1, 3.11.1 - 3.11.2

# Slotted Holes

Hole sizes for different diameter bolts can be found in this specification. For convenience, these sizes are provided in the following table:

AISC, 5-71, J3.2; Table J3.1

AISC, 5-268, Table 1
AISC, 5-271, para. 7 (b)
UBC, 2-698, J3.8.b

### Bolt Hole Dimensions in Inches

| Bolt Diameter | 1/2 in. | 5/8 in. | 3/4 in. | 7/8 in | 1 in |
|---|---|---|---|---|---|
| Standard | 9/16 | 11/16 | 13/16 | 15/16 | 1-1/16 |
| Oversize | 5/8 | 13/16 | 15/16 | 1-1/16 | 1-1/4 |
| Short Slot | 9/16 x 1 1/4 | 11/16 x 7/8 | 13/16 x 1 | 15/16 x 1-1/8 | 1-1/16 x 2-1/2 |
| Long Slot | 9/16 x 1-1/4 | 11/16 x 1-9/16 | 13/16 x 1-7/8 | 15/16 x 2-3/16 | 1-1/16 x 2-1/2 |

| Bolt Diameter | 1-1/8 | 1-1/4 | 1-3/8 | 1-1/2 |
|---|---|---|---|---|
| Standard | 1-3/16 | 1-5/16 | 1-7/16 | 1-9/16 |
| Oversize | 1-7/16 | 1-9/16 | 1-11/16 | 1-13/16 |
| Short Slot | 1-3/16 x 1-1/2 | 1-5/16 x 1-5/8 | 1-7/16 x 1-3/4 | 1-9/16 x 1-7/8 |
| Long Slot | 1-3/16 x 2-13/16 | 1-5/16 x 3-1/8 | 1-7/17 x 3-7/16 | 1-9-16 x 3-3/4 |

*(Continued)*

Slotted Holes *(Continued)*

*Bear in* mind that base plate holes for anchor bolts may be oversized as follows (also, see Anchor Bolts in Section A):

| Bolt Size, inches (mm) | Hole Size, inches (mm) |
|---|---|
| $3/4$ (19.1) | $5/16$ (7.9) oversized |
| $7/8$ (22.2) | $5/16$ (7.9) oversized |
| 1 < 2 (25.4 < 50.8) | $1/2$ (12.7) oversized |
| >2 (> 50.8) | 1 (25.4) > bolt diameter |

*Plate washers* or continuous bar washers with a minimum thickness of $5/16$ in. are required when long slotted holes are used in an outer ply. The plate or bar washer must be of a structural grade and need not be hardened.

*If a slotted* or oversized hole is used in an outer ply of the connection, and ASTM A490 bolts over 1 in. in diameter are used, a $5/16$ in. minimum thickness F436 washer must be used.

*Only with* the approval of the engineer does the specification allow the use of oversize, short, and long-slotted holes.

*The use* of oversize, short, and long-slotted holes is subject to the joint requirements found in the specification.

*Oversized holes* are permitted in any or all plies of slip-critical connections, but they shall not be used in bearing-type connections. Hardened washers shall be installed over oversized holes in an outer ply. . . .

*Short-slotted* holes are permitted in any or all plies of slip-critical connections. The slots are permitted without regard to direction of loading in slip-critical connections, but the length shall be normal to the direction of the load in bearing-type connections. Washers shall be installed over short-slotted holes in an outer ply. When high-strength bolts are used, such washers shall be hardened. . . .

*Long-slotted* holes are permitted in only one of the connected parts of either a slip-critical connection or bearing-type connection at an individual faying surface. Long-slotted holes are permitted without regard to direction of loading in slip-critical connections, but shall be normal to the direction of load in bearing-type connections.

*Plate washers* or continuous bar washers with a minimum thickness of $5/16$ in. are required when long-slotted holes are used in an outer ply. The plate or bar washer must be of a structural grade and need not be hardened.

*If a slotted* or oversized hole is used in an outer ply of the connection, and ASTM A490 bolts over 1 in. in diameter are used, a $5/16$ in. minimum thickness F436 washer must be used.

UBC, . . . The clear distance between holes shall not be less than one bolt diameter.

### Bolt Hole Dimensions in Inches

| Bolt Diameter | 1/2 in. | 5/8 in. | 3/4 in. | 7/8 in | 1 in |
|---|---|---|---|---|---|
| Standard | 9/16 | 11/16 | 13/16 | 15/16 | 1-1/16 |
| Oversize | 5/8 | 13/16 | 15/16 | 1-1/16 | 1-1/4 |
| Short Slot | 9/16 x 1 1/4 | 11/16 x 7/8 | 13/16 x 1 | 15/16 x 1-1/8 | 1-1/16 x 2-1/2 |
| Long Slot | 9/16 x 1-1/4 | 11/16 x 1-9/16 | 13/16 x 1-7/8 | 15/16 x 2-3/16 | 1-1/16 x 2-1/2 |

| Bolt Diameter | 1-1/8 | 1-1/4 | 1-3/8 | 1-1/2 |
|---|---|---|---|---|
| Standard | 1-3/16 | 1-5/16 | 1-7/16 | 1-9/16 |
| Oversize | 1-7/16 | 1-9/16 | 1-11/16 | 1-13/16 |
| Short Slot | 1-3/16 x 1-1/2 | 1-5/16 x 1-5/8 | 1-7/16 x 1-3/4 | 1-9/16 x 1-7/8 |
| Long Slot | 1-3/16 x 2-13/16 | 1-5/16 x 3-1/8 | 1-7/17 x 3-7/16 | 1-9-16 x 3-3/4 |

## Spacing of Longitudinal Fillet Welds

D1.1, 8.8.1

D1.1, . . . If longitudinal fillet welds are used alone in end connections of flat bar tension members, the length of each fillet weld shall be no less than the perpendicular distance between them. The transverse spacing of longitudinal fillet welds used in end connections shall not exceed 8 in. unless end transverse welds or intermediate plug or slot welds are used.

AISC 5-67, J2.2b

## Splices

UBC, 2-688, J1.7; 2-701, J7

UBC, . . . Groove welded splices in plate girders shall develop the full strength of the smaller spliced section.

AISC, 5-79, J7

## Special Moment Resisting Space Frames

UBC, 2-360, 2211.7

## Stiffeners

Bearing

AISC, 5-52, G4

D1.1, . . . The bearing ends of bearing stiffeners shall be square with the web and shall have at least 75% of the stiffener bearing cross sectional area in contact with the inner surfaces of the flanges. . . .

D1.1, 3.5.2 - 3.5.3.3

*The* out-of-straightness variation of bearing stiffeners shall not exceed $1/4$ in. up to 6 ft. deep or $1/2$ in. over 6 ft. deep. The actual centerline of the stiffener shall lie within the thickness of the stiffener as measured from the theoretical centerline location.

Intermediate

D1.1, 3.5.3.1-3.5.1.11

D1.1, . . . Where tight fit of intermediate stiffeners is specified, it shall be defined as allowing a gap of up to $1/16$ in. between stiffener and flange. . . .

UBC, 2-368, 2211.10.10

*The* out-of-straightness variation of intermediate stiffeners shall not exceed $1/2$ in. for girders up to 6 ft deep, and $3/4$ in. for girders over 6 ft. deep, with due regard for members that frame into them.

UBC, . . . For beams 24 in. in depth and greater, intermediate full-depth web stiffeners are required on both sides of the web. Such web stiffeners are required only on one side of the beam web for beams less than 24 in. in depth. . . .

Plates                                                                UBC, 2-704, K1.8

## Steel Construction (Conformance to)                                BOCA 1801.0

## Steel Identification                                               UBC, 2-356, 2202.2

UBC, . . . Structural steel shall be identified by         UBC, 2-356, 2202.2.2
the mill in accordance with approved national stan-        UBC, 2-555, A3.1
dards. When such steel is furnished to a specified         AISC, 5-92, M5.5
minimum yield point greater than 36,000 pounds             ASTM, A6/A
per square inch, the American Society for Testing
and Materials (ASTM) or other specification designation shall be so indicated. . . .

*Steel* that is not readily identifiable as to grade
from marking and test records shall be tested to determine conformity to such standards. . . .

*Certified* mill test reports or certified reports of
tests made by the fabricator or a testing laboratory
in accordance with ASTM A6 or A568, as applicable, shall constitute sufficient evidence of conformance with one of the above ASTM standards. If requested, the fabricator shall provide an affidavit
stating that the structural steel furnished meets
the requirements of the grade specified.

*Any identification* method used shall at least verify
and document the following items:

1. Material specification.
2. Heat number.
3. Material test reports for special requirements.

## Stop Work Order

BOCA 118.1 and 118.2

## Storage of Electrodes

D1.1, 4.5.2

D1.1, . . . All electrodes having low hydrogen coverings conforming to ANSI/AWS A5.1 shall be purchased in hermetically sealed containers or shall be dried for at least 2 hours between 500 and 800 deg. F before they are used. Electrodes having a low hydrogen covering conforming to ANSI/AWS A5.5 shall be purchased in hermetically sealed containers or shall be dried at least one hour at temperatures between 700 and 800 deg. F before being used. Electrodes shall be dried prior to use if the hermetically sealed container shows evidence of damage. Immediately after opening of the hermetically sealed container or removal of the electrodes from the drying ovens, electrodes shall be stored in ovens held at temperatures of at least 250 deg. F. After the opening of hermetically sealed containers or removal from drying or storage ovens, electrode exposure to the atmosphere shall not exceed the requirements of either 4.5.2.1 or 4.5.2.2.

D1.1, Table 4.6
D1.1 Comm., 4.5

### Table 4.6
### Permissible Atmospheric Exposure of
### Low Hydrogen Electrodes (see 4.5.2.1)

| Electrode | Column A (hours) | Column B (hours) |
|---|---|---|
| **A5.1** | | |
| E70XX | 4 max | Over 4 to 10 max |
| E70XXR | 9 max | |
| E70XXHZR | 9 max | |
| E7018M | 9 max | |
| **A5.5** | | |
| E70XX-X | 4 max | Over 4 to 10 max |
| E80XX-X | 2 max | Over 2 to 10 max |
| E90XX-X | 1 max | Over 1 to 5 max |
| E100XX-X | 1/2 max | Over 1/2 to 4 max |
| E110XX-X | 1/2 max | Over 1/2 to 4 max |

Notes:
1. Column A: Electrodes exposed to atmosphere for longer periods than shown shall be redried before use.
2. Column B: Electrodes exposed to atmosphere for longer periods than those established by testing shall be redried before use.
3. Entire table: Electrodes shall be issued and held in quivers, or other small open containers. Heated containers are not mandatory.
4. The optional supplemental designator, R, designates a low hydrogen electrode which has been tested for covering moisture content after exposure to a moist environment for 9 hours and has met the maximum level permitted in ANSI/AWS A5.1-91, *Specification for Carbon Steel Electrodes for Shielded Metal Arc Welding*.

## Straightening

D1.1, . . . *Members distorted* by welding shall be straightened by mechanical means or by application of a limited amount of localized heat. The temperature of heated areas as measured by approved methods shall not exceed 1100 deg. F (590 deg. C) for quenched and tempered steel nor 1200 deg. F (650 deg. C) for other steels. The part to be heated for straightening shall be substantially free of stress and from external forces, except those forces resulting from the mechanical straightening method used in conjunction with the application of heat.

D1.1, 3.7.3

D1.1 Comm., 3.7.3
AISC, 5-87, M2.1

## Stress in Fillet Welds

D1.1 Comm., 8.2, Part B

D1.1, . . . The stress on the effective throat of fillet welds is always considered to be shear. . . .

## Stress Relief

D1.1, 4.4

D1.1, . . . Where required by the contract drawings or specifications, welded assemblies shall be stress-relieved by heat treating. Finish machining shall preferably be done after stress relieving. . . .

D1.1 Comm., 4.4

*Consideration* must be given to possible distortion due to stress relief.

## Structural Steel Construction

SBCCI 1502

## Studs

Ambient Temperature

D1.1, 7.5.4

D1.1, . . . Welding shall not be done when the base metal temperature is below 0 deg. F or when the surface is wet or exposed to falling rain or snow. When the temperature of the base metal is below 32 deg. F, one additional stud in each 100 studs welded shall be tested by methods specified in 7.7.1.3 and 7.7.1.4, except that the angle of testing shall be approximately 15 degrees. This is in addition to the first two studs tested for each start of a new production period or change in set-up.

Amount to be Tested

D1.1, 7.7.1.1

D1.1, . . . Before production welding with a particular set-up and with a given size and type of stud, and at the beginning of each day's or shift's production, testing shall be performed on the first two studs that are welded. . . .

Areas to be Welded

D1.1, 7.4.3

D1.1, . . . The areas to which studs are to be welded shall be free from scale, rust, moisture, or other injurious material to the extent necessary to obtain satisfactory welds. These areas may be cleaned by wire brushing, scaling, prick-punching, or grinding.

Base Metal Requirements                                    D1.1, 7.4.2

D1.1, . . . The stud base shall not be painted, galvanized, or cadmium-plated prior to welding.

Bend Tests                                                 D1.1, 7.7.1.4

D1.1, . . . In addition to visual examination, the test shall consist of bending the studs after they are allowed to cool, to an angle of approximately 30 deg. from their original axis by either striking the studs with a hammer or placing a pipe or other suitable hollow device over the stud and manually or mechanically bending the stud. At temperatures below 50 deg. F, bending shall preferably be done by continuous slow application of load. For threaded studs, the torque test of Figure 7.3 shall be substituted for the bend test.

Diameter of                                                UBC, 2-684, I4, I5

UBC, . . . Unless located directly over the web, the diameter of studs shall not be greater than 2 $\frac{1}{2}$ times the thickness of the flange to which they are welded. . . .

*The* concrete slab shall be connected to the steel beam or girder with welded stud shear connectors $\frac{3}{4}$ in. or less in diameter. . . .

Ferrules for                                               D1.1, 7.4.4

D1.1, . . . The arc shields or ferrules shall be kept dry. Any arc shields that show signs of surface moisture from dew or rain shall be oven dried at 250 deg. F for 2 hours before use.

Minimum Weld Size                                          D1.1, 7.5.5.4

D1.1 . . . When fillet welds are used, the mini-    D1.1, Table 7.2
mum size shall be the larger of those required in Table 2.2 or Table 7.2.

*(Continued)*

Studs, Minimum Weld Size *(Continued)*

### Table 2.2
### Minimum Fillet Weld Size for
### Prequalified Joints (see 2.7.1.1)

| Base Metal Thickness (T)* | | Minimum Size of Fillet Weld** | | |
|---|---|---|---|---|
| in. | mm | in. | mm | |
| T≤1/4 | T≤ 6.4 | 1/8*** | 3 | Single-pass welds must be used |
| 1/4<T≤1/2 | 6.4<T≤12.7 | 3/16 | 5 | |
| 1/2<T≤3/4 | 12.7<T≤19.0 | 1/4 | 6 | |
| 3/4<T | 19.0<T | 5/16 | 8 | |

*For non-low hydrogen processes without preheat calculated in accordance with 4.2.2, T equals thickness of the thicker part joined. For non-low hydrogen processes using procedures established to prevent cracking in accordance with 4.2.2, and for low hydrogen processes, T equals thickness of the thinner part joined; single pass requirement does not apply.

**Except that the weld size need not exceed the thickness of the thinner part joined.

***Minimum size for dynamically loaded structures is 3/16 in. (5 mm).

### Table 7.2
### Minimum Fillet Weld Size for Small Diameter
### Studs (see 7.5.5.1)

| Stud Diameter | | Min. Size Fillet | |
|---|---|---|---|
| in. | mm | in. | mm |
| 1/4 thru 7/16 | 6.4 thru 11.1 | 3/16 | 5 |
| 1/2 | 12.7 | 1/4 | 6 |
| 5/8, 3/4, 7/8 | 15.9, 19, 22.2 | 5/16 | 8 |
| 1 | 25.4 | 3/8 | 10 |

Preparation of Surface                          D1.1, 7.5.5.1

D1.1, . . . Surfaces to be welded and surfaces
adjacent to a weld shall be free from loose or
thick scale, slag, rust, moisture, grease, and other
foreign material that would prevent proper welding or produce objectionable fumes.

Repair of                                       D1.1, 7.7.3 - 7.7.5

D1.1, . . . In production, studs on which a full
360 deg. flash is not obtained may, at the option
of the contractor, be repaired by adding the mini-

mum fillet weld as required by 7.5.5 in place of
the missing flash. The repair weld shall extend
at least $3/8$ in. beyond each end of the discontinuity being repaired.

Retesting of                                         D1.1, 7.7.1.5

D1.1, . . . If on visual examination the test
studs do not exhibit 360 deg. flash, or if on testing failure occurs in the weld zone of either stud,
the procedure shall be corrected, and two more
studs shall be welded to separate material or on
the production member and tested in accordance
with the provisions of 7.7.1.3 and 7.7.1.4. If either of the second two studs fails, additional
welding shall be continued on separate plates
until two consecutive studs are tested and found
to be satisfactory before any more production
studs are welded to the member.

Shall Conform to                                     D1.1, 7.3.1

D1.1, . . . Studs shall be made from cold drawn
bar stock conforming to the requirements of
ASTM A108, Specification for Steel Bars, Cold
Finished, Standard Quality, Grades 1010
through 1020, inclusive, either semi-killed or
killed deoxidation.

Shear Connectors                                     AISC, 5-58, I4

The maximum diameter of a welded shear con-       AISC, 5-60, I5.1
nector shall be 2 $1/2$ times the thickness of the   AISC, 2-246 and 2-255
flange to which it is welded. Larger studs may be   AISC, 5-29, A3.7
used if welded directly over the web of the         UBC, 2-684, I3
member.

*Composite* construction shall, where applicable,
be limited to decking having a 3 in. maximum rib
height and an average rib height of not less than
2 in.

*Welded shear* connectors shall be $3/4$ in. or less in
diameter.

*After installation* by welding, shear connectors
shall extend a minimum of 1 $1/2$ in. above the top
of the steel deck.

*(Continued)*

## Studs, Shear Connectors *(Continued)*

*The top* of the composite slab shall be a minimum of 2 in. above the top of the decking.

*The maximum* center-to-center spacing of shear connectors on a beam or girder shall not be greater than 36 in.

*Decking shall* be anchored, by welds or shear connectors, at a maximum center-to-center spacing of 16 in. Welds and studs are used as deck anchors to help resist uplifting of the deck.

### Size and Type Electrode

D1.1, 7.5.5.6

D1.1, . . . Shielded metal arc welding shall be performed using low hydrogen electrodes $5/32$ or $3/16$ in. in diameter, except that a smaller diameter electrode may be used on studs $7/16$ in. or less in diameter for out-of-position welds.

### Spacing of

D1.1, 7.4.5

D1.1, . . . Longitudinal and lateral spacing of stud shear connectors (Type B) with respect to each other and to edges of beam or girder flanges may vary a maximum of 1 in. from the location shown in the drawings. The minimum distance from the edge of a stud base to the edge of a flange shall be the diameter of the stud plus $1/8$ in., but preferably not less than 1-$1/2$ in.

### Tension Test

D1.1, 7.6.6.3

D1.1, Appendix IX, 7.2

D1.1, . . . Studs shall be tension tested to destruction using any machine capable of supplying the required force. A stud application shall be considered qualified if the test specimens do not fail in the weld.

### Torque

The specifications **do not permit** the use of torque tables for the installation, tensioning, or inspection of any bolt-up procedure.

*Bolt tensioning devices, such as a Skidmore,* must be used in all cases of high-strength bolt installation to determine proper bolt tensions and job torques.

The following tables are being provided as a convenience only:

**Average Torques for A325 and A490 Bolts**
(Foot-Pounds)

### A325 Bolts

| Nominal Bolt / Nut Diameter | Average Torque | Width of Head Across Flats |
|---|---|---|
| 5/8 | 200 | 1-1/16 |
| 3/4 | 355 | 1-1/4 |
| 7/8 | 570 | 1-7/16 |
| 1 | 850 | 1-5/8 |
| 1-1/8 | 1,060 | 1-13/16 |
| 1-1/4 | 1,495 | 2 |
| 1-3/8 | 1,960 | 2-3/16 |
| 1-1/2 | 2,600 | 2-3/8 |

### A490 Bolts

| Nominal Bolt/Nut Diameter | Average Torque | Width of Head Across Flats |
|---|---|---|
| 5/8 | 250 | 1-1/16 |
| 3/4 | 435 | 1-1/4 |
| 7/8 | 715 | 1-7/16 |
| 1 | 1,070 | 1-5/8 |
| 1-1/8 | 1,500 | 1-13/16 |
| 1-1/4 | 2,125 | 2 |
| 1-3/8 | 2,780 | 2-3/16 |
| 1-1/2 | 3,700 | 2-3/8 |

Studs, Torque *(Continued)*

### Nut and Bolt Head Dimensions
### A325 and A490

| Bolt Diameter | Width across the Flats (Nuts and Bolts) | Nut Height |
|---|---|---|
| 1/2 | 7/8 | 31/64 |
| 5/8 | 1-1/16 | 39/64 |
| 3/4 | 1-1/4 | 47/64 |
| 7/8 | 1-7/16 | 55/64 |
| 1 | 1-5/8 | 63/64 |
| 1-1/8 | 1-13/16 | 1-7/64 |
| 1-1/4 | 2 | 1-7/32 |
| 1-3/8 | 2-3/16 | 1-11/32 |
| 1-1-2 | 2-3/8 | 1-15/32 |

This information is useful in determining proper wrench sizes.

Torque Test                                    D1.1, 7.6.6.2

D1.1, . . . Studs shall be torque tested using a    D1.1, Fig. 7.3
torque test arrangement that is substantially in
accordance with Figure 7.3. A stud application
shall be considered qualified if all test specimens
are torqued to destruction without failure in the
weld.

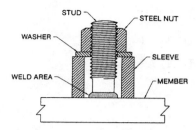

Note: The dimensions shall be appropriate to the size of the stud. The threads of the stud shall be clean and free of lubricant other than the residue of cutting oil.

| Required torque for testing threaded studs | | | |
|---|---|---|---|
| Nominal diameter of studs | | Threads per inch & series designated | Testing torque |
| in. | mm | | ft-lb | J |

| in. | mm | Threads per inch & series designated | ft-lb | J |
|---|---|---|---|---|
| 1/4 | 6.4 | 28 UNF | 5.0 | 6.8 |
| 1/4 | | 20 UNC | 4.2 | 5.7 |
| 5/16 | 7.9 | 24 UNF | 9.5 | 12.9 |
| 5/16 | | 18 UNC | 8.6 | 11.7 |
| 3/8 | 9.5 | 24 UNF | 17.0 | 23.0 |
| 3/8 | | 16 UNC | 15.0 | 20.3 |
| 7/16 | 11.1 | 20 UNF | 27.0 | 36.6 |
| 7/16 | | 14 UNC | 24.0 | 32.5 |
| 1/2 | 12.7 | 20 UNF | 42.0 | 57.0 |
| 1/2 | | 13 UNC | 37.0 | 50.2 |
| 9/16 | 14.3 | 18 UNF | 60.0 | 81.4 |
| 9/16 | | 12 UNC | 54.0 | 73.2 |
| 5/8 | 15.9 | 18 UNF | 84.0 | 114.0 |
| 5/8 | | 11 UNC | 74.0 | 100.0 |
| 3/4 | 19.0 | 16 UNF | 147.0 | 200.0 |
| 3/4 | | 10 UNC | 132.0 | 180.0 |
| 7/8 | 22.2 | 14 UNF | 234.0 | 320.0 |
| 7/8 | | 9 UNC | 212.0 | 285.0 |
| 1 | 25.4 | 12 UNF | 348.0 | 470.0 |
| 1 | | 8 UNC | 318.0 | 430.0 |

**Figure 7.3 — Torque Testing Arrangement and Table of Testing Torques (see 7.6.6.2)**

### Type "B" Studs

D1.1, 7.4.5; Table 7.1

D1.1, . . . Type B studs shall be studs that are headed, bent, or of other configuration in ½ in., ⅝ in., ¾ in., ⅞ in., and 1 in. in diameter that are used as an essential component in composite beam design and construction.

### Visual Examinations

D1.1, 7.7.1.3

D1.1, . . . The studs shall be visually examined. They shall exhibit full 360 deg. flash.

### Welding Below 50 Deg. Ambient

D1.1, 7.7.1.4

D1.1, . . . At temperatures below 50 deg. F, bending shall preferably be done by continuous slow application of load. . . .

*(Continued)*

Studs *(Continued)*

| | |
|---|---|
| Welding of | D1.1, 7.1 |
| With or Without Flux | D1.1, 7.2.3 |

D1.1, . . . A suitable deoxidizing and arc stabilizing flux shall be furnished with each stud $^5/_{16}$ in. in diameter or larger. Studs less than $^5/_{16}$ in. in diameter may be furnished with or without flux.

| | |
|---|---|
| Workmanship | D1.1, 7.4 |

## Surfaces Adjacent to Field Welds

AISC, 5-90, M3.5

D1.1, . . . Surfaces on which weld metal is to be deposited shall be smooth, uniform, and free from fins, tears, cracks, and other discontinuities that would adversely affect the quality or strength of the weld. Surfaces to be welded, and surfaces adjacent to a weld, shall also be free from loose or thick scale, slag, rust, moisture, grease, and other foreign material that would prevent proper welding or produce objectionable fumes. Mill scale that can withstand vigorous wire brushing, a thin rust-inhibitive coating, or antispatter compound may remain with the following exception: for girders in dynamically loaded structures, all mill scale shall be removed from the surfaces on which flange-to-web welds are to be made by submerged arc welding or by shielded arc welding with low hydrogen electrodes.

D1.1, 3.2.1

## Tack Welds

D1.1, 3.3.7

D1.1, . . . Tack welds shall be subject to the same quality requirements as the final welds, with the following exceptions:

D1.1 Comm., 3.3.7

1. Preheat is not mandatory for single pass tack welds that are remelted and incorporated into continuous submerged arc welds.

2. Discontinuities, such as undercut, unfilled craters, and porosity need not be removed before the final submerged arc welding. . . .

*Tack* welds not incorporated into the final welds shall be removed, except that, for statically loaded structures, they need not be removed unless required by the engineer.

## Temperature of Heated Areas

D1.1, 3.7.3

D1.1, . . . Members distorted by welding shall be straightened by mechanical means or by application of a limited amount of localized heat. The temperature of heated areas as measured by approved methods shall not exceed 1100 deg. F for quenched and tempered steel nor 1200 deg. F for other steels. The part to be heated shall be substantially free of stress and from external forces, except those stresses resulting from the mechanical straightening method used in conjunction with the application of heat.

## Tension Members

AISC, 5-40, D2

D1.1, . . . If longitudinal fillet welds are used alone in end connections of flat bar tension members, the length of each fillet weld shall be no less than the perpendicular distance between them. The transverse spacing of longitudinal fillet welds used in end connections shall not exceed 8 in. unless end transverse welds or intermediate plug or slot welds are used.

D1.1, 8.8.1
D1.1, 8.12.3
D1.1, 9.17
UBC, 2-671, D2

*In built-up* tension members, the longitudinal spacing of stitch welds connecting a plate component to other components, or connecting two plate components to each other, shall not exceed 12 in. or 24 times the thickness of the thinner plate.

## Thermal Cutting Repair Tolerance

AISC, 5-87, M2.2

D1.1, . . . Roughness exceeding these values and notches or gouges not more than $3/16$ in. deep on otherwise satisfactory surfaces shall be removed by machining or grinding. Notches or gouges exceeding $3/16$ in. deep may be repaired by grinding if the nom-

D1.1, 3.2.2.1
D1.1, Table 3.1

*(Continued)*

Thermal Cutting Repair Tolerance *(Continued)*

inal cross sectional area is not reduced by more than 2%. Ground or machined surfaces shall be faired to the original surface with a slope not exceeding one in ten. Cut surfaces and adjacent edges shall be left free of slag. In thermal cut surfaces, occasional notches or gouges may, with the approval of the engineer, be repaired by welding.

*Reentrant corners* of cut material shall be formed to provide a gradual transition with a radius of not less than 1 in. Adjacent surfaces shall meet without offset or cutting past the point of tangency. The reentrant corners may be formed by thermal cutting, followed by grinding, if necessary, to meet the surface requirements of 3.2.2.

## Thickness Limitations

D1.1, 1.2.3

D1.1, . . . The provisions of this code are not intended to apply to welding base metals less than $1/8$ in. thick.

## Transition in Thickness or Width

D1.1, 8.10

D1.1, . . . Tension butt joints between axially aligned members of different thicknesses or widths, or both, and subject to tensile stress greater than $1/3$ the allowable design tensile stress shall be made in such a manner that the slope in the transition does not exceed 1 in 2 $1/2$ (see Figures 8.3 and 8.4). The transition shall be accomplished by chamfering the thicker part, tapering the wider part, sloping the weld metal, or by any combination of these.

## Turn of Nut Method   *(See Bolt Section)*

## Unacceptable Welds or Work

D1.1, 3.7.6

D1.1, . . . If, after an unacceptable weld has been made, work has been performed that has rendered that weld inaccessible or has created new conditions that make corrections of the unacceptable weld dangerous or ineffectual, then the original conditions shall be restored by removing welds or members, or both, before the corrections are made. If this is not done, the deficiency shall be compensated for by additional work performed according to an approved revised design.

## Undercut in

### Dynamically Loaded Structures                          D1.1, 9.25.1.5

D1.1, . . . In primary members, undercut shall be no more than 0.01 in. (0.25 mm) deep when the weld is transverse to tensile stress under any design loading condition. Undercut shall be no more than $1/32$ in. deep for all other cases.

### Statically Loaded Structures                          D1.1, 8.15.1.5

D1.1, . . . For material less than 1 in. thick, undercut shall not exceed $1/32$ in., except that a maximum $1/16$ in. is permitted for an accumulated length of 2 in. in any 12 in. For material equal to or greater than 1 in. thick, undercut shall not exceed $1/16$ in. for any length of weld.

## Underrun in Fillet Welds                          D1.1, 8.15.1.7

D1.1, . . . A fillet weld in any single continuous weld shall be permitted to underrun the nominal fillet size specified by $1/16$ in. without correction, provided that the undersize portion of the weld does not exceed 10% of the length of the weld. On web-to-flange welds on girders, no underrun is permitted at the ends for a length equal to twice the width of the flange.

## Undersize Welds *(See Underrun)*                          D1.1, 3.7.2.2

D1.1, . . . The surfaces shall be prepared and additional weld metal deposited.                          D1.1, 8.15.1.7

## Unsafe Construction Equipment

BOCA 3003.3

## Unsafe Conditions

BOCA 3003.7

## Verification Inspector

D1.1, 6.1.2

D1.1, . . . The verification inspector is the duly designated person who acts for and in behalf of the owner or engineer on all inspection and quality matters within the scope of the contract documents.

## Visual Inspection (Studs)

D1.1, 7.7.1.3

D1.1, . . . The test studs shall be visually examined. They shall exhibit full 360 deg. flash.

## Welder Requalification

D1.1, 6.4.3

D1.1, . . . The inspector shall require requalification of any qualified welder, welding operator, or tack welder who has for a period exceeding 6 months not used the process for which the welder, welding operator, or tack welder was qualified.

## Welding of Structural Steel

D1.1-94
SBCCI 1507

## Welds in Combinations with Rivets and Bolts

D1.1, 8.7

D1.1, . . . Rivets or bolts used in bearing type connections shall not be considered as sharing the stress in combination with welds. Welds, if used, shall be provided to carry the entire stress in the connection. However, connections that are welded to one member and riveted or bolted to the other member are permitted. High-strength bolts properly installed as a friction type connection prior to welding may be considered as sharing the stress with the welds.

UBC, 2-688, J1.10; 2-689, J1.12
AISC 5-64, J1.10

## Welds (Details of) *(See Specific Weld)*

UBC, 2-689, J2

## Weld Termination

D1.1, . . . Welds shall be terminated at the ends of a joint in a manner that will insure sound welds. Whenever necessary, this shall be done by the use of weld tabs aligned in such a manner to provide an extension of the joint preparation. . . .

*For* statically loaded structures, weld tabs need not be removed unless required by the engineer. . . .

*For* dynamically loaded structures, weld tabs shall be removed upon completion and cooling of the weld, and the ends of the weld shall be made smooth and flush with the edges of abutting parts. . . .

*Ends* of welded butt joints required to be flush shall be finished so as not to reduce the width beyond the detailed width or the actual width furnished, whichever is greater, by more than $\frac{1}{8}$ in. or so as not to leave reinforcement at each end that exceeds $\frac{1}{8}$ in. Ends of welded butt joints shall be faired at a slope not to exceed 1 in 10.

## Weld Washers

UBC, . . . Weld washers shall be used when the thickness of the sheet is less than 0.028 in. Weld washers shall have a thickness between 0.05 and 0.08 in. with a minimum prepunched hole of $\frac{3}{8}$ in. diameter.

D1.3, . . . Weld washers are to be used in containing the arc spot welds in sheet steel thinner than 0.028 in. (See Figure 2.1.) Weld washers shall be made of one of the sheet steels listed in 1.2.1 and shall have a thickness between 0.05 in. and 0.08 in., with a minimum prepunched hole of $\frac{3}{8}$ in.

*(Continued)*

Weld Washers *(Continued)*

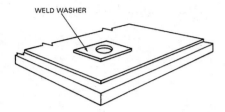

WELD WASHER

Figure 2.1 — Typical Weld Washer
(See 2.2.2.2)

## Welding Precautions

BOCA 3020.1

## Wind Velocity

D1.1, 4.14.3

D1.1, . . . Gas metal arc or flux cored welding with
external gas shielding shall not be done in a draft
or wind unless the weld is protected by a shelter.
Such shelter shall be of a material and shape appropriate to reduce wind velocity in the vicinity of
the weld to a maximum of 5 miles per hour.

## Workmanship (Studs) *(See Studs)*

D1.1, 7.4

# C  Tables

## Anchor Bolts

AISC 4-4, Table 1-C

### Anchor Bolt and Threaded Rod Material

| ASTM Specification | Diameter in inches | Material |
|---|---|---|
| A307 | 4 | Carbon |
| A325 | 1/2 to 1-1/2 | Carbon, Quenched and Tempered |
| A354 Grades B,C & D | 1/4 to 4 | Alloy, Quenched and Tempered |
| A449 | 1/4 ot 3 | Carbon, Quenched and Tempered |
| A490 | 1/2 to 1-1/2 | Alloy, Quenched and Tempered |
| A687 | 5/8 to 3 | Alloy, Quenched and Tempered, Notch Tough |
| A36 | 8 | Carbon |
| A572 Grade 50 | 2 | High-Strength Low Alloy |
| A572 Grade 42 | 6 | High-Strength Low Alloy |
| A588 | 4 to 8 | High-Strength Low Alloy, Weather Resistant |

Quenched and Tempered Anchor Bolts should not be heated or welded without the expressed approval of the Engineer.

Oversized Holes for

UBC, 2-358, 2210

| Bolt Size, inches (mm) | Hole Size, inches (mm) |
|---|---|
| $3/4$ (19.1) | $5/16$ (7.9) oversized |
| $7/8$ (22.2) | $5/16$ (7.9) oversized |
| $1 < 2$ (25.4 < 50.8) | $1/2$ (12.7) oversized |
| >2 (> 50.8) | 1 (25.4) > bolt diameter |

# Bolt Dimensions

AISC, 5-289, Table C1

**Dimensions for A325 and A490 Bolts**

| Diameter in Inches | Width across Flats (Wrench Size) | Head Height | Thread Length |
|---|---|---|---|
| 1/2 | 7/8 | 5/16 | 1 |
| 5/8 | 1-1/16 | 25/64 | 1-1/4 |
| 3/4 | 1-1/4 | 15/32 | 1-3/8 |
| 7/8 | 1-7/16 | 35/64 | 1-1/2 |
| 1 | 1-5/8 | 39/64 | 1-3/4 |
| 1-1/8 | 1-12/16 | 11/16 | 2 |
| 1-1-4 | 2 | 25/32 | 2 |
| 1-3/8 | 2-3/16 | 27/32 | 2-1/4 |
| 1-1/2 | 2-3/8 | 15/16 | 2-1/4 |

# Bolt Markings

AISC 5-293, Fig. C2

(1) ADDITIONAL OPTIONAL 3 RADIAL LINES AT 120° MAY BE ADDED.
(2) TYPE 3 ALSO ACCEPTABLE.
(3) ADDITIONAL OPTIONAL MARK INDICATING WEATHERING GRADE MAY BE ADDED.

*Required marking for acceptable bolt and nut assemblies*

# Boxing (End Returns)

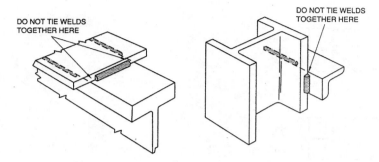

Figure 8.2 — Fillet Welds on Opposite Side of a Common Plane of Contact (see 8.8.5)

# Composite Construction with Formed Steel Decking

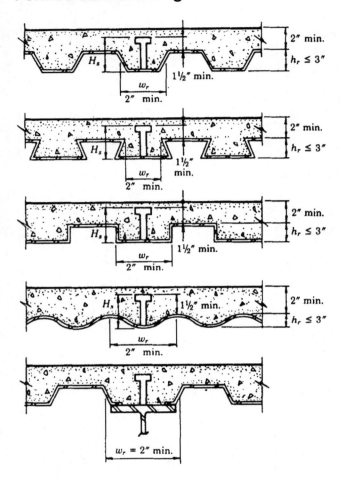

*Figure C-I5.1*

# Cope Holes

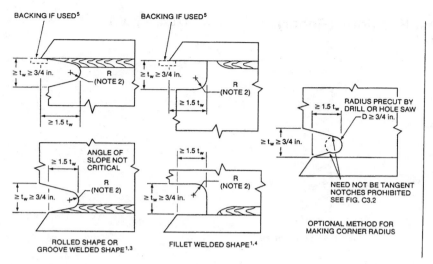

Notes:
1. For ASTM A6 Group 4 and 5 shapes and welded built-up shapes with web thickness more than 1-1/2 in. (38.1 mm), preheat to 150°F (66°C) prior to thermal cutting, grind and inspect thermally cut edges of access hole using magnetic particle or dye penetration methods prior to making web and flange splice groove welds.
2. Radius shall provide smooth notch-free transition; R ≥ 3/8 in. (9 mm) [Typical 1/2 in. (13 mm)].
3. Access opening made after welding web to flange.
4. Access opening made before welding web to flange. Weld not returned through opening.
5. These are typical details for joints welded from one side against steel backing. Alternative joint designs should be considered.

Figure 3.2 — Weld Access Hole Geometry (see 3.2.2, 3.2.2.1, 3.2.5 and 3.2.5.1)

# Cut Edge Discontinuities

## Table 3.1
### Limits on Acceptability and Repair of Mill Induced Laminar Discontinuities in Cut Surfaces (see 3.2.3)

| Description of Discontinuity | Repair Required |
|---|---|
| Any discontinuity 1 in. (25 mm) in length or less | None, need not be explored. |
| Any discontinuity over 1 in. (25 mm) in length and 1/8 in. (3 mm) maximum depth | None, but the depth should be explored.* |
| Any discontinuity over 1 in. (25 mm) in length with depth over 1/8 in. (3 mm) but not greater then 1/4 in. (6 mm) | Remove, need not weld. |
| Any discontinuity over 1 in. (25 mm) in length with depth over 1/4 in. (6 mm) but not greater than 1 in. | Completely remove and weld. |
| Any discontinuity over 1 in. (25 mm) in length with depth greater than 1 in. | See 3.2.3.2. |

*A spot check of 10% of the discontinuities on the cut surface in question should be explored by grinding to determine depth. If the depth of any one of the discontinuities explored exceeds 1/8 in. (3 mm), then all of the discontinuities over 1 in. (25 mm) in length remaining on that cut surface shall be explored by grinding to determine depth. If none of the discontinuities explored in the 10% spot check have a depth exceeding 1/8 in. (3 mm), then the remainder of the discontinuities on that cut surface need not be explored.

# Direct Butt Joint Test Positions (Rebar)
## (See Reinforcement)

D1.4, Fig. 6.1

BARS HORIZONTAL

BARS VERTICAL

A - TEST POSITION 1G

B - TEST POSITION 2G

BARS HORIZONTAL

C - TEST POSITION 3G

D - TEST POSITION 4G

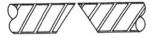

NOTE: SEE FIGURE 6.3 FOR DEFINITION OF POSITIONS FOR
GROOVE WELDS

Figure 6.1 — Direct Butt Joint Test Positions for Groove Welds (See 6.2.4.1)

## Effective Throats of Flare Groove Welds

AISC 5-66, Table J2.2
UBC, 2-690, Table J2.2
D1.1, Table 2.1

### TABLE J2.2
### Effective Throat Thickness of Flare Groove Welds

| Type of Weld | Radius (*R*) of Bar or Bend | Effective Throat Thickness |
|---|---|---|
| | | × 25.4 for mm |
| Flare bevel groove | All | $5/16R$ |
| Flare V-groove | All | $1/2R$[a] |

[a]Use $3/8R$ for Gas Metal Arc Welding (except short circuiting transfer process) when $R \geq 1/2$-in. (13 mm).

### Table 2.1
### Effective Weld Sizes of Flare Groove Welds
### (see 2.3.1.4)

| Flare-Bevel-Groove Welds | Flare-V-Groove Welds |
|---|---|
| 5/16 R | 1/2 R* |

NOTE:  R = radius of outside surface

*Use 3/8 R for GMAW (except short circuiting transfer) process when R is 1/2 in. (13 mm) or greater.

## Effective Throats of Partial Penetration Welds

AISC, 5-66, Table J2.1
UBC, 2-690, Table J2.1
D1.1, Table 2.3

### Table 2.3
### Minimum Weld Size for Partial Joint
### Penetration Groove Welds (see 2.10.3)

| Base Metal Thickness of Thicker Part Joined | | Minimum Weld Size* | |
|---|---|---|---|
| in. | mm | in. | mm |
| 1/8 (3.2) to 3/16 (4.8) incl. | | 1/16 | 2 |
| Over 3/16 (4.8) to 1/4 (6.4) incl. | | 1/8 | 3 |
| Over 1/4 (6.4) to 1/2 (12.7) incl. | | 3/16 | 5 |
| Over 1/2 (12.7) to 3/4 (19.0) incl. | | 1/4 | 6 |
| Over 3/4 (19.0) to 1-1/2 (38.1) incl. | | 5/16 | 8 |
| Over 1-1/2 (38.1) to 2-1/4 (57.1) incl. | | 3/8 | 10 |
| Over 2-1/4 (57.1) to 6 (152) incl. | | 1/2 | 13 |
| Over 6 (152) | | 5/8 | 16 |

*Except the weld size need not exceed the thickness of the thinner part.

# Electrode Storage

### Table 4.6
### Permissible Atmospheric Exposure of
### Low Hydrogen Electrodes (see 4.5.2.1)

| Electrode | Column A (hours) | Column B (hours) |
|---|---|---|
| **A5.1** | | |
| E70XX | 4 max | Over 4 to 10 max |
| E70XXR | 9 max | |
| E70XXHZR | 9 max | |
| E7018M | 9 max | |
| **A5.5** | | |
| E70XX-X | 4 max | Over 4 to 10 max |
| E80XX-X | 2 max | Over 2 to 10 max |
| E90XX-X | 1 max | Over 1 to 5 max |
| E100XX-X | 1/2 max | Over 1/2 to 4 max |
| E110XX-X | 1/2 max | Over 1/2 to 4 max |

Notes:
1. Column A: Electrodes exposed to atmosphere for longer periods than shown shall be redried before use.
2. Column B: Electrodes exposed to atmosphere for longer periods than those established by testing shall be redried before use.
3. Entire table: Electrodes shall be issued and held in quivers, or other small open containers. Heated containers are not mandatory.
4. The optional supplemental designator, R, designates a low hydrogen electrode which has been tested for covering moisture content after exposure to a moist environment for 9 hours and has met the maximum level permitted in ANSI/AWS A5.1-91, *Specification for Carbon Steel Electrodes for Shielded Metal Arc Welding.*

Reinforcing Steel                                   D1.4, Table 5.3

**Table 5.3**
**Permissible Atmospheric Exposure of Low-Hydrogen Electrodes**
**(See 5.7.2.1)**

| Electrode | Column A (hours) | Column B (hours) |
|---|---|---|
| A5.1 | | |
| E70XX | 4 max | over 4 to 10 max |
| A5.5 | | |
| E70XX-X | 4 max | over 4 to 10 max |
| E80XX-X | 2 max | over 2 to 10 max |
| E90XX-X | 1 max | over 1 to 5 max |
| E100XX-X | 1/2 max | over 1/2 to 4 max |
| E110XX-X | 1/2 max | over 1/2 to 4 max |

Notes:

1. Column A: Electrodes exposed to the atmosphere for longer periods than shown shall be redried before use.

2. Column B: Electrode exposure to the atmosphere for longer periods than those established by test shall be redried before use.

3. Entire Table: Electrodes shall be issued and held in quivers, or other small containers which may be open. Heated containers are not mandatory.

## End Returns *(See Boxing)*                      D1.4, Table 5.1

## Filler Metal Requirements

Reinforcing Steel                                   D1.4, Table 5.1

                                                    D1.4, 5.1

## Table 5.1
## Matching Filler Metals Requirements
### (See 5.1)

| Group | Steel Specification Requirements | | | | | Filler Metal Requirements | | | | |
|---|---|---|---|---|---|---|---|---|---|---|
| | Steel Specification | | Minimum Yield Point/Strength ksi | MPa | Minimum Tensile Strength ksi | MPa | Electrode Specification[4] | Yield Point/Strength[1] ksi | MPa | Tensile Strength[1] ksi | MPa |
| I | ASTM A615 | Grade 40 | 40 | — | 70 | — | SMAW AWS A5.1 and A5.5 E7015, E7016, E7018, E7028, E7015-X, E7016-X, E7018-X | 60 | 415 | 72 | 495 |
| | ASTM A615M | Grade 300 | — | 300 | — | 500 | | 57 | 390 | 70 | 480 |
| | ASTM A617 | Grade 40 | 40 | — | 70 | — | GMAW AWS A5.18 ER70S-X | 60 | 415 | 72 | 495 |
| | ASTM A617M | Grade 300 | — | 300 | — | 500 | FCAW AWS A5.20 E7XT-X (Except -2, -3, -10, -GS) | 60 | 415 | 72 | 495 |
| II | ASTM A616 | Grade 50 | 50 | — | 80 | — | SMAW AWS A5.5 E8015-X, E8016-X, E8018-X | 67 | 460 | 80 | 550 |
| | ASTM A616M | Grade 350 | — | 350 | — | 550 | | | | | |
| | ASTM A706 | Grade 60 | 60 | — | 80 | — | GMAW AWS A5.28 ER80S-X | 68 | 470 | 80 | 550 |
| | ASTM A706M | Grade 400 | — | 400 | — | 550 | FCAW AWS A5.29 E8XTX-X | 68 | 470 | 80-100 | 550-690 |
| III | ASTM A615 | Grade 60 | 60 | — | 90 | — | SMAW AWS A5.5 E9015-X, E9016-X, E9018-X | 77 | 530 | 90 | 620 |
| | ASTM A615M | Grade 400 | — | 400 | — | 600 | | | | | |
| | ASTM A616 | Grade 60 | 60 | — | 90 | — | GMAW AWS A5.28 ER90S-X | 78 | 540 | 90 | 620 |
| | ASTM A616M | Grade 400 | — | 400 | — | 600 | | | | | |
| | ASTM A617 | Grade 60 | 60 | — | 90 | — | FCAW AWS A5.29 E9XTX-X | 78 | 540 | 90-110 | 620-760 |
| | ASTM A617M | Grade 400 | — | 400 | — | 600 | | | | | |

(Continued)

## Table 5.1
### (Continued)

| | Steel Specification Requirements | | | | | Filler Metal Requirements | | | | |
|---|---|---|---|---|---|---|---|---|---|---|
| | | Minimum Yield Point/Strength | | Minimum Tensile Strength | | | Yield Point/Strength[1] | | Tensile Strength[1] | |
| Group | Steel Specification | ksi | MPa | ksi | MPa | Electrode Specification[4] | ksi | MPa | ksi | MPa |
| | | | | | | SMAW AWS A5.5 | | | | |
| | | | | | | E10015-X, E10016-X, E10018-X | 87 | 600 | 100 | 690 |
| | | | | | | E10018-M | 88-100 | 610-690 | 100 | 690 |
| IV | ASTM A615 Grade 75[2] | 75 | — | 100 | — | GMAW AWS A5.28 | | | | |
| | | | | | | ER100S-X | 88-102 | 610-700 | 100 | 690 |
| | ASTM A615M Grade 500[3] | — | 500 | — | 700 | FCAW AWS A5.29 | | | | |
| | | | | | | E10XTX-X | 88 | 610 | 100-120 | 690-830 |

Notes:

1. This table is based on filler metal as-welded properties. Single values are minimums. Hyphenated values indicated minimum and maximum.

2. Applicable to bar sizes Nos. 11, 14, and 18.

3. Applicable to bar sizes Nos. 35, 45, and 55.

4. Filler metals classified in the postweld heat treated (PWHT) condition by the AWS filler metal specification may be used when given prior approval by the Engineer. Consideration shall be made of the differences in tensile strength, ductility and hardness between the PWHT versus as-welded condition.

255

Filler Metal Requirements *(Continued)*

Restrictions for A242 and A588              D1.1, Table 4.2
                                            D1.1, 4.1.4

### Table 4.2
### Filler Metal Requirements for Exposed Bare Applications of ASTM A242 and A588 Steel (see 4.1.4)

| | Welding Processes | | |
|---|---|---|---|
| Shielded Metal Arc | Submerged Arc | Gas Metal Arc or Gas Tungsten Arc | Flux Cored Arc |
| A5.5 | A5.23[1,4] | A5.28[4] | A5.29 |
| E7018-W | F7AX-EXXX-W | | |
| E8018-W | | | E8XT1-W |
| E8016-C3 or E8018-C3 | F7AX-EXXX-Ni1[2] | ER80S-Ni1 | E8XTX-Ni1 |
| E8016-C1 or E8018-C1 | F7AX-EXXX-Ni4[2] | | |
| E8016-C2 or E8018-C2 | | | |
| E7016-C1L or E7018-C1L | F7AX-EXXX-Ni2[2] | ER80S-Ni2 | E8XTX-Ni2 |
| E7016-C2L or E7018-C2L | F7AX-EXXX-Ni3[2] | ER80S-Ni3 | E80T5-Ni3 |
| E8018-B2L[1] | | ER80S-B2L[1] | E80T5-B2L[1] |
| | | ER80S-G[1,3] | |
| | | | E71T8-Ni1 |
| | | | E71T8-Ni2 |
| | | | E7XTX-K2 |

Restrictions for A514 and A517              D1.1, 4.5.3

When used for welding ASTM A514 or A517 steels, electrodes of any classification lower than E100XX-X, except for E7018M and E70XXH4R, shall be dried at least 1 hour at temperatures between 700 and 800 deg. F before being used, whether furnished in hermetically sealed containers or otherwise.

Sheet Steel                                        D1.3, Table 5.1

## Table 5.1
## Matching Filler Metal Requirements
## (see 5.1.1)

| Steel Specification | Minimum Yield Point | | Minimum Tensile Strength | | Filler Metal Requirements |
|---|---|---|---|---|---|
| | ksi | MPa | ksi | MPa | |
| A446 Gr A | 33 | 230 | 45 | 310 | SMAW AWS A5.1 or A5.5 |
| Gr B | 37 | 255 | 52 | 360 | E60XX, E70XX or E70XX-X |
| Gr C | 40 | 275 | 55 | 380 | |
| A570 Gr 30 | 30 | 205 | 49 | 340 | SAW AWS A5.17 or A5.23 |
| Gr 33 | 33 | 230 | 52 | 360 | F6X-EXXX or F7AX-EXXX |
| Gr 36 | 36 | 250 | 53 | 365 | |
| Gr 40 | 40 | 275 | 55 | 380 | GMAW AWS A5.18 |
| Gr 45 | 45 | 310 | 60 | 415 | ER70S-X |
| Gr 50 | 50 | 345 | 65 | 450 | |
| A606 | 45 | 310 | 65 | 450 | |
| A607 Gr 45 | 45 | 310 | 60 | 415 | FCAW AWS A5.20 |
| Gr 50 | 50 | 345 | 65 | 450 | E60T-X or E70T-X (except 2 and 3) |
| A611 Gr A | 25 | 175 | 42 | 290 | |
| Gr B | 30 | 205 | 45 | 310 | |
| Gr C | 33 | 230 | 48 | 335 | |
| Gr D | 40 | 275 | ·52 | 360 | |
| A607 Gr 55 | 55 | 380 | 70 | 485 | SMAW AWS A5.1 or A5.5 E70XX or E70XX-X SAW AWS A5.17 or A5.23 F7AX-EXXX GMAW AWS A5.18 ER70S-X FCAW AWS A5.20 E70T-X (except 2 and 3) |
| A446 Gr E | 80 | 550 | 82 | 565 | SMAW AWS A5.5 E80XX-X |
| A607 Gr 60 | 60 | 415 | 75 | 515 | SAW AWS A5.23 F8AX-EXXX |
| Gr 70 | 70 | 485 | 85 | 585 | GMAW AWS A5.28 ER80S |
| A611 Gr E | 80 | 550 | 82 | 565 | FCAW AWS A5.29 E8XT |

Note: Low hydrogen electrodes must be used when required by ANSI/AWS D1.1. See 1.1.1.

Structural Steel                                   D1.1, Table 4.1
                                                   AISC, 5-28, A3.6

*(Continued)*

# Table 4.1
## Matching Filler Metal Requirements (see 4.1.1)

**Group I**

### Steel Specification Requirements

| Steel Specification[1,2] | | Minimum Yield Point/Strength ksi | MPa | Tensile Range ksi | MPa |
|---|---|---|---|---|---|
| ASTM A36[5] | | 36 | 250 | 58-80 | 400-550 |
| ASTM A53 | Grade B | 35 | 240 | 60 min | 415 min |
| ASTM A106 | Grade B | 35 | 240 | 60 min | 415 min |
| ASTM A131 | Grades A, B, CS, D, DS, E | 34 | 235 | 58-71 | 400-490 |
| ASTM A139 | Grade B | 35 | 241 | 60 min | 414 min |
| ASTM A381 | Grade Y35 | 35 | 240 | 60 min | 415 min |
| ASTM A500 | Grade A | 33 | 228 | 45 min | 310 min |
| | Grade B | 42 | 290 | 58 min | 400 min |
| ASTM A501 | | 36 | 250 | 58 min | 400 min |
| ASTM A516 | Grade 55 | 30 | 205 | 55-75 | 380-515 |
| | Grade 60 | 32 | 220 | 60-80 | 415-550 |
| ASTM A524 | Grade I | 35 | 240 | 60-85 | 415-586 |
| | Grade II | 30 | 205 | 55-80 | 380-550 |
| ASTM A529 | | 42 | 290 | 60-85 | 415-585 |
| ASTM A570 | Grade 30 | 30 | 205 | 49 min | 340 min |
| | Grade 33 | 33 | 230 | 52 min | 360 min |
| | Grade 36 | 36 | 250 | 53 min | 365 min |
| | Grade 40 | 40 | 275 | 55 min | 380 min |
| | Grade 45 | 45 | 310 | 60 min | 415 min |
| | Grade 50 | 50 | 345 | 65 min | 450 min |
| ASTM A573 | Grade 65 | 35 | 240 | 65-77 | 450-530 |
| | Grade 58 | 32 | 220 | 58-71 | 400-490 |
| ASTM A709 | Grade 36[5] | 36 | 250 | 58-80 | 400-550 |
| API 5L | Grade B | 35 | 240 | 60 | 415 |
| | Grade X42 | 42 | 290 | 60 | 415 |
| ABS | Grades A, B, D, CS, DS | | | 58-71 | 400-490 |
| | Grade E[6] | | | 58-71 | 400-490 |

### Filler Metal Requirements

| Electrode Specification[3,4] | Minimum Yield Point/Strength ksi | MPa | Tensile Strength Range ksi | MPa |
|---|---|---|---|---|
| **SMAW** AWS A5.1 or A5.5[7,9] | | | | |
| E60XX | 50 | 345 | 62 min | 425 |
| E70XX | 60 | 415 | 72 min | 495 |
| E70XX-X | 57 | 390 | 70 min | 480 |
| **SAW** AWS A5.17 or A5.23[7,9] | | | | |
| F6XX-EXXX | 48 | 330 | 60-80 | 415-550 |
| F7XX-EXXX or F7XX-EXX-XX | 58 | 400 | 70-95 | 485-660 |
| **GMAW, GTAW** AWS A5.18 | | | | |
| ER70S-X | 60 | 415 | 72 min | 495 |
| **FCAW** AWS A5.20 | | | | |
| E6XT-X | 50 | 345 | 62 min | 425 |
| E7XT-X | 60 | 415 | 72 min | 495 |
| (Except -2, -3, -10, -GS) | | | | |

(continued)

Table 4.1 (continued)

<table>
<tr><th rowspan="3">Group</th><th colspan="5">Steel Specification Requirements</th></tr>
<tr><th rowspan="2">Steel Specification[1,2]</th><th colspan="2">Minimum Yield Point/Strength</th><th colspan="2">Tensile Range</th></tr>
<tr><th>ksi</th><th>MPa</th><th>ksi</th><th>MPa</th></tr>
<tr><td rowspan="10">II</td><td>ASTM A131   Grades AH32, DH32, EH32</td><td>46</td><td>315</td><td>68-85</td><td>470-585</td></tr>
<tr><td>            Grades AH36, DH36, EH36</td><td>51</td><td>350</td><td>71-90</td><td>490-620</td></tr>
<tr><td>ASTM A242[6]</td><td>42-50</td><td>290-345</td><td>63-70</td><td>435-485</td></tr>
<tr><td>ASTM A441</td><td>40-50</td><td>275-345</td><td>60-70</td><td>415-485</td></tr>
<tr><td>ASTM A516   Grade 65</td><td>35</td><td>240</td><td>65-85</td><td>450-585</td></tr>
<tr><td>            Grade 70</td><td>38</td><td>260</td><td>70-90</td><td>485-620</td></tr>
<tr><td>ASTM A537   Class 1</td><td>45-50</td><td>310-345</td><td>65-90</td><td>450-620</td></tr>
<tr><td>ASTM A572   Grade 42</td><td>42</td><td>290</td><td>60 min</td><td>415 min</td></tr>
<tr><td>ASTM A572   Grade 50</td><td>50</td><td>345</td><td>65 min</td><td>450 min</td></tr>
<tr><td>ASTM A588[6]   (4 in. and under)</td><td>50</td><td>345</td><td>70 min</td><td>485 min</td></tr>
<tr><td></td><td>ASTM A595   Grade A</td><td>55</td><td>380</td><td>65 min</td><td>450 min</td></tr>
<tr><td></td><td>            Grades B and C</td><td>60</td><td>415</td><td>70 min</td><td>480 min</td></tr>
<tr><td></td><td>ASTM A606[6]</td><td>45-50</td><td>310-340</td><td>65 min</td><td>450 min</td></tr>
<tr><td></td><td>ASTM A607   Grade 45</td><td>45</td><td>310</td><td>60 min</td><td>410 min</td></tr>
<tr><td></td><td>            Grade 50</td><td>50</td><td>345</td><td>65 min</td><td>450 min</td></tr>
<tr><td></td><td>            Grade 55</td><td>55</td><td>380</td><td>70 min</td><td>480 min</td></tr>
<tr><td></td><td>ASTM A618</td><td>46-50</td><td>315-345</td><td>65 min</td><td>450 min</td></tr>
<tr><td></td><td>ASTM A633   Grade A</td><td>42</td><td>290</td><td>63-83</td><td>430-570</td></tr>
<tr><td></td><td>            Grades C, D (2-1/2 in. and under)</td><td>50</td><td>345</td><td>70-90</td><td>485-620</td></tr>
<tr><td></td><td>ASTM A709   Grade 50</td><td>50</td><td>345</td><td>65 min</td><td>450 min</td></tr>
<tr><td></td><td>            Grade 50W</td><td>50</td><td>345</td><td>70 min</td><td>485 min</td></tr>
<tr><td></td><td>ASTM A710   Grade A. Class 2   >2 in.</td><td>55</td><td>380</td><td>65 min</td><td>450 min</td></tr>
<tr><td></td><td>ASTM A808   (2-1/2 in. and under)</td><td>42</td><td>290</td><td>60 min</td><td>415 min</td></tr>
<tr><td></td><td>API 2H[6]   Grade 42</td><td>42</td><td>290</td><td>62-80</td><td>430-550</td></tr>
<tr><td></td><td>            Grade 50</td><td>50</td><td>345</td><td>70 min</td><td>485 min</td></tr>
<tr><td></td><td>API 5L   Grade X52</td><td>52</td><td>360</td><td>66-72</td><td>455-495</td></tr>
<tr><td></td><td>ABS   Grades AH32, DH32, EH32</td><td>45.5</td><td>315</td><td>71-90</td><td>490-620</td></tr>
<tr><td></td><td>        Grades AH36, DH36, EH36[6]</td><td>51</td><td>350</td><td>71-90</td><td>490-620</td></tr>
</table>

Filler Metal Requirements

<table>
<tr><th rowspan="2">Electrode Specification[3,4]</th><th colspan="2">Minimum Yield Point/Strength</th><th colspan="2">Tensile Strength Range</th></tr>
<tr><th>ksi</th><th>MPa</th><th>ksi</th><th>MPa</th></tr>
<tr><td>SMAW<br>AWS A5.1 or A5.5[7,9]<br>E7015, E7016<br>E7018, E7028</td><td>60</td><td>415</td><td>72 min</td><td>495</td></tr>
<tr><td>E7015-X, E7016-X<br>E7018-X</td><td>57</td><td>390</td><td>70 min</td><td>480</td></tr>
<tr><td>SAW<br>AWS A5.17 or A5.23[7,9]<br>F7XX-EXXX or F7XX-EXX-XX</td><td>58</td><td>400</td><td>70-95</td><td>485-660</td></tr>
<tr><td>GMAW, GTAW<br>AWS A5.18<br>ER70S-X</td><td>60</td><td>415</td><td>72 min</td><td>495</td></tr>
<tr><td>FCAW<br>AWS A5.20<br>E7XT-X<br>(Except -2, -3, -10, -GS)</td><td>60</td><td>415</td><td>72 min</td><td>495</td></tr>
</table>

(continued)

**Table 4.1 (continued)**

| Group | Steel Specification[1,2] | Min Yield Point/Strength ksi | MPa | Tensile Range ksi | MPa | Electrode Specification[3,4] | Min Yield Point/Strength ksi | MPa | Tensile Strength Range ksi | MPa |
|---|---|---|---|---|---|---|---|---|---|---|
| III | ASTM A572 Grade 60 | 60 | 415 | 75 min | 515 min | SMAW AWS A5.5[7,9] E8015-X, E8016-X E8018-X | 67 | 460 | 80 min | 550 |
|  | Grade 65 | 65 | 450 | 80 min | 550 min |  |  |  |  |  |
|  | ASTM A537 Class 2[6] | 46-60 | 315-415 | 80-100 | 550-690 | SAW AWS A5.23[7,9] F8XX-EXX-XX | 68 | 470 | 80-100 | 550-690 |
|  | ASTM A633 Grade E[6] | 55-60 | 380-415 | 75-100 | 515-690 |  |  |  |  |  |
|  | ASTM A710 Grade A. Class 2 ≤2 in. | 60-65 | 415-450 | 72 min | 495 min | GMAW, GTAW AWS A5.28[7,9] ER80S-X | 68 | 470 | 80 min | 550 |
|  | ASTM A710 Grade A. Class 3 >2 in. | 60-65 | 415-450 | 70 min | 485 min | FCAW AWS A5.29[7,9] E8XTX-X | 68 | 470 | 80-100 | 550-690 |
| IV | ASTM A514 Over 2-1/2 in. (63.5 mm) ASTM A709 Grades 100, 100W 2-1/2 in. to 4 in. (63.5 to 102 mm) | 90 | 620 | 100-130 | 690-895 | SMAW AWS A5.5[7] E10015-X, E10016-X E10018-X | 87 | 600 | 100 min | 690 |
|  | ASTM A710 Grade A. Class 1 ≤3/4 in. | 90 | 620 | 100-130 | 690-895 | SAW AWS A5.23[7] F10XX-EXX-XX | 88 | 610 | 100-120 | 690-830 |
|  | ASTM A710 Grade A. Class 3 ≤2 in. | 80 | 550 | 90 min | 620 min | GMAW, GTAW AWS A5.28[7] ER100S-X | 88-102 | 610-700 | 100 min | 690 |
|  |  | 75 | 515 | 85 min | 585 min | FCAW AWS A5.29[7] E10XTX-X | 88 | 605 | 100-120 | 690-830 |

(continued)

**Table 4.1 (continued)**

| Group | Steel Specification[1,2] | | Minimum Yield Point/Strength ksi | MPa | Tensile Range ksi | MPa | Electrode Specification[3,4] | Minimum Yield Point/Strength ksi | MPa | Tensile Strength Range ksi | MPa |
|---|---|---|---|---|---|---|---|---|---|---|---|
| | | | | | | | **SMAW** AWS A5.5[7] E11015-X, E11016-X E11018-X | 97 | 670 | 110 min | 760 |
| V | ASTM A514 | 2-1/2 in. (63.5 mm) and under | 100 | 690 | 110-130 | 760-895 | **SAW** AWS A5.23[7] F11XX-EXX-XX | 98 | 680 | 110-130 | 760-895 |
| | ASTM A517 | Grades 100, 100W | 90-100 | 620-690 | 105-135 | 725-930 | **GMAW, GTAW** AWS A5.28[7] ER110S-X | 95-107 | 660-740 | 110 min | 760 |
| | ASTM A709 | 2-1/2 in. (63.5 mm) and under | 100 | 690 | 110-130 | 760-895 | **FCAW** AWS A5.29[7] E11XTX-X | 98 | 675 | 110-130 | 760-900 |

Steel Specification Requirements — Filler Metal Requirements

Notes:
1. In joints involving base metals of different groups, low-hydrogen filler metal requirements applicable to the lower strength group may be used. The low-hydrogen processes shall be subject to the technique requirements applicable to the higher strength group.
2. Match API Standard 2B (fabricated tubes) according to steel used.
3. When welds are to be stress-relieved, the deposited weld metal shall not exceed 0.05 percent vanadium.
4. See 4.16 for electrogas and electroslag weld metal requirements.
5. Only low hydrogen electrodes shall be used when welding A36 or A709 Grade 36 steel more than 1 in. (25.4 mm) thick for dynamically loaded structures.
6. Special welding materials and procedures (e.g., E80XX-X low alloy electrodes) may be required to match the notch toughness of base metal (for applications involving impact loading or low temperature), or for atmospheric corrosion and weathering characteristics (see 4.1.4).
7. Deposited weld metal shall have a minimum impact strength of 20 ft • lbs (27.1 J) at 0° F (−18° C) when Charpy V-notch specimens are required.
8. The designation of ER70S-1B has been reclassified as ER80S-D2 in A5.28-79. Prequalified joint welding procedures prepared prior to 1981 and specifying AWS A5.18, ER70S-1B, may now use AWS A5.28-79 ER80S-D2 when welding steels in Groups I and II.
9. Filler metals of alloy groups B3, B3L, B4, B4L, B5, B5L, B6, B6L, B7, B7L, B8, B8L, or B9, in ANSI/AWS A5.5, A5.23, A5.28, or A5.29, are not prequalified for use in the as-welded condition.

# Fillers

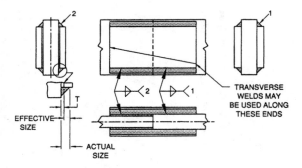

Note: The effective area of weld 2 shall equal that
of weld 1, but its size shall be its effective size
plus the thickness of the filler T.

**Figure 2.1 — Fillers Less Than 1/4 in. (6.4 mm) Thick (see 2.4.2)**

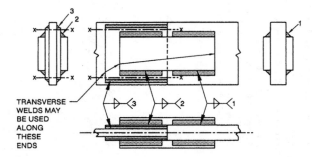

Notes:

1.  The effective area of weld 2 shall equal that of weld 1.
    The length of weld 2 shall be sufficient to avoid overstressing
    the filler in shear along planes x-x.

2.  The effective area of weld 3 shall equal that of weld 1.
    and there shall be no overstress of the ends of weld 3 resulting
    from the eccentricity of the forces acting on the filler.

**Figure 2.2 — Fillers 1/4 in. (6.4 mm) or Thicker (see 2.4.3)**

# Fillet Weld (Maximum Size)

D1.1, Fig. 2.3

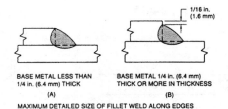

BASE METAL LESS THAN
1/4 in. (6.4 mm) THICK

(A)

BASE METAL 1/4 in. (6.4 mm)
THICK OR MORE IN THICKNESS

(B)

MAXIMUM DETAILED SIZE OF FILLET WELD ALONG EDGES

**Figure 2.3 — Details for Prequalified
Fillet Welds (see 2.7.1.2)**

# Friction Connections (Classes of)

AISC 5-271, Table 3

**Allowable Loads for Different Classes of Slip-Critical Connections**

| Faying Surfaces | Direction of Load for Different Hole Types | | | | | | | |
|---|---|---|---|---|---|---|---|---|
| | All Directions | | | | Transverse Direction | | Parallel Direction | |
| | Standard Holes | | Oversize and Short Slots | | Long Slotted Holes | | Long Slotted Holes | |
| | A325 | A490 | A325 | A490 | A325 | A490 | A325 | A490 |
| Class A | 17 | 21 | 15 | 18 | 12 | 15 | 10 | 13 |
| Class B | 28 | 34 | 24 | 29 | 20 | 24 | 17 | 20 |
| Class C | 22 | 27 | 19 | 23 | 16 | 19 | 14 | 16 |

Class A:    1. Clean mill scale is allowed.
            2. Blast cleaned surfaces with Class A protective coatings.

Class B:    1. Blast cleaned surfaces.
            2. Blast cleaned surfaces with Class B protective coatings.

Class C:    1. Hot dipped galvanized and roughened surfaces.

# Joists (Open Web, Steel)

## Bridging

UBC, 2-385,

**Number Of Rows Of Bridging\*\*\***
**Distances are Span Lengths**
**(See "Definition of Span" in front of load table)**

| *Section Number | 1 Row | 2 Rows | 3 Rows | 4 Rows** | 5 Rows** |
|---|---|---|---|---|---|
| | Multiply table span lengths × 304.8 for mm | | | | |
| #1 | Up thru 16' | Over 16' thru 24' | Over 24' thru 28' | | |
| #2 | Up thru 17' | Over 17' thru 25' | Over 25' thru 32' | | |
| #3 | Up thru 18' | Over 18' thru 28' | Over 28' thru 38' | Over 38' thru 40' | |
| #4 | Up thru 19' | Over 19' thru 28' | Over 28' thru 38' | Over 38' thru 48' | |
| #5 | Up thru 19' | Over 19' thru 29' | Over 29' thru 39' | Over 39' thru 50' | Over 50' thru 52' |
| #6 | Up thru 19' | Over 19' thru 29' | Over 29' thru 39' | Over 39' thru 51' | Over 51' thru 56' |
| #7 | Up thru 20' | Over 20' thru 33' | Over 33' thru 45' | Over 45' thru 58' | Over 58' thru 60' |
| #8 | Up thru 20' | Over 20' thru 33' | Over 33' thru 45' | Over 45' thru 58' | Over 58' thru 60' |
| #9 | Up thru 20' | Over 20' thru 33' | Over 33' thru 46' | Over 46' thru 59' | Over 59' thru 60' |
| #10 | Up thru 20' | Over 20' thru 37' | Over 37' thru 51' | Over 51' thru 60' | |
| #11 | Up thru 20' | Over 20' thru 38' | Over 38' thru 53' | Over 53' thru 60' | |
| #12 | Up thru 20' | Over 20' thru 39' | Over 39' thru 53' | Over 53' thru 60' | |

\*Last digit(s) of joist designation shown in Load Tables.
\*\*Where 4 or 5 rows of bridging are required, a row nearest the mid-span of the joist shall be diagonal bridging with bolted connections at chords and intersection.
\*\*\*See Section 5.11 for additional bridging required for uplift design.

## Camber

UBC, 2-383, 4.7

### 4.7 CAMBER

Camber is optional with the manufacturer but, when provided, recommended approximate camber is as follows:

| Top Chord Length | Approximate Camber |
|---|---|
| 20 feet (6096 mm) | $1/4$ inch (7 mm) |
| 30 feet (9144 mm) | $3/8$ inch (10 mm) |
| 40 feet (12 192 mm) | $5/8$ inch (16 mm) |
| 50 feet (15 240 mm) | 1 inch (26 mm) |
| 60 feet (18 288 mm) | $1^1/2$ inches (38 mm) |

In no case will joists be manufactured with negative camber.

# Maximum Hole Size

### Bolt Hole Dimensions in Inches

| Bolt Diameter | 1/2 in. | 5/8 in. | 3/4 in. | 7/8 in | 1 in |
|---|---|---|---|---|---|
| Standard | 9/16 | 11/16 | 13/16 | 15/16 | 1-1/16 |
| Oversize | 5/8 | 13/16 | 15/16 | 1-1/16 | 1-1/4 |
| Short Slot | 9/16 x 1 1/4 | 11/16 x 7/8 | 13/16 x 1 | 15/16 x 1-1/8 | 1-1/16 x 2-1/2 |
| Long Slot | 9/16 x 1-1/4 | 11/16 x 1-9/16 | 13/16 x 1-7/8 | 15/16 x 2-3/16 | 1-1/16 x 2-1/2 |

| Bolt Diameter | 1-1/8 | 1-1/4 | 1-3/8 | 1-1/2 |
|---|---|---|---|---|
| Standard | 1-3/16 | 1-5/16 | 1-7/16 | 1-9/16 |
| Oversize | 1-7/16 | 1-9/16 | 1-11/16 | 1-13/16 |
| Short Slot | 1-3/16 x 1-1/2 | 1-5/16 x 1-5/8 | 1-7/16 x 1-3/4 | 1-9/16 x 1-7/8 |
| Long Slot | 1-3/16 x 2-13/16 | 1-5/16 x 3-1/8 | 1-7/17 x 3-7/16 | 1-9-16 x 3-3/4 |

# Minimum Bolt Tension in kips

AISC, 5-77, Table J3.7
AISC, 5-274, Table 4
UBC, 2-591, Table J3.1

### Minimum Bolt Tension in Kips

| Bolt Diameters in inches | A325 | A490 |
|:---:|:---:|:---:|
| 1/2 | 12 | 15 |
| 5/8 | 19 | 24 |
| 3/4 | 28 | 35 |
| 7/8 | 39 | 49 |
| 1 | 51 | 64 |
| 1-1/8 | 56 | 80 |
| 1-1/4 | 71 | 102 |
| 1-3/8 | 85 | 121 |
| 1-1/2 | 103 | 148 |

# Minimum Edge Distance

UBC, 2-699, Table J3.5
AISC, 5-76, Table J3.5

### Minimum Edge Distance

| Bolt Diameter in inches | From a sheared Edge | From a Rolled Edge |
|:---:|:---:|:---:|
| 1/2 | 7/8 | 3/4 |
| 5/8 | 1-1/8 | 7/8 |
| 3/4 | 1-1/4 | 1 |
| 7/8 | 1-1/2 | 1-1/8 |
| 1 | 1-3/4 | 1-1/4 |
| 1-1/8 | 2 | 1-1/2 |
| 1-1/4 | 2-1/4 | 1-5/8 |
| 1-1/2 | 2-5/8 | 1-7/8 |

Edge distances are measured from the edge of the connected material to the centerline of the hole.

# Minimum Effective Throat of Partial Penetration Groove Weld

AISC, 5-66, Table J2.3
UBC, 2-691, Table J2.3
D1.1, Table 2.3

### Table 2.3
### Minimum Weld Size for Partial Joint Penetration Groove Welds (see 2.10.3)

| Base Metal Thickness of Thicker Part Joined | | Minimum Weld Size* | |
|---|---|---|---|
| in. | mm | in. | mm |
| 1/8 (3.2) to 3/16 (4.8) incl. | | 1/16 | 2 |
| Over 3/16 (4.8) to 1/4 (6.4) incl. | | 1/8 | 3 |
| Over 1/4 (6.4) to 1/2 (12.7) incl. | | 3/16 | 5 |
| Over 1/2 (12.7) to 3/4 (19.0) incl. | | 1/4 | 6 |
| Over 3/4 (19.0) to 1-1/2 (38.1) incl. | | 5/16 | 8 |
| Over 1-1/2 (38.1) to 2-1/4 (57.1) incl. | | 3/8 | 10 |
| Over 2-1/4 (57.1) to 6 (152) incl. | | 1/2 | 13 |
| Over 6 (152) | | 5/8 | 16 |

*Except the weld size need not exceed the thickness of the thinner part.

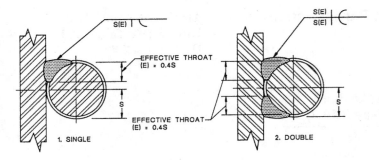

**A - FLARE-BEVEL-GROOVE WELD**

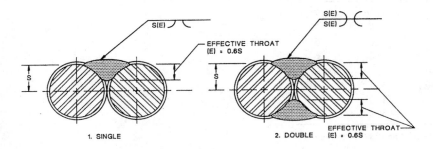

**B - FLARE-V-GROOVE WELD**

NOTE:   1.   RADIUS OF REINFORCING BAR = S

2.   THESE ARE SECTIONAL VIEWS.  BAR
     DEFORMATIONS ARE SHOWN ONLY
     FOR ILLUSTRATIVE PURPOSES.

**Figure 2.1 — Effective Weld Sizes for Flare-Groove Welds (See 2.3.2.3)**

## Minimum Fillet Weld Size for Small Studs

D1.1, Table 7.2

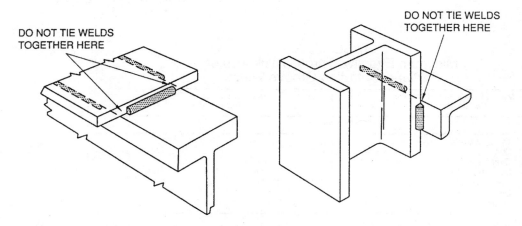

Figure 8.2 — Fillet Welds on Opposite Side of a Common Plane of Contact (see 8.8.5)

## Minimum Size Backing Bar

D1.1, 3.13.3

| Process | Thickness, min | |
|---|---|---|
| | in. | mm |
| SMAW | 3/16 | 4.8 |
| GMAW | 1/4 | 6.4 |
| FCAW-SS | 1/4 | 6.4 |
| FCAW-G | 3/8 | 9.5 |
| SAW | 3/8 | 9.5 |

Note: Commercially available steel backing for pipe and tubing is acceptable, provided there is no evidence of melting on exposed interior surfaces.

## Minimum Size Fillet Weld

UBC, 2-691, Table J2.4
AISC, 5-67, Table J2.4
D1.1, Table 2.2

TABLE J2.3
Minimum Effective Throat Thickness of
Partial-penetration Groove Welds

| Material Thickness of Thicker Part Joined (in.) | Minimum Effective Throat Thickness[a] (in.) |
|---|---|
| × 25.4 for mm | |
| To $1/4$ inclusive | $1/8$ |
| Over $1/4$ to $1/2$ | $3/16$ |
| Over $1/2$ to $3/4$ | $1/4$ |
| Over $3/4$ to $1 1/2$ | $5/16$ |
| Over $1/2$ to $2 1/4$ | $3/8$ |
| Over $2 1/4$ to 6 | $1/2$ |
| Over 6 | $5/8$ |

[a]See Sect. J2.

### Table 2.2
### Minimum Fillet Weld Size for
### Prequalified Joints (see 2.7.1.1)

| Base Metal Thickness (T)* | | Minimum Size of Fillet Weld** | | |
|---|---|---|---|---|
| in. | mm | in. | mm | |
| T≤1/4 | T≤ 6.4 | 1/8*** | 3 | Single-pass welds must be used |
| 1/4<T≤1/2 | 6.4<T≤12.7 | 3/16 | 5 | |
| 1/2<T≤3/4 | 12.7<T≤19.0 | 1/4 | 6 | |
| 3/4<T | 19.0<T | 5/16 | 8 | |

*For non-low hydrogen processes without preheat calculated in accordance with 4.2.2, T equals thickness of the thicker part joined. For non-low hydrogen processes using procedures established to prevent cracking in accordance with 4.2.2, and for low hydrogen processes, T equals thickness of the thinner part joined; single pass requirement does not apply.

**Except that the weld size need not exceed the thickness of the thinner part joined.

***Minimum size for dynamically loaded structures is 3/16 in. (5 mm).

# Nominal Hole Dimensions

AISC, 5-71, Table J3.1
AISC, 5-268, Table 1

### Bolt Hole Dimensions in Inches

| Bolt Diameter | 1/2 in. | 5/8 in. | 3/4 in. | 7/8 in | 1 in |
|---|---|---|---|---|---|
| Standard | 9/16 | 11/16 | 13/16 | 15/16 | 1-1/16 |
| Oversize | 5/8 | 13/16 | 15/16 | 1-1/16 | 1-1/4 |
| Short Slot | 9/16 x 1 1/4 | 11/16 x 7/8 | 13/16 x 1 | 15/16 x 1-1/8 | 1-1/16 x 2-1/2 |
| Long Slot | 9/16 x 1-1/4 | 11/16 x 1-9/16 | 13/16 x 1-7/8 | 15/16 x 2-3/16 | 1-1/16 x 2-1/2 |

| Bolt Diameter | 1-1/8 | 1-1/4 | 1-3/8 | 1-1/2 |
|---|---|---|---|---|
| Standard | 1-3/16 | 1-5/16 | 1-7/16 | 1-9/16 |
| Oversize | 1-7/16 | 1-9/16 | 1-11/16 | 1-13/16 |
| Short Slot | 1-3/16 x 1-1/2 | 1-5/16 x 1-5/8 | 1-7/16 x 1-3/4 | 1-9/16 x 1-7/8 |
| Long Slot | 1-3/16 x 2-13/16 | 1-5/16 x 3-1/8 | 1-7/17 x 3-7/16 | 1-9-16 x 3-3/4 |

# Nondestructive Testing

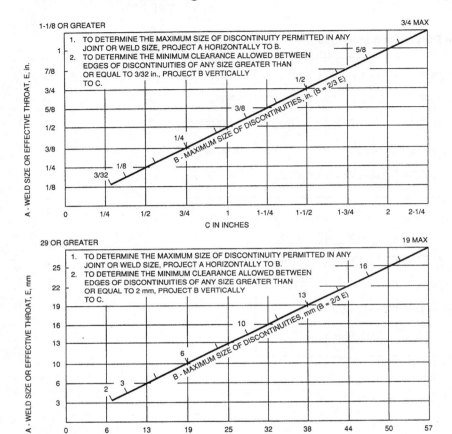

C - MINIMUM CLEARANCE MEASURED ALONG THE LONGITUDINAL AXIS OF THE WELD BETWEEN EDGES OF POROSITY OR FUSION TYPE DISCONTINUITIES ( LARGER OF ADJACENT DISCONTINUITIES GOVERNS), OR TO AN EDGE OR AN END OF AN INTERSECTING WELD.

(C = 3B = 2E)

**Figure 8.5 — Weld Quality Requirements for Elongated Discontinuities as Determined by Radiography for Statically Loaded Structures (see 8.15.3.2)**

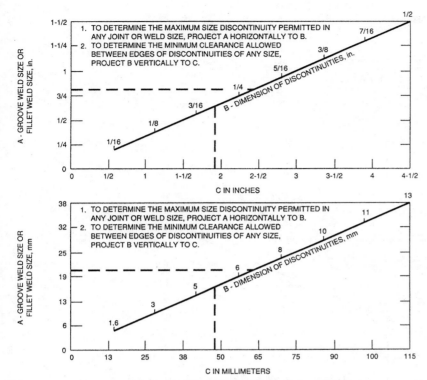

C - MINIMUM CLEARANCE MEASURED ALONG THE LONGITUDINAL AXIS OF THE WELD
BETWEEN EDGES OF POROSITY OR FUSION-TYPE DISCONTINUITIES.
(LARGER OF ADJACENT DISCONTINUITIES GOVERNS)

Note: Adjacent discontinuities, spaced less than the minimum spacing required by Figure 9.7,
shall be measured as one length equal to the sum of the total length of the discontinuities
plus the length of the space between them and evaluated as a single discontinuity.

**Figure 9.7 — Weld Quality Requirements for Discontinuities Occurring in Tension
Welds (Limitations of Porosity and Fusion Discontinuities) (see 9.25.2.1)**

*(Continued)*

## Nondestructive Testing *(Continued)*

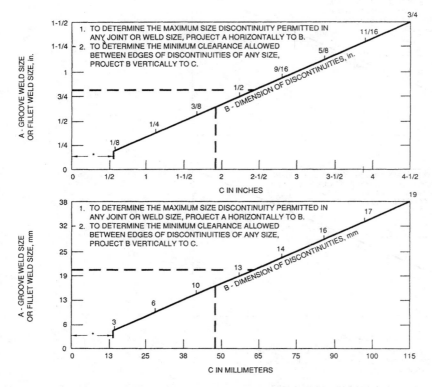

C - MINIMUM CLEARANCE MEASURED ALONG THE LONGITUDINAL AXIS OF THE WELD
BETWEEN EDGES OF POROSITY OR FUSION-TYPE DISCONTINUITIES.
(LARGER OF ADJACENT DISCONTINUITIES GOVERNS)

* THE MAXIMUM SIZE OF A DISCONTINUITY LOCATED WITHIN THIS DISTANCE FROM AN EDGE OF
PLATE SHALL BE 1/8 in.(3 mm), BUT A 1/8 in. DISCONTINUITY MUST BE 1/4 in.(6 mm)
OR MORE AWAY FROM THE EDGE.  THE SUM OF DISCONTINUITIES LESS THAN 1/8 in. IN SIZE
AND LOCATED WITHIN THIS DISTANCE FROM THE EDGE SHALL NOT EXCEED 3/16 in.(5 mm).
DISCONTINUITIES 1/16 in.(1.6 mm) TO LESS THAN 1/8 in. WILL NOT BE RESTRICTED IN OTHER LOCATIONS
UNLESS THEY ARE SEPARATED BY LESS THAN 2 L (L BEING THE LENGTH OF THE LARGER
DISCONTINUITY); IN WHICH CASE, THE DISCONTINUITIES SHALL BE MEASURED AS ONE LENGTH
EQUAL TO THE TOTAL LENGTH OF THE DISCONTINUITIES AND SPACE AND EVALUATED AS SHOWN
IN FIGURE 9.8.

**Figure 9.8 — Weld Quality Requirements for Discontinuities Occurring in Compression Welds
(Limitations of Porosity or Fusion Type Discontinuities) (see 9.25.2.2)**

## Table 9.3
## Ultrasonic Acceptance-Rejection Criteria (see 9.25.3.1)

| Discontinuity Severity Class | Weld Thickness* in in. (mm) and Search Unit Angle | | | | | | | | | | |
|---|---|---|---|---|---|---|---|---|---|---|---|
| | 5/16 (8) thru 3/4 (19) | > 3/4 thru 1-1/2 (38) | > 1-1/2 thru 2-1/2 (64) | | | > 2-1/2 thru 4 (100) | | | > 4 thru 8 (200) | | |
| | 70° | 70° | 70° | 60° | 45° | 70° | 60° | 45° | 70° | 60° | 45° |
| Class A | +10 & lower | +8 & lower | +4 & lower | +7 & lower | +9 & lower | +1 & lower | +4 & lower | +6 & lower | −2 & lower | +1 & lower | +3 & lower |
| Class B | +11 | +9 | +5 +6 | +8 +9 | +10 +11 | +2 +3 | +5 +6 | +7 +8 | −1 0 | +2 +3 | +4 +5 |
| Class C | +12 | +10 | +7 +8 | +10 +11 | +12 +13 | +4 +5 | +7 +8 | +9 +10 | +1 +2 | +4 +5 | +6 +7 |
| Class D | +13 & up | +11 & up | +9 & up | +12 & up | +14 & up | +6 & up | +9 & up | +11 & up | +3 & up | +6 & up | +8 & up |

Notes:

1. Class B and C discontinuities shall be separated by at least 2L, L being the length of the longer discontinuity, except that when two or more such discontinuities are not separated by at least 2L, but the combined length of discontinuities and their separation distance is equal to or less than the maximum allowable length under the provisions of Class B or C, the discontinuity shall be considered a single acceptable discontinuity.

2. Class B and C discontinuities shall not begin at a distance less than 2L from the end of the weld, L being the discontinuity length.

3. Discontinuities detected at "scanning level" in the root face area of complete joint penetration double groove weld joints shall be evaluated using an indication rating 4 dB more sensitive than that described in 6.19.6.5 when such welds are designated as "tension welds" on the drawing (subtract 4 dB from the indication rating "d").

4. For indications that remain on the display as the search unit is moved, refer to 9.25.3.2.

*Weld thickness shall be defined as the nominal thickness of the thinner of the two parts being joined.

**Class A (large discontinuities)**
Any indication in this category shall be rejected (regardless of length).

**Class B (medium discontinuities)**
Any indication in this category having a length greater than 3/4 inch (19 mm) shall be rejected.

**Class C (small discontinuities)**
Any indication in this category having a length greater than 2 in. (51 mm) in the middle half or 3/4 inch (19 mm) length in the top or bottom quarter of weld thickness shall be rejected.

**Class D (minor discontinuities)**
Any indication in this category shall be accepted regardless of length or location in the weld.

| Scanning Levels | |
|---|---|
| Sound Path** in in. (mm) | Above Zero Reference, dB |
| through 2-1/2 (64 mm) | 20 |
| > 2-1/2 through 5 (64-127 mm) | 25 |
| > 5 through 10 (127-254 mm) | 35 |
| > 10 through 15 (254-381 mm) | 45 |

**This column refers to sound path distance; NOT material thickness.

*(Continued)*

Nondestructive Testing *(Continued)*

### Table 8.2
### Ultrasonic Acceptance-Rejection Criteria (see 8.15.4)

| Discontinuity Severity Class | Weld Thickness* in in. (mm) and Search Unit Angle | | | | | | | | | | | |
|---|---|---|---|---|---|---|---|---|---|---|---|
| | 5/16(8) thru 3/4(19) | > 3/4 thru 1-1/2(38) | > 1-1/2 thru 2-1/2(64) | | | > 2-1/2 thru 4(100) | | | > 4 thru 8(200) | | |
| | 70° | 70° | 70° | 60° | 45° | 70° | 60° | 45° | 70° | 60° | 45° |
| Class A | +5 & lower | +2 & lower | −2 & lower | +1 & lower | +3 & lower | −5 & lower | −2 & lower | 0 & lower | −7 & lower | −4 & lower | −1 & lower |
| Class B | +6 | +3 | −1 0 | +2 +3 | +4 +5 | −4 −3 | −1 0 | +1 +2 | −6 −5 | −3 −2 | 0 +1 |
| Class C | +7 | +4 | +1 +2 | +4 +5 | +6 +7 | −2 to +2 | +1 +2 | +3 +4 | −4 to +2 | −1 to +2 | +2 +3 |
| Class D | +8 & up | +5 & up | +3 & up | +6 & up | +8 & up | +3 & up | +3 & up | +5 & up | +3 & up | +3 & up | +4 & up |

Notes:
1. Class B and C discontinuities shall be separated by at least 2L, L being the length of the longer discontinuity, except that when two or more such discontinuities are not separated by at least 2L, but the combined length of discontinuities and their separation distance is equal to or less than the maximum allowable length under the provisions of Class B or C, the discontinuity shall be considered a single acceptable discontinuity.
2. Class B and C discontinuities shall not begin at a distance less than 2L from weld ends carrying primary tensile stress, L being the discontinuity length.
3. Discontinuities detected at "scanning level" in the root face area of complete joint penetration double groove weld joints shall be evaluated using an indication rating 4 dB more sensitive than described in 6.19.6.5 when such welds are designated as "tension welds" on the drawing (subtract 4 dB from the indication rating "d").
4. Electroslag or electrogas welds: discontinuities detected at "scanning level" which exceed 2 in. (51 mm) in length shall be suspected as being piping porosity and shall be further evaluated with radiography.
5. For indications that remain on the display as the search unit is moved, refer to 8.15.4.

*Weld thickness shall be defined as the nominal thickness of the thinner of the two parts being joined.

**Class A (large discontinuities)**
Any indication in this category shall be rejected (regardless of length).

**Class B (medium discontinuities)**
Any indication in this category having a length greater than 3/4 inch (19 mm) shall be rejected.

**Class C (small discontinuities)**
Any indication in this category having a length greater than 2 inches (51 mm) shall be rejected.

**Class D (minor discontinuities)**
Any indication in this category shall be accepted regardless of length or location in the weld.

**Scanning Levels**

| Sound path** in in. (mm) | Above Zero Reference, dB |
|---|---|
| through 2-1/2 (64 mm) | 14 |
| > 2-1/2 through 5 (64-127 mm) | 19 |
| > 5 through 10 (127-254 mm) | 29 |
| > 10 through 15 (254-381 mm) | 39 |

**This column refers to sound path distance; NOT material thickness.

# Nut and Bolt Head Dimensions

### Nut and Bolt Head Dimensions
### A325 and A490

| Bolt Diameter | Width across the Flats (Nuts and Bolts) | Nut Height |
|---|---|---|
| 1/2 | 7/8 | 31/64 |
| 5/8 | 1-1/16 | 39/64 |
| 3/4 | 1-1/4 | 47/64 |
| 7/8 | 1-7/16 | 55/64 |
| 1 | 1-5/8 | 63/64 |
| 1-1/8 | 1-13/16 | 1-7/64 |
| 1-1/4 | 2 | 1-7/32 |
| 1-3/8 | 2-3/16 | 1-11/32 |
| 1-1-2 | 2-3/8 | 1-15/32 |

This information is useful in determining proper wrench sizes.

# Nut Rotation from Snug Tight

AISC, 5-275, Table 5
UBC, 2-454, Table 22-lV-E

| **Nut Rotation from Snug Tight**<br>(Required amount of nut rotation when using the Turn-of-Nut Method of Tightening) | | | |
|---|---|---|---|
| **Bolt Length in Relationship to Bolt Diameter** | | | |
| Degree of slope under Bolt Head | Bolt length up to and including 4 diameters | Bolt length over 4 diameters and up to 8 diameters | Bolt length over 8 diameters and up to, but not over 12 diameters |
| Both surfaces normal | 1/3 Turn | 1/2 Turn | 2/3 Turn |
| One surface normal, one surface sloped no more than 1:20 | 1/2 Turn | 2/3 Turn | 5/6 Turn |
| Both surfaces sloped no more than 1:20 | 2/3 Turn | 5/6 Turn | 1 Turn |

The Specification allows a rotational tolerance of plus or minus 30 degrees for nuts rotated 1/2 turn and less, and plus or minus 45 degrees for nuts rotated 2/3 or more turns.

# Nut Specification

UBC 2-445, 2221.3
AISC, 5-266, 2(c)

### Nuts for A325 and A490 Bolts

| A325 Bolts | Grade of Nuts | | | | | | |
|---|---|---|---|---|---|---|---|
| | A563 | | | | | A194 | |
| | C | C3 | D | DH | DH3 | 2 | 2H |
| Type 1 Plain | C | C3 | D | DH | DH3 | 2 | 2H |
| Type 1 Galvanized | | | | DH$^g$ | | | 2H$^g$ |
| Type 2 Discontinued | | | | | | | |
| Type 3 Plain | | C3 | | | DH3 | | |

| A490 Bolts | Grade of Nuts | | | | |
|---|---|---|---|---|---|
| | A563 | | | | A194 |
| | | | DH | DH3 | 2H |
| Type 1 Plain | | | DH | DH3 | 2H |
| Type 2 Plain | | | DH | DH3 | 2H |
| Type 3 Plain | | | | DH3 | |

# Nut Marking *(See Bolts)*

# Plate Flatness

AISC, 1-157, 1-158

### Variations From Flatness in
### CARBON STEEL PLATE

| Material Thickness in Inches | Flatness Variations for Different Width Materials in Inches | | | | | | | |
|---|---|---|---|---|---|---|---|---|
| | < 36 | 36 to < 48 | 48 to < 60 | 60 to < 72 | 72 to < 84 | 84 to < 96 | 96 to < 108 | 108 to < 120 |
| < 1/4 | 9/16 | 3/4 | 15/16 | 1-1/4 | 1-3/8 | 1-1/2 | 1-5/8 | 1-3/4 |
| 1/4 to < 3/8 | 1/2 | 5/8 | 3/4 | 15/16 | 1-1/8 | 1-1/4 | 1-3/8 | 1-1/2 |
| 3/8 to < 1/2 | 1/2 | 9/16 | 5/8 | 5/8 | 3/4 | 7/8 | 1 | 1-1/8 |
| 1/2 to < 3/4 | 7/16 | 1/2 | 9/16 | 5/8 | 5/8 | 3/4 | 1 | 1 |
| 3/4 to < 1 | 7/16 | 1/2 | 9/16 | 5/8 | 5/8 | 5/8 | 3/4 | 7/8 |
| 1 to < 2 | 3/8 | 1/2 | 1/2 | 9/16 | 9/16 | 5/8 | 5/8 | 5/8 |
| 2 to < 4 | 5/16 | 3/8 | 7/16 | 1/2 | 1/2 | 1/2 | 1/2 | 9/16 |
| 4 to < 6 | 3/8 | 7/16 | 1/2 | 1/2 | 9/16 | 9/16 | 5/8 | 3/4 |
| 6 to < 8 | 7/16 | 1/2 | 1/2 | 5/8 | 11/16 | 3/4 | 7/8 | 7/8 |

**Variations From Flatness in
High-Strength Low Alloy Steel,
Hot Rolled or Heat Treated**

| Material Thickness in Inches | Flatness Variations for Different Width Materials in Inches | | | | | | | |
|:---:|:---:|:---:|:---:|:---:|:---:|:---:|:---:|:---:|
| | < 36 | 36 to < 48 | 48 to < 60 | 60 to < 72 | 72 to < 84 | 84 to < 96 | 96 to < 108 | 108 to < 120 |
| < 1/4 | 13/16 | 1-1/8 | 1-3/8 | 1-7/8 | 2 | 2-1/4 | 2-3/8 | 2-5/8 |
| 1/4 to < 3/8 | 3/4 | 15/16 | 1-1/8 | 1-3/3 | 1-3/4 | 1-7/8 | 2 | 2-1/4 |
| 3/8 to < 1/2 | 3/4 | 7/8 | 15/16 | 15/16 | 1-1/8 | 1-5/16 | 1-1/2 | 1-5/8 |
| 1/2 to < 3/4 | 5/8 | 3/4 | 13/16 | 7/8 | 1 | 1-1/8 | 1-1/4 | 1-3/8 |
| 3/4 to < 1 | 5/8 | 3/4 | 7/8 | 7/8 | 15/16 | 1 | 1-1/8 | 1-5/16 |
| 1 to < 2 | 9/16 | 5/8 | 3/4 | 13/16 | 7/8 | 15/16 | 1 | 1 |
| 2 to < 4 | 1/2 | 9/16 | 11/16 | 3/4 | 3/4 | 3/4 | 3/4 | 7/8 |
| 4 to < 6 | 9/16 | 11/16 | 3/4 | 3/4 | 7/8 | 7/8 | 15/16 | 1-1/8 |
| 6 to < 8 | 5/8 | 3/4 | 3/4 | 15/16 | 1 | 1-1/8 | 1-1/4 | 1-5/16 |

# Preheat and Interpass Temperatures

Reinforcing Steel                                                    D1.4, Table 5.2

**Table 5.2**
**Minimum Preheat and Interpass Temperature[1,2]**
**(see 5.2.1)**

| Carbon Equivalent[3,4] (C.E.) Range, % | Size of Reinforcing Bar | Shielded Metal Arc Welding with Low Hydrogen Electrodes, Gas Metal Arc Welding, or Flux Cored Arc Welding | |
|---|---|---|---|
| | | Minimum Temperature | |
| | | °F | °C |
| Up to 0.40 | up to 11 inclusive | none[5] | none[5] |
| | 14 and 18 | 50 | 10 |
| Over 0.40 to 0.45 inclusive | up to 11 inclusive | none[5] | none[5] |
| | 14 and 18 | 100 | 40 |
| Over 0.45 to 0.55 inclusive | up to 6 inclusive | none[5] | none[5] |
| | 7 to 11 inclusive | 50 | 10 |
| | 14 to 18 | 200 | 90 |
| Over 0.55 to 0.65 inclusive | up to 6 inclusive | 100 | 40 |
| | 7 to 11 inclusive | 200 | 90 |
| | 14 to 18 | 300 | 150 |
| Over 0.65 to 0.75 | up to 6 inclusive | 300 | 150 |
| | 7 to 18 inclusive | 400 | 200 |
| Over 0.75 | 7 to 18 inclusive | 500 | 260 |

1. When reinforcing steel is to be welded to main structural steel, the preheat requirements of the structural steel shall also be considered (see ANSI/AWS D1.1, table titled "Minimum Preheat and Interpass Temperature." The minimum preheat requirement to apply in this situation shall be the higher requirement of the two tables. However, extreme caution shall be exercised in the case of welding reinforcing steel to quenched and tempered steels, and such measures shall be taken as to satisfy the preheat requirements for both. If not possible, welding shall not be used to join the two base metals.

2. Welding shall not be done when the ambient temperature is lower than 0°F (-18°C). When the base metal is below the temperature listed for the welding process being used and the size and carbon equivalent range of the bar being welded, it shall be preheated (except as otherwise provided) in such a manner that the cross section of the bar for not less than 6 in. (150 mm) on each side of the joint shall be at or above the specified minimum temperature. Preheat and interpass temperatures shall be sufficient to prevent crack formation.

3. After welding is complete, bars shall be allowed to cool naturally to ambient temperature. Accelerated cooling is prohibited.

4. Where it is impractical to obtain chemical analysis, the carbon equivalent shall be assumed to be above 0.75%. See also 1.3.4.

5. When the base metal is below 32°F (0°C), the base metal shall be preheated to at least 70°F (20°C), or above, and maintained at this minimum temperature during welding.

Structural Steel                                                    D1.1, Table 4.3

## Table 4.3
## Minimum Preheat and Interpass Temperature[1,2] (see 4.2)

### Category A

Welding Process: Shielded metal arc welding with other than low hydrogen electrodes

Steel Specification:

| Steel Specification | Grade |
|---|---|
| ASTM A36[3] | |
| ASTM A53 | Grade B |
| ASTM A106 | Grade B |
| ASTM A131 | Grades A, B, CS, D, DS, E |
| ASTM A139 | Grade B |
| ASTM A381 | Grade Y35 |
| ASTM A500 | Grade A |
| ASTM A500 | Grade B |
| ASTM A501 | |
| | Grades I & II |
| | All grades |
| | Grade 65 |
| | Grade 36[3] |
| | Grade B |
| | Grades X42 |
| ABS | Grades A, B, D, CS, DS |
| | Grade E |

| Thickness of Thickest Part at Point of Welding, in. | mm | Minimum Temperature, °F | °C |
|---|---|---|---|
| Up to 3/4 | 19 incl. | None[4] | |
| Over 3/4 thru 1-1/2 | 19 / 38.1 incl. | 150 | 66 |
| Over 1-1/2 thru 2-1/2 | 38.1 / 63.5 | 225 | 107 |
| Over 2-1/2 | 63.5 | 300 | 150 |

### Category B

Welding Process: Shielded metal arc welding with low hydrogen electrodes, submerged arc welding,[5] gas metal arc welding, gas tungsten arc welding, flux cored arc welding

Steel Specification:

| Steel Specification | Grade |
|---|---|
| ASTM A36[3] | |
| ASTM A53 | Grade B |
| ASTM A106 | Grade B |
| ASTM A131 | Grades A, B, CS, D, DS, E; AH 32 & 36; DH 32 & 36; EH 32 & 36 |
| ASTM A139 | Grade B |
| ASTM A242 | |
| ASTM A381 | Grade Y35 |
| ASTM A441 | |
| ASTM A500 | Grade A |
| ASTM A500 | Grade B |
| ASTM A501 | |
| ASTM A516 | Grades 55 & 60, 65 & 70 |
| ASTM A524 | Grades I & II |
| ASTM A529 | |
| ASTM A537 | Classes 1 & 2 |
| ASTM A570 | All grades |
| ASTM A572 | Grades 42, 50 |
| ASTM A573 | Grade 65 |
| ASTM A588 | Grades A, B, C |
| ASTM A595 | |
| ASTM A606 | |
| ASTM A607 | Grades 45, 50, 55 |
| ASTM A618 | |
| ASTM A633 | Grades A, B; Grades C, D |
| ASTM A709 | Grades 36, 50, 50W |
| ASTM A808 | |
| API 5L | Grade B; Grade X42 |
| API Spec. 2H | Grades 42, 50 |
| ABS | Grades AH 32 & 36, DH 32 & 36, EH 32 & 36 |
| ABS | Grades A, B, D, CS, DS; Grade E |

| Thickness of Thickest Part at Point of Welding, in. | mm | Minimum Temperature, °F | °C |
|---|---|---|---|
| Up to 3/4 | 19 incl. | None[4] | |
| Over 3/4 thru 1-1/2 | 19 / 38.1 incl. | 50 | 10 |
| Over 1-1/2 thru 2-1/2 | 38.1 / 63.5 incl. | 150 | 66 |
| Over 2-1/2 | 63.5 | 225 | 107 |

(continued)

## Table 4.3 (continued)

| Category | Steel Specification | Welding Process | Thickness of Thickest Part at Point of Welding, in. | Thickness of Thickest Part at Point of Welding, mm | Minimum Temperature, °F | Minimum Temperature, °C |
|---|---|---|---|---|---|---|
| C | ASTM A572 Grades 60 & 65<br>ASTM A633 Grade E<br>API 5L Grade X52 | Shielded metal arc welding with low hydrogen electrodes, submerged arc welding,[5] gas metal arc welding, gas tungsten arc welding, flux cored arc welding | Up to 3/4 | 19 incl. | 50 | 10 |
| | | | Over 3/4 thru 1-1/2 | 19, 38.1 incl. | 150 | 66 |
| | | | Over 1-1/2 thru 2-1/2 | 38.1, 63.5 incl. | 225 | 107 |
| | | | Over 2-1/2 | 63.5 | 300 | 150 |
| D | ASTM A514<br>ASTM A517<br>ASTM A709 Grades 100 & 100W | Shielded metal arc welding with low hydrogen electrodes, submerged arc welding[5] with carbon or alloy steel wire, neutral flux, gas metal arc welding, gas tungsten arc welding, or flux cored arc welding | Up to 3/4 | 19 incl. | 50 | 10 |
| | | | Over 3/4 thru 1-1/2 | 19, 38.1 incl. | 125 | 50 |
| | | | Over 1-1/2 thru 2-1/2 | 38.1, 63.5 incl. | 175 | 80 |
| | | | Over 2-1/2 | 63.5 | 225 | 107 |
| E | ASTM A710 Grade A (All classes) | | no preheat is required[6] | | | |

Notes:

1. Welding shall not be done when the ambient temperature is lower than 0° F (–18° C). Zero ° F (–18° C) does not mean the ambient environmental temperature but the temperature in the immediate vicinity of the weld. The ambient environmental temperature may be below 0° F, but a heated structure or shelter around the area being welded could maintain the temperature adjacent to the weldment at 0° F or higher. When the base metal is below the temperature listed for the welding process being used and the thickness of material being welded, it shall be preheated (except as otherwise provided) in such manner that the surfaces of the parts on which weld metal is being deposited are at or above the specified minimum temperature for a distance equal to the thickness of the part being welded, but not less than 3 in. (75 mm) in all directions from the point of welding. Preheat and interpass temperatures must be sufficient to prevent crack formation. Temperature above the minimum shown may be required for highly restrained welds. For A514, A517, and A709 Grades 100 and 100W steel, the maximum preheat and interpass temperature shall not exceed 400° F (205° C) for thickness up to 1-1/2 in. (38.1 mm) inclusive, and 450° F (230° C) for greater thickness. Heat input when welding A514, A517, and A709 Grades 100 and 100W steel shall not exceed the steel producer's recommendations.

2. In joints involving combinations of base metals, preheat shall be as specified for the higher strength steel being welded.

3. Only low hydrogen electrodes shall be used when welding A36 or A709 Grade 36 steel more than 1 in. (25.4 mm) thick for dynamically loaded structures.

4. When the base metal temperature is below 32° F (0° C), the base metal shall be preheated to at least 70° F (21° C) and this minimum temperature maintained during welding.

5. For modification of preheat requirements for submerged arc welding with parallel or multiple electrodes, see 4.10.6 or 4.11.6.

6. Pr̲ ᵗ is not required for the base metal. Preheat for E80XX-X filler metal shall be as for Gr ᵗ; for higher strength filler metal treat as Group D.

## Prequalified Welds

Full Penetration Groove Welded Joints          D1.1, pp. 8–33, see A3–A46

Partial Penetration Groove Welded Joints       D1.1, pp. 35–53, see A3–A46

Weld Joints                                    AISC, 4-155 through 4-175


## Qualification                               D1.1, Table 5.6

Welder                                         D1.1, Table 5.6

---

**Table 5.6**
**Number and Type of Specimens and Range of Thickness Qualified —**
**Welder and Welding Operator Qualification (see 5.26.1)**
**(Dimensions in Inches)**

**(1)  Test on Plate**

| Type of Weld | Thickness of Test Plate (T) as Welded, in. | Visual Inspection | Number of Specimens | | | T-Joint Break | Macro-etch Test | Plate Thickness Qualified, in. |
| | | | Bend Tests* | | | | | |
| | | | Face | Root | Side | | | |
|---|---|---|---|---|---|---|---|---|
| Groove | 3/8 | Yes | 1 | 1 | — | — | — | 1/8 to 3/4 max[3] |
| Groove | 3/8 ≤T<1 | Yes | — | — | 2 | — | — | 1/8 to 2T max[3] |
| Groove | 1 or over | Yes | — | — | 2 | — | — | 1/8 to Unlimited[3] |
| Fillet Option No. 1[1] | 1/2 | Yes | — | — | — | 1 | 1 | 1/8 to Unlimited |
| Fillet Option No. 2[2] | 3/8 | Yes | — | 2 | — | — | — | 1/8 to Unlimited |
| Plug | 3/8 | Yes | — | — | — | — | 2 | 1/8 to Unlimited |

Notes:
1. See Figure 5.27 or 5.36 as applicable.
2. See Figure 5.28 or 5.37 as applicable.
3. Also qualifies for welding fillet welds on material of unlimited thickness.
*Radiographic examination of the welder or welding operator test plate may be made in lieu of the bend test. (See 5.3.2.)

*(Continued)*

Qualification *(Continued)*

Procedure                                          D1.1, Table 5.4

## Table 5.4
## Procedure Qualification — Type and Position Limitations (see 5.10.5)

| Qualification Test | | Type of Weld and Position of Welding Qualified* | | | |
| | | Plate[1] | | Pipe[1] | |
| Weld | Plate or Pipe Positions** | Groove | Fillet | Groove | Fillet |
|---|---|---|---|---|---|
| Plate-groove | 1G | F | F | F | F |
| | 2G | H | F,H | F,H | F,H |
| Complete joint penetration | 3G | V | V | | |
| | 4G | OH | OH | | |
| Plate-groove | 1G | F | F | F | F |
| | 2G | H | F,H | F,H | F,H |
| Partial joint penetration | 3G | V | V | | |
| | 4G | OH | OH | | |
| Plate-fillet | 1F | | F | | F |
| | 2F | | F,H | | F,H |
| | 3F | | V | | V |
| | 4F | | OH | | OH |
| Pipe-groove | 1G Rotated | F | F | F | F |
| | 2G | F,H | F,H | F,H | F,H |
| Complete joint penetration | 5G | F,V,OH | F,V,OH | F,V,OH | F,V,OH |
| | 6G | F,H,V,OH (Note 2) | F,H,V,OH | F,H,V,OH (Note 2) | F,H,V,OH (Note 2) |
| | 6GR Only | All[3] | All | All[4, 5] | All (see 5.10.3.3 and 10.12) |
| | 6GR plus sample joints or mock-up per 10.12.3.3 | All[3] | All | All[4, 5] | All |
| Pipe-fillet | 1F Rotated | | F | | F |
| | 2F | | F,H | | F,H |
| | 2F Rotated | | F,H | | F,H (Note 6) |
| | 4F | | F,H,OH | | F,H,OH |
| | 5F | | All | | All |

Notes:

1. Qualifies for a welding axis with an essentially straight line and specifically includes plates, wrought shapes, fabricated sections, and rectangular fabricated sections and rectangular tubing, and pipe or tubing over 24 in. (610 mm) in diameter, except for complete joint penetration welds in tubular T-, Y-, and K-connections. This includes welding along a line parallel to the axis of round pipe.

2. Qualifies for fillet and groove welds in all positions except for complete joint penetration groove welding of T-, Y-, and K-connections.

3. Limited to prequalified joint details. See 2.9.1 and Figure 2.4; also 2.10.1 and Figure 2.5.

4. Qualifies for T-, Y-, and K-connections subject to limitations of 10.12, and any prequalified joint detail, see 2.9.1 and Figure 2.4; also 2.10.1 and Figure 2.5.

5. Qualification limited to groove angles 30° or greater. Does not qualify for butt joints welded from one side without backing. See 10.12.3.1, 2.10.1 and Figure 2.5.

6. Qualifies for horizontal fillet welds on rotated pipes only.

 *Positions of welding: F = flat, H = horizontal, V = vertical, OH = overhead.

**See Figs. 5.8.1.1, 5.8.1.2, and 5.8.1.3.

# Reinforcing Steel

Anchorages and Inserts                                    D1.4, Fig. 3.5

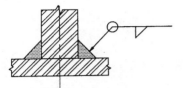

A - EXTERNAL FILLET WELD

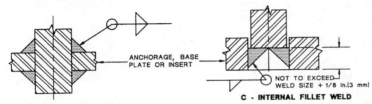

B - EXTERNAL FILLET WELD

ANCHORAGE, BASE
PLATE OR INSERT

NOT TO EXCEED
WELD SIZE + 1/8 in.(3 mm)

C - INTERNAL FILLET WELD

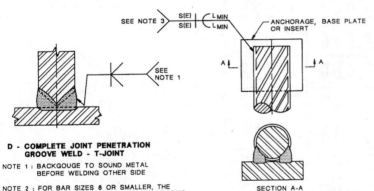

SEE NOTE 3

SEE
NOTE 1

ANCHORAGE, BASE PLATE
OR INSERT

D - COMPLETE JOINT PENETRATION
    GROOVE WELD - T-JOINT

NOTE 1 : BACKGOUGE TO SOUND METAL
         BEFORE WELDING OTHER SIDE

NOTE 2 : FOR BAR SIZES 8 OR SMALLER, THE
         SINGLE-BEVEL WELD WITH BACKGOUGING
         AND BACK WELDING IS RECOMMENDED

NOTE 3 : $L_{MIN}$ = 2 X BAR DIAMETER

SECTION A-A

E - LAP JOINTS IN AN ANCHORAGE
    USING FLARE-BEVEL-GROOVE WELDS

*(Continued)*

Reinforcing Steel *(Continued)*

Dimensions of                                    D1.4, Appendix B

| ASTM STANDARD REINFORCING BARS | | | |
|---|---|---|---|
| BAR SIZE DESIGNATION | NOMINAL AREA SQ. INCHES | NOMINAL WEIGHT POUNDS PER FT. | NOMINAL DIAMETER INCHES |
| #3 | 0.11 | 0.376 | 0.375 |
| #4 | 0.20 | 0.668 | 0.500 |
| #5 | 0.31 | 1.043 | 0.625 |
| #6 | 0.44 | 1.502 | 0.750 |
| #7 | 0.60 | 2.044 | 0.875 |
| #8 | 0.79 | 2.670 | 1.000 |
| #9 | 1.00 | 3.400 | 1.128 |
| #10 | 1.27 | 4.303 | 1.270 |
| #11 | 1.56 | 5.313 | 1.410 |
| #14 | 2.25 | 7.65 | 1.693 |
| #18 | 4.00 | 13.60 | 2.257 |

Direct Butt Joint Test Positions                    D1.4, Fig. 6.1

BARS HORIZONTAL

A - TEST POSITION 1G

BARS VERTICAL

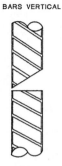

B - TEST POSITION 2G

BARS HORIZONTAL

C - TEST POSITION 3G        D - TEST POSITION 4G

NOTE: SEE FIGURE 6.3 FOR DEFINITION OF POSITIONS FOR
GROOVE WELDS

**Figure 6.1 — Direct Butt Joint Test Positions for Groove Welds (See 6.2.4.1)**

*(Continued)*

Reinforcing Steel, Direct Butt Joint
Test Positions *(Continued)*

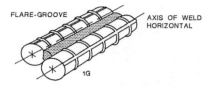

**A - TEST POSITION : FLAT**

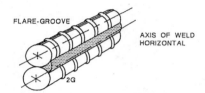

**B - TEST POSITION : HORIZONTAL**

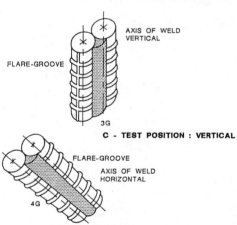

**C - TEST POSITION : VERTICAL**

**D - TEST POSITION : OVERHEAD**

NOTE: SEE FIGURES 6.3 AND 6.4 FOR DEFINITIONS OF
POSITIONS FOR FLARE-GROOVE AND FILLET WELDS.

**Figure 6.2 — Indirect Butt Joint Test Positions for Flare-Groove Welds
or Fillet Welds (See 6.2.4.2)**

Flare Groove Welds                                    D1.4, Fig. 2.1

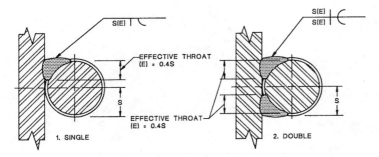

**A - FLARE-BEVEL-GROOVE WELD**

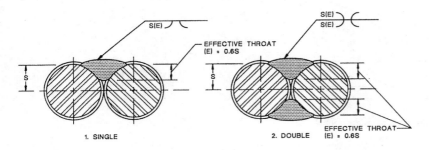

**B - FLARE-V-GROOVE WELD**

NOTE:   1.   RADIUS OF REINFORCING BAR = S

2.   THESE ARE SECTIONAL VIEWS.  BAR
DEFORMATIONS ARE SHOWN ONLY
FOR ILLUSTRATIVE PURPOSES.

**Figure 2.1 — Effective Weld Sizes for Flare-Groove Welds (See 2.3.2.3)**

*(Continued)*

Reinforcing Steel *(Continued)*

Indirect Butt Joints                                                D1.4, Fig. 3.3

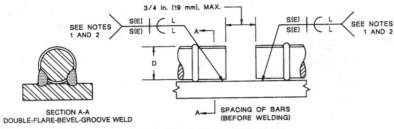

A - INDIRECT BUTT JOINT WITH A SPLICE PLATE

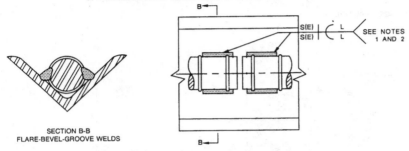

B - INDIRECT BUTT JOINT WITH A SPLICE ANGLE

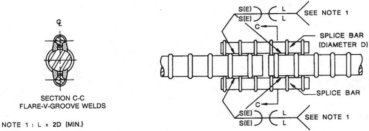

NOTE 1 : L = 2D (MIN.)

NOTE 2 : VARIATION OF THIS WELD
USING SINGLE FLARE-V
WELDS IS PERMITTED PROVIDED
ECCENTRICITY IS CONSIDERED IN DESIGN

NOTE 3 : GAPS BETWEEN BARS
OR BARS AND PLATES WILL VARY DEPENDING
ON HEIGHT OF DEFORMATIONS

NOTE 4   DEFORMATIONS SHOWN ON SECTIONAL VIEWS
ARE FOR ILLUSTRATIVE PURPOSES ONLY.

C - INDIRECT BUTT JOINT WITH TWO SPLICE BARS

**Figure 3.3 — Indirect Butt Joints (See 3.5)**

Lap Joints                                        D1.4, Fig. 3.4

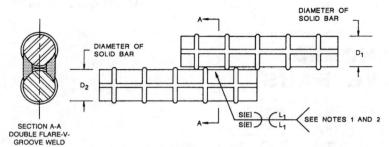

SECTION A-A
DOUBLE FLARE-V-
GROOVE WELD

**A - DIRECT LAP JOINT
WITH BARS IN CONTACT**

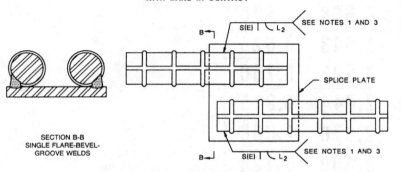

SECTION B-B
SINGLE FLARE-BEVEL-
GROOVE WELDS

**B - INDIRECT LAP JOINTS
WITH BARS SEPARATED**

NOTES :

1. THE EFFECTS OF ECCENTRICITY SHALL BE
   CONSIDERED OR RESTRAINT PROVIDED IN
   THE DESIGN OF THE JOINT

2. $L_1$  = 2 $D_1$ (MIN.) ; $D_1 \le D_2$

3. $L_2$ MIN = 2 X DIAMETER OF BAR

4. GAPS BETWEEN BARS AND PLATES WILL VARY BASED
   ON HEIGHT OF DEFORMATIONS.

**Figure 3.4 — Lap Joints (See 3.2.1)**

*(Continued)*

Reinforcing Steel *(Continued)*

Size

## ASTM STANDARD REINFORCING BARS

| BAR SIZE DESIGNATION | NOMINAL AREA SQ. INCHES | NOMINAL WEIGHT POUNDS PER FT. | NOMINAL DIAMETER INCHES |
|---|---|---|---|
| #3 | 0.11 | 0.376 | 0.375 |
| #4 | 0.20 | 0.668 | 0.500 |
| #5 | 0.31 | 1.043 | 0.625 |
| #6 | 0.44 | 1.502 | 0.750 |
| #7 | 0.60 | 2.044 | 0.875 |
| #8 | 0.79 | 2.670 | 1.000 |
| #9 | 1.00 | 3.400 | 1.128 |
| #10 | 1.27 | 4.303 | 1.270 |
| #11 | 1.56 | 5.313 | 1.410 |
| #14 | 2.25 | 7.65 | 1.693 |
| #18 | 4.00 | 13.60 | 2.257 |

## V-Groove and Bevel-Groove Welds            D1.4, Fig. 3.2

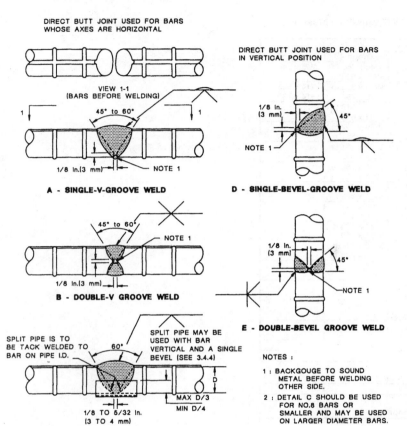

DIRECT BUTT JOINT USED FOR BARS
WHOSE AXES ARE HORIZONTAL

VIEW 1-1
(BARS BEFORE WELDING)

45° to 60°

1/8 In.(3 mm)              NOTE 1

**A - SINGLE-V-GROOVE WELD**

DIRECT BUTT JOINT USED FOR BARS
IN VERTICAL POSITION

1/8 In.
(3 mm)                45°

NOTE 1

**D - SINGLE-BEVEL-GROOVE WELD**

45° to 60°

NOTE 1

1/8 In.(3 mm)

**B - DOUBLE-V GROOVE WELD**

1/8 In.
(3 mm)                45°

NOTE 1

**E - DOUBLE-BEVEL GROOVE WELD**

SPLIT PIPE IS TO
BE TACK WELDED TO
BAR ON PIPE I.D.               60°

SPLIT PIPE MAY BE
USED WITH BAR
VERTICAL AND A SINGLE
BEVEL (SEE 3.4.4)

MAX D/3
MIN D/4

1/8 TO 5/32 In.
(3 TO 4 mm)

**C - SINGLE-V GROOVE WELD WITH SPLIT PIPE BACKING
(NOTE 2)**

NOTES :

1 : BACKGOUGE TO SOUND
    METAL BEFORE WELDING
    OTHER SIDE.

2 : DETAIL C SHOULD BE USED
    FOR NO.8 BARS OR
    SMALLER AND MAY BE USED
    ON LARGER DIAMETER BARS.

# Structural Shape Size Groupings

AISC, 1-8, Table 2

**Structural Shape Sizes Per Tensile Group Classifications**

| W Shapes | | | | |
|---|---|---|---|---|
| Group 1 | Group 2 | Group 3 | Group 4 | Group5 |
| W24 x 55 | W44 x 198 | W44 x 248 | W40 x 362 to W40 x 655 | W36 x 848 |
| W24 x 62 | W44 x 224 | W44 x 285 | W36 x 328 to W36 x 798 | W14 x 605 to W14 x 730 |
| W21 x 44 to W21 x 57 | W40 x 149 to W40 x 268 | W40 x 277 to W40 x 328 | W33 x 318 to W33 x 619 | |
| W18 x 35 to W 18 x 71 | W36 x 135 to W36 x 210 | W36 x 230 to W36 x 300 | W30 x 292 to W30 x 581 | |
| W16 x 26 to W16 x 57 | W33 x 118 to W33 x 152 | W33 x 201 to W33 x 291 | W27 x 281 to W27 x 539 | |
| W14 x 22 to W14 x 53 | W30 x 90 to W30 x 211 | W30 x 235 to W30 x 261 | W24 x 250 to W24 x 492 | |
| W12 x 14 to W12 x 58 | W27 x 84 to W27 x 178 | W27 x 194 to W27 x 258 | W21 x 248 to W21 x 402 | |
| W10 x 12 to W10 x 45 | W24 x 68 to W24 x 162 | W24 x 176 to W24 x 229 | W18 x 211 to W18 x 311 | |
| W8 x 10 to W8 x 48 | W21 x 62 to W21 x 147 | W21 x 166 to W21 x 223 | W14 x 233 to W14 x 550 | |
| W6 x 9 to W6 x 25 | W18 x 76 to W18 x 143 | W18 x 158 to W18 x 192 | W12 x 21- to W12 x 336 | |
| W5 x 16 W5 x 19 | W16 x 67 to W16 x 100 | W14 x 145 to W14 x 211 | | |
| W4 x 13 | W14 x 61 to W14 x 132 | W12 x 120 to W12 x 190 | | |
| | W12 x 65 to W12 x 106 | | | |
| | W10 x 49 to W10 x 112 | | | |
| | W8 x 58 | | | |
| | W8 x 67 | | | |

**M Shapes**

| to 37.7 lb./ft. | | | | |
|---|---|---|---|---|

**S Shapes**

| to 35 lb./ft. | | | | |
|---|---|---|---|---|

**HP Shapes**

| | to 102 lb./ft | > 102 lb./ft. | | |
|---|---|---|---|---|

**Standard Channel**

| to 20.7 lb./ft. | > 20.7 lb./ft. | | | |
|---|---|---|---|---|

**Miscellaneous Channel**

| to 28.5 lb./ft. | >28.5 lb./ft. | | | |
|---|---|---|---|---|

**Angle Iron**

| to 1/2 in. | >1/2 to 3/4 in. | >3/4 in. | | |
|---|---|---|---|---|

# Sheet Steel

Gage Numbers and Thickness Equivalents          D1.3, Appendix D

| Table D1 Gage Numbers and Equivalent Thicknesses Hot-Rolled and Cold-Rolled Sheet | | | Table D2 Gage Numbers and Equivalent Thicknesses Galvanized Sheet | | |
|---|---|---|---|---|---|
| Manufacturers' Standard Gage Number | Thickness Equivalent in. | mm | Galvanized Sheet Gage Number | Thickness Equivalent in. | mm |
| 3 | 0.2391 | 6.073 | | | |
| 4 | 0.2242 | 5.695 | | | |
| 5 | 0.2092 | 5.314 | | | |
| 6 | 0.1943 | 4.935 | | | |
| 7 | 0.1793 | 4.554 | | | |
| 8 | 0.1644 | 4.176 | 8 | 0.1681 | 4.270 |
| 9 | 0.1495 | 3.800 | 9 | 0.1532 | 3.891 |
| 10 | 0.1345 | 3.416 | 10 | 0.1382 | 3.510 |
| 11 | 0.1196 | 3.038 | 11 | 0.1233 | 3.132 |
| 12 | 0.1046 | 2.657 | 12 | 0.1084 | 2.753 |
| 13 | 0.0897 | 2.278 | 13 | 0.0934 | 2.372 |
| 14 | 0.0747 | 1.900 | 14 | 0.0785 | 1.993 |
| 15 | 0.0673 | 1.709 | 15 | 0.0710 | 1.803 |
| 16 | 0.0598 | 1.519 | 16 | 0.0635 | 1.613 |
| 17 | 0.0538 | 1.366 | 17 | 0.0575 | 1.460 |
| 18 | 0.0478 | 1.214 | 18 | 0.0516 | 1.311 |
| 19 | 0.0418 | 1.062 | 19 | 0.0456 | 1.158 |
| 20 | 0.0359 | 0.912 | 20 | 0.0396 | 1.006 |
| 21 | 0.0329 | 0.836 | 21 | 0.0366 | 0.930 |
| 22 | 0.0299 | 0.759 | 22 | 0.0336 | 0.853 |
| 23 | 0.0269 | 0.660 | 23 | 0.0306 | 0.777 |
| 24 | 0.0239 | 0.607 | 24 | 0.0276 | 0.701 |
| 25 | 0.0209 | 0.531 | 25 | 0.0247 | 0.627 |
| 26 | 0.0179 | 0.455 | 26 | 0.0217 | 0.551 |
| 27 | 0.0164 | 0.417 | 27 | 0.0202 | 0.513 |
| 28 | 0.0149 | 0.378 | 28 | 0.0187 | 0.475 |
| | | | 29 | 0.0172 | 0.437 |
| | | | 30 | 0.0157 | 0.399 |
| | | | 31 | 0.0142 | 0.361 |
| | | | 32 | 0.0134 | 0.340 |

Note: Table D1 is for information only. This product is commonly specified to decimal thickness, not to gage number.

Note: Table D2 is for information only. This product is commonly specified to decimal thickness, not to gage number.

*(Continued)*

Sheet Steel *(Continued)*

Welding Positions                                UBC, 2-518, Table E2

**TABLE E2**

| Connection | Welding Position | | | | | |
|---|---|---|---|---|---|---|
| | Square Groove Butt Weld | Arc Spot Weld | Arc Seam Weld | Fillet Weld, Lap or T | Flare-Bevel Groove | Flare-V Groove Weld |
| Sheet to Sheet | F<br>H<br>V<br>OH | —<br>—<br>—<br>— | F<br>H<br>—<br>— | F<br>H<br>V<br>OH | F<br>H<br>V<br>OH | F<br>H<br>V<br>OH |
| Sheet to Supporting Member | —<br>—<br>—<br>— | F<br>—<br>—<br>— | F<br>—<br>—<br>— | F<br>H<br>V<br>OH | F<br>H<br>V<br>OH | —<br>—<br>—<br>— |

(F = flat, H = horizontal, V = vertical, OH = overhead)

Square Groove Welds in Butt Joints                    D1.3, Fig. 3.1

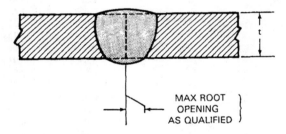

Figure 3.1 — Square-Groove Welds in Butt Joints (See 3.2)

## Symbols

AWS A2.4

# AMERICAN WELDING SOCIETY

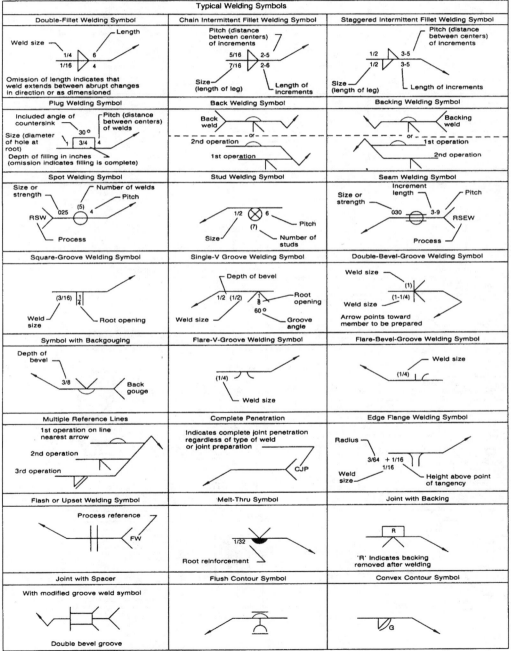

Typical Welding Symbols

**Double-Fillet Welding Symbol** — Length, Weld size, 1/4, 6, 1/16, 4. Omission of length indicates that weld extends between abrupt changes in direction or as dimensioned

**Chain Intermittent Fillet Welding Symbol** — Pitch (distance between centers) of increments, 5/16, 2-5, 7/16, 2-6, Size (length of leg), Length of increments

**Staggered Intermittent Fillet Welding Symbol** — Pitch (distance between centers) of increments, 1/2, 3-5, 1/2, 3-5, Size (length of leg), Length of increments

**Plug Welding Symbol** — Included angle of countersink, 30°, Pitch (distance between centers) of welds, Size (diameter of hole at root), 3/4, 4, Depth of filling in inches (omission indicates filling is complete)

**Back Welding Symbol** — Back weld, or, 2nd operation, 1st operation

**Backing Welding Symbol** — Backing weld, or, 1st operation, 2nd operation

**Spot Welding Symbol** — Size or strength, Number of welds, Pitch, RSW, 025, (5), 4, Process

**Stud Welding Symbol** — 1/2, 6, (7), Size, Pitch, Number of studs

**Seam Welding Symbol** — Increment length, Pitch, Size or strength, 030, 3-9, RSEW, Process

**Square-Groove Welding Symbol** — (3/16), 1/4, Weld size, Root opening

**Single-V Groove Welding Symbol** — Depth of bevel, 1/2 (1/2), 1/8, Root opening, 60°, Weld size, Groove angle

**Double-Bevel-Groove Welding Symbol** — Weld size, (1), (1-1/4), Weld size, Arrow points toward member to be prepared

**Symbol with Backgouging** — Depth of bevel, 3/8, Back gouge

**Flare-V-Groove Welding Symbol** — (1/4), Weld size

**Flare-Bevel-Groove Welding Symbol** — Weld size, (1/4)

**Multiple Reference Lines** — 1st operation on line nearest arrow, 2nd operation, 3rd operation

**Complete Penetration** — Indicates complete joint penetration regardless of type of weld or joint preparation, CJP

**Edge Flange Welding Symbol** — Radius, 3/64, + 1/16, 1/16, Weld size, Height above point of tangency

**Flash or Upset Welding Symbol** — Process reference, FW

**Melt-Thru Symbol** — 1/32, Root reinforcement

**Joint with Backing** — R, 'R' indicates backing removed after welding

**Joint with Spacer** — With modified groove weld symbol, Double bevel groove

**Flush Contour Symbol**

**Convex Contour Symbol** — G

* It should be understood that these charts are intended only as shop aids. The only complete and official presentation of the standard welding symbols is in A2.4.

# AMERICAN WELDING SOCIETY

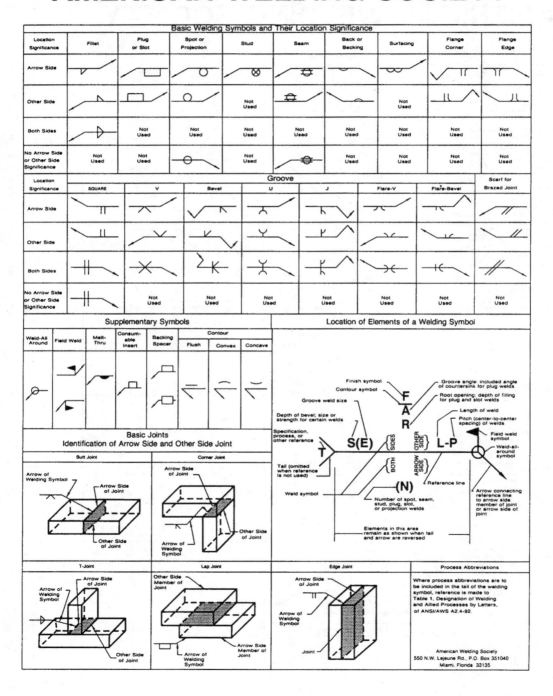

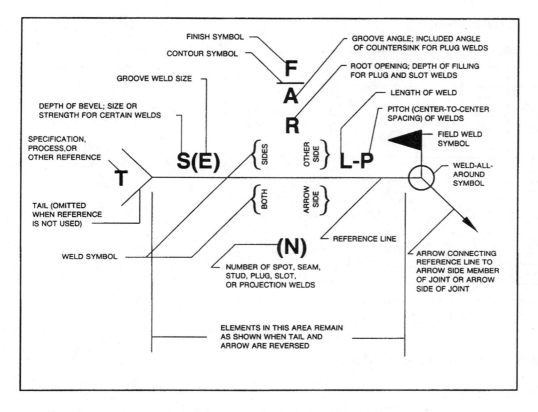

FINISH SYMBOL

CONTOUR SYMBOL

GROOVE WELD SIZE

DEPTH OF BEVEL; SIZE OR
STRENGTH FOR CERTAIN WELDS

SPECIFICATION,
PROCESS, OR
OTHER REFERENCE

TAIL (OMITTED
WHEN REFERENCE
IS NOT USED)

WELD SYMBOL

GROOVE ANGLE; INCLUDED ANGLE
OF COUNTERSINK FOR PLUG WELDS

ROOT OPENING; DEPTH OF FILLING
FOR PLUG AND SLOT WELDS

LENGTH OF WELD

PITCH (CENTER-TO-CENTER
SPACING) OF WELDS

FIELD WELD
SYMBOL

WELD-ALL-
AROUND
SYMBOL

REFERENCE LINE

ARROW CONNECTING
REFERENCE LINE TO
ARROW SIDE MEMBER
OF JOINT OR ARROW
SIDE OF JOINT

NUMBER OF SPOT, SEAM,
STUD, PLUG, SLOT,
OR PROJECTION WELDS

ELEMENTS IN THIS AREA REMAIN
AS SHOWN WHEN TAIL AND
ARROW ARE REVERSED

$F$
$A$
$R$

$T$   $S(E)$   SIDES   OTHER SIDE   $L\text{-}P$

BOTH   ARROW SIDE

$(N)$

**Standard Location of Elements of a Welding Symbol**

| WELD ALL AROUND | FIELD WELD | MELT THROUGH | CONSUMABLE INSERT (SQUARE) | BACKING OR SPACER (RECTANGLE) | CONTOUR | | |
|---|---|---|---|---|---|---|---|
| | | | | | FLUSH OR FLAT | CONVEX | CONCAVE |
| | | | | | | | |

**Supplementary Symbols**

*(Continued)*

Symbols *(Continued)*

| GROOVE | | | | | | | |
|---|---|---|---|---|---|---|---|
| SQUARE | SCARF | V | BEVEL | U | J | FLARE-V | FLARE-BEVEL |
| ⊥⊥ | ∕∕ | ⋎ | ⋎ | ⋎ | ⋎ | ⊃⊂ | ⊥⊂ |

| FILLET | PLUG OR SLOT | STUD | SPOT OR PROJECTION | SEAM | BACK OR BACKING | SURFACING | FLANGE | |
|---|---|---|---|---|---|---|---|---|
| | | | | | | | EDGE | CORNER |
| ▷ | ▭ | ⊗ | ○ | ⊖ | ⌣ | ◡◡ | ⊥∟ | ⊥⊥ |

NOTE: THE REFERENCE LINE IS SHOWN DASHED FOR ILLUSTRATIVE PURPOSES.

**Weld Symbols**

# Transition of Thickness or Width

Reinforcing                                                   D1.4, Fig. 3.1

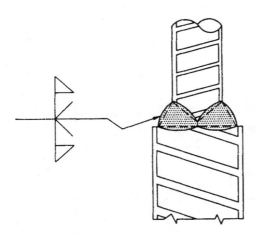

**Figure 3.1 — Direct Butt Joint  Showing
Transition Between Bars of Different
Sizes (See 3.1)**

Structural                                        D1.1, Fig. 8 3

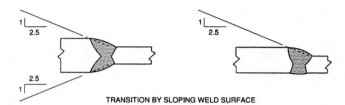

TRANSITION BY SLOPING WELD SURFACE

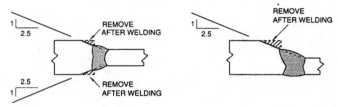

TRANSITION BY SLOPING WELD SURFACE AND CHAMFERING

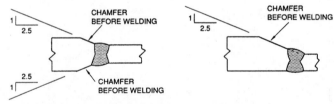

TRANSITION BY CHAMFERING THICKER PART

CENTERLINE ALIGNMENT
(PARTICULARLY APPLICABLE
TO WEB PLATES)

OFFSET ALIGNMENT
(PARTICULARLY APPLICABLE
TO FLANGE PLATES)

Notes:

1.  Groove may be of any permitted or qualified type and detail.

2.  Transition slopes shown are the maximum permitted.

**Figure 8.3 — Transition of Butt Joints in Parts of Unequal Thickness (see 8.10)**

# Torque

## Average Torques for A325 and A490 Bolts
### (Foot-Pounds)

### A325 Bolts

| Nominal Bolt / Nut Diameter | Average Torque | Width of Head Across Flats |
|---|---|---|
| 5/8 | 200 | 1-1/16 |
| 3/4 | 355 | 1-1/4 |
| 7/8 | 570 | 1-7/16 |
| 1 | 850 | 1-5/8 |
| 1-1/8 | 1,060 | 1-13/16 |
| 1-1/4 | 1,495 | 2 |
| 1-3/8 | 1,960 | 2-3/16 |
| 1-1/2 | 2,600 | 2-3/8 |

### A490 Bolts

| Nominal Bolt/Nut Diameter | Average Torque | Width of Head Across Flats |
|---|---|---|
| 5/8 | 250 | 1-1/16 |
| 3/4 | 435 | 1-1/4 |
| 7/8 | 715 | 1-7/16 |
| 1 | 1,070 | 1-5/8 |
| 1-1/8 | 1,500 | 1-13/16 |
| 1-1/4 | 2,125 | 2 |
| 1-3/8 | 2,780 | 2-3/16 |
| 1-1/2 | 3,700 | 2-3/8 |

### Nut and Bolt Head Dimensions
### A325 and A490

| Bolt Diameter | Width across the Flats (Nuts and Bolts) | Nut Height |
|---|---|---|
| 1/2 | 7/8 | 31/64 |
| 5/8 | 1-1/16 | 39/64 |
| 3/4 | 1-1/4 | 47/64 |
| 7/8 | 1-7/16 | 55/64 |
| 1 | 1-5/8 | 63/64 |
| 1-1/8 | 1-13/16 | 1-7/64 |
| 1-1/4 | 2 | 1-7/32 |
| 1-3/8 | 2-3/16 | 1-11/32 |
| 1-1-2 | 2-3/8 | 1-15/32 |

This information is useful in determining proper wrench sizes.

# Torque Test for Studs

D1.1, Fig. 7.3

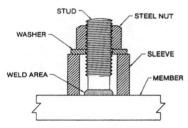

Note: The dimensions shall be appropriate to the size of the stud. The threads of the stud shall be clean and free of lubricant other than the residue of cutting oil.

| Required torque for testing threaded studs | | | | |
|---|---|---|---|---|
| Nominal diameter of studs | | Threads per inch & series designated | Testing torque | |
| in. | mm | | ft-lb | J |
| 1/4 | 6.4 | 28 UNF | 5.0 | 6.8 |
| 1/4 | | 20 UNC | 4.2 | 5.7 |
| 5/16 | 7.9 | 24 UNF | 9.5 | 12.9 |
| 5/16 | | 18 UNC | 8.6 | 11.7 |
| 3/8 | 9.5 | 24 UNF | 17.0 | 23.0 |
| 3/8 | | 16 UNC | 15.0 | 20.3 |
| 7/16 | 11.1 | 20 UNF | 27.0 | 36.6 |
| 7/16 | | 14 UNC | 24.0 | 32.5 |
| 1/2 | 12.7 | 20 UNF | 42.0 | 57.0 |
| 1/2 | | 13 UNC | 37.0 | 50.2 |
| 9/16 | 14.3 | 18 UNF | 60.0 | 81.4 |
| 9/16 | | 12 UNC | 54.0 | 73.2 |
| 5/8 | 15.9 | 18 UNF | 84.0 | 114.0 |
| 5/8 | | 11 UNC | 74.0 | 100.0 |
| 3/4 | 19.0 | 16 UNF | 147.0 | 200.0 |
| 3/4 | | 10 UNC | 132.0 | 180.0 |
| 7/8 | 22.2 | 14 UNF | 234.0 | 320.0 |
| 7/8 | | 9 UNC | 212.0 | 285.0 |
| 1 | 25.4 | 12 UNF | 348.0 | 470.0 |
| 1 | | 8 UNC | 318.0 | 430.0 |

**Figure 7.3 — Torque Testing Arrangement and Table of Testing Torques (see 7.6.6.2)**

# Variance in Groove Welded Joints

D1.1, Fig. 3.3

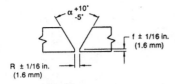

(A) GROOVE WELD WITHOUT BACKING -
ROOT NOT BACKGOUGED

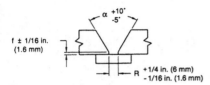

(B) GROOVE WELD WITH BACKING -
ROOT NOT BACKGOUGED

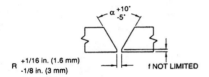

(C) GROOVE WELD WITHOUT BACKING -
ROOT BACKGOUGED

|  | Root not back gouged* | | Root back gouged | |
|---|---|---|---|---|
|  | in. | mm | in. | mm |
| (1) Root face of joint | ±1/16 | 1.6 | Not limited | |
| (2) Root opening of joints without backing | ±1/16 | 1.6 | +1/16 −1/8 | 1.6 3 |
| Root opening of joints with backing | +1/4 −1/16 | 6 1.6 | Not applicable | |
| (3) Groove angle of joint | +10° −5° | | +10° −5° | |

*see 10.13.1 for tolerances for complete joint penetration tubular groove welds made from one side without backing.

**Figure 3.3 — Workmanship Tolerances in Assembly of Groove Welded Joints (see 3.3.4)**

# Washers

**TABLE 22-IV-L—HARDENED CIRCULAR AND CLIPPED CIRCULAR WASHERS[1]**

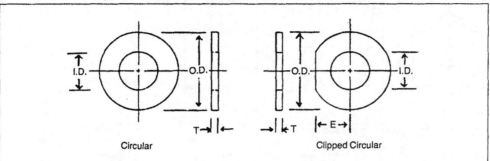

Circular                                    Clipped Circular

| BOLT SIZE (inches) | CIRCULAR AND CLIPPED CIRCULAR | | | | CLIPPED |
|---|---|---|---|---|---|
| | Nominal Outside Diameter (O.D.) (inches) | Normal Inside Diameter (I.D.) (inches) | Thickness (T) (inches) | | Minimum Edge Distance (E)[2] (inches) |
| | | | min. | max. | |
| | | × 25.4 for mm | | | |
| $1/4$ | $5/8$ | $9/32$ | 0.051 | 0.080 | $7/32$ |
| $5/16$ | $11/16$ | $11/32$ | 0.051 | 0.080 | $9/32$ |
| $3/8$ | $13/16$ | $13/32$ | 0.051 | 0.080 | $11/32$ |
| $7/16$ | $59/64$ | $15/32$ | 0.051 | 0.080 | $13/32$ |
| $1/2$ | $1^1/16$ | $17/32$ | 0.097 | 0.177 | $7/16$ |
| $5/8$ | $1^5/16$ | $11/16$ | 0.122 | 0.177 | $9/16$ |
| $3/4$ | $1^{15}/32$ | $13/16$ | 0.122 | 0.177 | $21/32$ |
| $7/8$ | $1^3/4$ | $15/16$ | 0.136 | 0.177 | $25/32$ |
| 1 | 2 | $1^1/8$ | 0.136 | 0.177 | $7/8$ |
| $1^1/8$ | $2^1/4$ | $1^1/4$ | 0.136 | 0.177 | 1 |
| $1^1/4$ | $2^1/2$ | $1^3/8$ | 0.136 | 0.177 | $1^3/32$ |
| $1^3/8$ | $2^3/4$ | $1^1/2$ | 0.136 | 0.177 | $1^7/32$ |
| $1^1/2$ | 3 | $1^5/8$ | 0.136 | 0.177 | $1^5/16$ |
| $1^3/4$ | $3^3/8$ | $1^7/8$ | 0.178[3] | 0.28[3] | $1^{17}/32$ |
| 2 | $3^3/4$ | $2^1/8$ | 0.178[3] | 0.28[3] | $1^3/4$ |
| $2^1/4$ | 4 | $2^3/8$ | 0.24[4] | 0.34[4] | 2 |
| $2^1/2$ | $4^1/2$ | $2^5/8$ | 0.24[4] | 0.34[4] | $2^3/16$ |
| $2^3/4$ | 5 | $2^7/8$ | 0.24[4] | 0.34[4] | $2^{13}/32$ |
| 3 | $5^1/2$ | $3^1/8$ | 0.24[4] | 0.34[4] | $2^5/8$ |
| $3^1/4$ | 6 | $3^3/8$ | 0.24[4] | 0.34[4] | $2^7/8$ |
| $3^1/2$ | $6^1/2$ | $3^5/8$ | 0.24[4] | 0.34[4] | $3^1/16$ |
| $3^3/4$ | 7 | $3^7/8$ | 0.24[4] | 0.34[4] | $3^5/16$ |
| 4 | $7^1/2$ | $4^1/8$ | 0.24[4] | 0.34[4] | $3^1/2$ |

[1] Tolerances are as noted in Table 22-IV-N.
[2] Clipped edge, $E$, shall not be closer than $7/8$ of the bolt diameter from the center of the washer.
[3] $3/16$ inch (4.8 mm) nominal.
[4] $1/4$ inch (6.4 mm) nominal.

*(Continued)*

Washers *(Continued)*

### TABLE 22-IV-M—HARDENED BEVELED WASHERS[1]

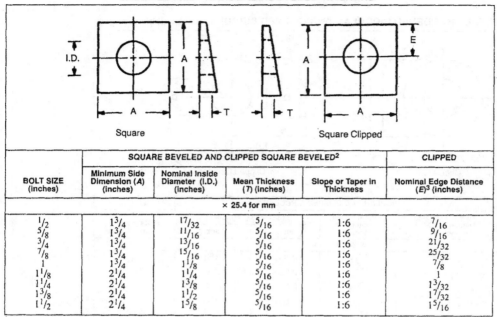

Square                              Square Clipped

| BOLT SIZE (inches) | SQUARE BEVELED AND CLIPPED SQUARE BEVELED[2] | | | | CLIPPED |
| | Minimum Side Dimension (A) (inches) | Nominal Inside Diameter (I.D.) (inches) | Mean Thickness (T) (inches) | Slope or Taper in Thickness | Nominal Edge Distance (E)[3] (inches) |
| --- | --- | --- | --- | --- | --- |
| | | | × 25.4 for mm | | |
| $1/2$ | $1^3/4$ | $17/32$ | $5/16$ | 1:6 | $7/16$ |
| $5/8$ | $1^3/4$ | $11/16$ | $5/16$ | 1:6 | $9/16$ |
| $3/4$ | $1^3/4$ | $13/16$ | $5/16$ | 1:6 | $21/32$ |
| $7/8$ | $1^3/4$ | $15/16$ | $5/16$ | 1:6 | $25/32$ |
| 1 | $1^3/4$ | $1^1/8$ | $5/16$ | 1:6 | $7/8$ |
| $1^1/8$ | $2^1/4$ | $1^1/4$ | $5/16$ | 1:6 | 1 |
| $1^1/4$ | $2^1/4$ | $1^3/8$ | $5/16$ | 1:6 | $1^3/32$ |
| $1^3/8$ | $2^1/4$ | $1^1/2$ | $5/16$ | 1:6 | $1^7/32$ |
| $1^1/2$ | $2^1/4$ | $1^5/8$ | $5/16$ | 1:6 | $1^5/16$ |

[1]Tolerances are as noted in Table 22-IV-N.
[2]Rectangular beveled washers shall conform to the dimensions shown above, except that one side may be longer than that shown for the *A* dimension.
[3]Clipped edge *E* shall not be closer than $7/8$ of the bolt diameter from the center of the washer.

## Weld Washers                                                    D1.3, Fig. 2.1

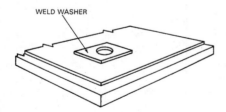

WELD WASHER

Figure 2.1 — Typical Weld Washer
(See 2.2.2.2)

# Weld Profiles
## Reinforcing

D1.4, Fig. 4.1

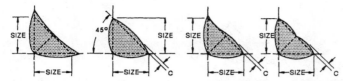

**(A) DESIRABLE FILLET WELD PROFILES**   **(B) ACCEPTABLE FILLET WELD PROFILES**

NOTE : CONVEXITY, C, OF A WELD OR INDIVIDUAL SURFACE BEAD SHALL NOT EXCEED THE VALUE OF THE FOLLOWING TABLE:

| MEASURED LEG SIZE OR WIDTH OF INDIVIDUAL SURFACE BEAD, L | MAX. CONVEXITY |
|---|---|
| L ≤ 5/16 In. (8mm) | 1/16 In. (1.6mm) |
| 5/16 In. < L < 1 In. (25mm) | 1/8 In. (3mm) |
| L ≥ 1 In. | 3/16 In. (5mm) |

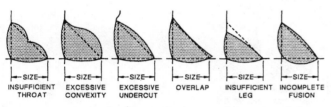

INSUFFICIENT THROAT   EXCESSIVE CONVEXITY   EXCESSIVE UNDERCUT   OVERLAP   INSUFFICIENT LEG   INCOMPLETE FUSION

**(C) UNACCEPTABLE FILLET WELD PROFILES**

Figure 4.1 — Acceptable and Unacceptable Weld Profiles (See 4.4)

*(Continued)*

Weld Profiles *(Continued)*

Structural                                                    D1.1, Fig. 3.4

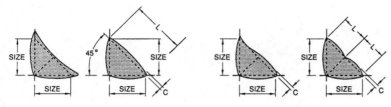

(A) DESIRABLE FILLET WELD PROFILES          (B) ACCEPTABLE FILLET WELD PROFILES

Note: Convexity, C, of a weld or individual surface bead shall not exceed the value of the following table:

| Measured Leg Size or Width of Individual Surface Bead, L | Max. Convexity |
|---|---|
| L ≤ 5/16 in. (8 mm) | 1/16 in. (1.6 mm) |
| L > 5/16 in. to L < 1 in. (25 mm) | 1/8 in. (3 mm) |
| L ≥ 1 in. | 3/16 in. (5 mm) |

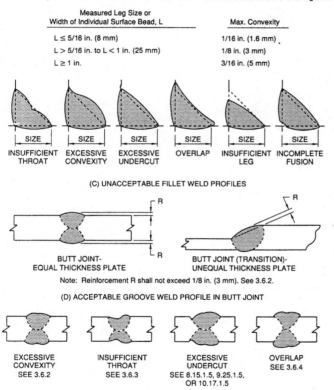

| INSUFFICIENT THROAT | EXCESSIVE CONVEXITY | EXCESSIVE UNDERCUT | OVERLAP | INSUFFICIENT LEG | INCOMPLETE FUSION |

(C) UNACCEPTABLE FILLET WELD PROFILES

BUTT JOINT-
EQUAL THICKNESS PLATE

BUTT JOINT (TRANSITION)-
UNEQUAL THICKNESS PLATE

Note:  Reinforcement R shall not exceed 1/8 in. (3 mm). See 3.6.2.

(D) ACCEPTABLE GROOVE WELD PROFILE IN BUTT JOINT

| EXCESSIVE CONVEXITY SEE 3.6.2 | INSUFFICIENT THROAT SEE 3.6.3 | EXCESSIVE UNDERCUT SEE 8.15.1.5, 9.25.1.5, OR 10.17.1.5 | OVERLAP SEE 3.6.4 |

(E) UNACCEPTABLE GROOVE WELD PROFILES IN BUTT JOINTS

**Figure 3.4 — Acceptable and Unacceptable Weld Profiles (see 3.6)**

# Weld Washers *(See Washers)*

# D  HIGH-STRENGTH BOLTS

## Anchor Bolts

AISC, 4-4, Table 1-C

Shear between the column base plate and the anchor bolts is very rarely, if ever critical. Anchor bolts normally are in tensile stress only. This is due to the fact that the vertical load on a column is usually more than sufficient to transfer any amount of shear from the column base plate into the foundation.

AISC, 5-79, J10
AISC, 5-172, J10
UBC, 2-358, 2210

The use of oversize holes in column base plates is permitted by the specification and is not detrimental to the integrity of the structure.

UBC, . . . Anchor bolts shall be set accurately to the pattern and dimensions called for on the plans. The protrusion of the threaded ends through the connected material shall be sufficient to fully engage the threads of the nuts, but shall not be greater than the length of the threads on the bolts. Base plate holes for anchor bolts may be oversized as follows:

| Bolt Size, inches (mm) | Hole Size, inches (mm) |
|---|---|
| $3/4$ (19.1) | $5/16$ (7.9) oversized |
| $7/8$ (22.2) | $5/16$ (7.9) oversized |
| $1 < 2$ (25.4 < 50.8) | $1/2$ (12.7) oversized |
| >2 (> 50.8) | 1 (25.4) > bolt diameter |

## Application (SC, N, X)

AISC, 4-9

The specification allows for the following three different types of applications for high-strength bolts:

Application (SC, N, X) *(Continued)*

SC   Slip-Critical Connections

N    Bearing-Type Connection/Threads may be included in the shear plane

X    Bearing-Type Connection/Threads must be excluded from the shear plane

It is the responsibility of the design engineer to provide this information on the drawings.

## A325 and A490

AISC, 5-263, RCSC Spec.

## A449

AISC 5-27, A3.4

ASTM A449 bolts are permitted only in non-slip-
critical connections requiring bolt diameters over
$1\frac{1}{2}$ inches in diameter. Any diameter of
ASTM A449 may be used in an anchor bolt or
threaded rod application.

## Bolt Tension in kips

AISC, 5-77, Table J3.7
UBC, 2-454, Table 22-IV-D

### Minimum Bolt Tension in Kips

| Bolt Diameters in inches | A325 | A490 |
|:---:|:---:|:---:|
| 1/2 | 12 | 15 |
| 5/8 | 19 | 24 |
| 3/4 | 28 | 35 |
| 7/8 | 39 | 49 |
| 1 | 51 | 64 |
| 1-1/8 | 56 | 80 |
| 1-1/4 | 71 | 102 |
| 1-3/8 | 85 | 121 |
| 1-1/2 | 103 | 148 |

## Bolts and Threaded Parts

UBC, 2-695, J3.4

UBC, . . . Allowable tension and shear stresses on
bolts, threaded parts, and rivets shall be as given in
Table J3.2, in ksi of the nominal body area of rivets

(before driving) or the unthreaded nominal body area of bolts and threaded parts other than upset rods. . . .

*Design* for bolts, threaded parts, and rivets subject to fatigue loading shall be in accordance with Appendix K4.3.

## Bolted Parts (Slope of)

AISC, 5-267, para. 3 (a)

No foreign material is permitted in the faying surface of a connection. Nothing but the steel of the connected member is permitted.

UBC, 2-446, 2222.1

*The slope* of the contact surfaces in the connection are not allowed to be greater than 1:20 without installing a hardened beveled washer to compensate for the slope.

*While paint* is permitted on the faying surfaces of bearing-type connections, coatings are only allowed on the faying surfaces of slip-critical connections with the approval of the engineer.

## Connection Slip

AISC, 5-297 para. 1

In certain cases, and with engineering approval, the specification permits so designated joints to be tightened only to a snug tight condition.

## Calibrated Wrench Method

AISC, 5-305, para. 3

The calibrated wrench method of tightening, when used, should be performed with regard to any variables that affect torque. The most commonly found variables are:

1. The finish and tolerance of the nut and bolt threads.
2. Different manufacturers of nuts vs. bolts.
3. Amount of, and type of lubrication.
4. Storage conditions at the job site.
5. Dirt or burrs on the threads.

*(Continued)*

Calibrated Wrench Method *(Continued)*

6.  Condition and capacity of the impact wrench.

7.  Air supply, hose diameter, etc.

*When the calibrated* wrench method of tightening is used, hard washers must be installed, fasteners must be protected from dirt and moisture, and wrenches must be calibrated daily.

*Recognition* of the calibrated wrench method of tightening was removed from the specification with the 1980 edition. This action was taken because it is the least reliable of all methods of installation and many costly controversies had occurred. It is to be suspected that short cut procedures in the use of the calibrated wrench method of installation, not in accordance with the specification provisions, were being used. These incorrect procedures, plus others, had a compounding effect upon the uncertainty of the installed bolt tension, and were responsible for many of the controversies. It is recognized, however, that if the calibrated wrench method is implemented without short cuts, as intended by the specification, that there will be a 90% assurance that the tensions specified in Table 4 will be equaled or exceeded. Because the specification should not prohibit any method that will give acceptable results when used as specified, the calibrated wrench method of installation is reinstated in this edition of the Council Specification.

## Edge Distance

Maximum                                                          UBC, 2-699, J3.10

UBC, . . . The maximum distance from the cen-                    AISC, 5-76, J3.10
ter of any rivet or bolt hole to the nearest edge of
parts in contact shall be 12 times the thickness
of the connected part under consideration, but
shall not exceed 6 in. . . .

*When fasteners* are used to connect a plate with a
shape, or two plates together, the center-to-cen-

ter spacing shall not exceed a maximum of 14 times the thickness of the thinner part, nor 7 in.

*For unpainted* built-up members made of weather resistant steel, the maximum edge distance shall not be greater than 8 times the thickness of the thinner part, or 5 in.

*The maximum* edge distance as measured from the center of a bolt hole to the nearest edge of the connection shall not be greater than 12 times the thickness of the connected part and shall not be greater than 6 in.

Minimum

UBC, 2-698, J3.9

UBC, . . . The distance from the center of a standard hole to an edge of a connected part shall not be less than the applicable value from Table J3.5 or the value from Equation (J3-6), as applicable.

UBC, 2-699, Table J3.5
AISC, 5-75, J3.9
AISC, 5-76, Table J3.5

### Minimum Edge Distance

| Bolt Diameter in inches | From a sheared Edge | From a Rolled Edge |
|:---:|:---:|:---:|
| 1/2 | 7/8 | 3/4 |
| 5/8 | 1-1/8 | 7/8 |
| 3/4 | 1-1/4 | 1 |
| 7/8 | 1-1/2 | 1-1/8 |
| 1 | 1-3/4 | 1-1/4 |
| 1-1/8 | 2 | 1-1/2 |
| 1-1/4 | 2-1/4 | 1-5/8 |
| 1-1/2 | 2-5/8 | 1-7/8 |

Edge distances are measured from the edge of the connected material to the centerline of the hole.

## Faying Surface (Paint in)

AISC, 5-295, para. 3

In slip-critical connections that have been designated by the engineer to be free of paint, even the smallest amount of overspray is prohibited and must be removed.

AISC, 5-293, para. 3
UBC, 2-446, 2221.7; 2222.2

## Grades (Nuts)

AISC, 5-266, para. 2.(c)

Suitable nuts for high-strength bolts are listed in the following table:

**Nuts for A325 and A490 Bolts**

| A325 Bolts | Grade of Nuts | | | | | | |
|---|---|---|---|---|---|---|---|
| | A563 | | | | | A194 | |
| | C | C3 | D | DH | DH3 | 2 | 2H |
| Type 1 Plain | C | C3 | D | DH | DH3 | 2 | 2H |
| Type 1 Galvanized | | | | DH[G] | | | 2H[G] |
| Type 2 Discontinued | | | | | | | |
| Type 3 Plain | | C3 | | | DH3 | | |

| A490 Bolts | Grade of Nuts | | | | | | |
|---|---|---|---|---|---|---|---|
| | A563 | | | | | A194 | |
| | | | | DH | DH3 | | 2H |
| Type 1 Plain | | | | DH | DH3 | | 2H |
| Type 2 Plain | | | | DH | DH3 | | 2H |
| Type 3 Plain | | | | | DH3 | | |

## Galvanized High-Strength Bolts

AISC, 5-291, para. 5

Galvanized high-strength bolts must be manufactured and shipped as a matched assembly.

*Some of the areas* of concern that must be considered when deciding whether or not to use galvanized bolts are:

1. The effect the galvanizing process might have on the high-strength bolt.

2. Whether the nut stripping strength will be reduced due to the galvanized coatings.

3. How will the galvanizing effect the torque induced in tensioning.

4. Shipping requirements.

*This specification* allows A325 bolts to be galvanized. A490 bolts are not permitted by the specification to be galvanized.

# High Tension Bolts

AISC, 5-71, J3

The specification specifies that the use of high-strength bolts shall conform to the requirements of the *Specification for Structural Joints Using ASTM A325 or A490 Bolts.*

AISC, 5-263, RCSC Spec.
UBC, 2-605, M2.2.5
UBC, 2-445, Division IV

When ASTM A449 bolts are used in bearing-type and tension-type connections, and are to be tightened to a condition exceeding 50% of their minimum specified tensile strength, an ASTM F436 hardened washer shall be installed under the bolt head and the nuts shall be ASTM A563.

UBC, . . . The use of high-strength bolts shall conform to the requirements of the *RCSC Load and Resistance Factor Design Specification for Structural Joints Using ASTM A325 or A490 Bolts.*

In Combination with Rivets or Welds

AISC, 5-302, para. 2

UBC, . . . In new work, A307 bolts or high-strength bolts used in bearing-type connections shall not be considered as sharing the stress in combination with welds. Welds, if used, shall be provided to carry the entire stress in the connection.

AISC, 5-64, J1.11
UBC, 2-688, J1.10
UBC, 2-689, J1.12
D1.1, 8.7

D1.1, . . . In new work, rivets or bolts in combination with welds shall not be considered as sharing the stress, and the welds shall be provided to carry the entire stress for which the connection is designed. If bolts are to be removed, the plans should indicate whether holes should be filled and in what manner.

D1.1, 9.13

In Construction

AISC, 5-88, M2.5

*During erection,* the use of drift pins to align bolt holes in members is allowed, provided their use does not distort, oblong, or otherwise enlarge the holes.

AISC, 5-273, 8.(a)-(e)

*Improper* or poor matching of bolt holes shall be cause for rejection of the member.

*If the slope* of the faying surfaces in the connection exceeds 1:20, with regard to the bolt or nut

*(Continued)*

### High Tension Bolts *(Continued)*

face, a single hardened beveled washer shall be installed to compensate for the slope.

### Inspection of

AISC, 5-276, para. 9 (a) - (c)

Before work begins, and when required, the inspector shall observe the calibration procedure. The inspector shall inspect materials to verify compliance with the approved plans, specifications, and contract documents.

AISC, 5-307, C9
UBC, 2-451, 2228

*While work* is in progress, the inspector shall assure that the areas within the faying surfaces of the connections are in firm contact with one another and that the installation procedures are being met.

UBC, . . . While work is in progress, the special inspector shall determine that the requirements of Sections 2221, 2222, and 2227 are met.

### Installation and Tightening

AISC, 5-272, para. 8 (a) - (e)

The RCSC Specification requires that fastener components be stored in suitable containers and protected from the elements. Plastic coverings are not recommended as they most likely will produce condensation that in turn will be deposited onto the fasteners. If the lubricant on the bolts is water soluble, and many bolt manufacturers use water soluble oil, it will be washed off. Also, being water soluble, this particular lubricant will eventually evaporate and invite corrosion. For these and other reasons it is of great importance to institute a sound, clean and dry storage program.

AISC, 5-274, Table 4;
AISC, 5-275, Table 5
UBC, 2-449, 2227

*If bolts* are to be tightened to a snug tight condition only, the engineer must clearly state so on the drawings.

*Snug tight* can be defined as the condition that exists when the faying surfaces in the connection are in firm contact. This condition can be accomplished by making a few hits with an impact wrench or by a person using full effort with a spud wrench.

*Care should* be observed when tightening galvanized bolts. Effective lubricants have, in the past, made it possible for a person using full effort with a spud wrench, to achieve full tension.

*Tightening* may be done by turning the bolt while the nut is prevented from rotating when it is impractical to turn the nut. If impact wrenches are used to tighten the bolts in the connection, the specification dictates they be capable of tightening the bolts in approximately 10 seconds.

*When using* the turn-of-nut method, three bolts of each diameter, length, grade, and lot number, and three nuts of each diameter, grade, and lot number, along with three washers of the grade and size to be used, shall, at the beginning of each shift, be installed in a bolt tensioning device, such as a Skidmore Wilhelm, and brought to a degree of tension no less than 5% of the required **minimum bolt tension in kips.**

*When tightening* is to be performed using the calibrated wrench method, hardened washers conforming to F436 must be installed under the component being turned and installation procedures must be calibrated daily or at the start of each shift.

*When torque* wrenches are to be used, they are to set at least 5% above the **minimum bolt tension in kips.**

*It shall* be verified during the tightening process that the calibrated torque wrench does not produce nut rotation from the snug tight condition greater than allowed in the following table:

*(Continued)*

High Tension Bolts *(Continued)*

| Nut Rotation from Snug Tight | | |
|---|---|---|
| (Required amount of nut rotation when using the Turn-of-Nut Method of Tightening) | | |
| **Bolt Length in Relationship to Bolt Diameter** | | |
| Degree of slope under Bolt Head | Bolt length up to and including 4 diameters | Bolt length over 4 diameters and up to 8 diameters | Bolt length over 8 diameters and up to, but not over 12 diameters |
| Both surfaces normal | 1/3 Turn | 1/2 Turn | 2/3 Turn |
| One surface normal, one surface sloped no more than 1:20 | 1/2 Turn | 2/3 Turn | 5/6 Turn |
| Both surfaces sloped no more than 1:20 | 2/3 Turn | 5/6 Turn | 1 Turn |

The Specification allows a rotational tolerance of plus or minus 30 degrees for nuts rotated 1/2 turn and less, and plus or minus 45 degrees for nuts rotated 2/3 or more turns.

*Galvanized A325* bolts and A490 bolts cannot be reused. Retightening of bolts that were loosened when adjacent bolts were tightened does not constitute reuse and may be retightened.

Minimum Tension in kips

AISC, 5-274, Table 4
AISC, 5-77, Table J3.7
UBC, 2-454, Table 22-IV-D

### Minimum Bolt Tension in Kips

| Bolt Diameters in inches | A325 | A490 |
|---|---|---|
| 1/2 | 12 | 15 |
| 5/8 | 19 | 24 |
| 3/4 | 28 | 35 |
| 7/8 | 39 | 49 |
| 1 | 51 | 64 |
| 1-1/8 | 56 | 80 |
| 1-1/4 | 71 | 102 |
| 1-3/8 | 85 | 121 |
| 1-1/2 | 103 | 148 |

Reuse of

UBC, 2-451, 2227.5

UBC, . . . A490 and galvanized A325 bolts shall not be reused. Other A325 bolts may be reused if approved by the building official. . . .

AISC, 5-276, (e)
AISC, 5-276, para. 8 (e)

## Inspection of

AISC, 5-276, para. (a) - (c)

Before work begins, and when required, the inspector shall observe the calibration procedure. The inspector shall inspect materials to verify compliance with the approved plans, specifications, and contract documents.

AISC, 5-307, C9
UBC, 2-451, 2228

*While work* is in progress, the inspector shall assure that the areas within the faying surfaces of the connections are in firm contact with one another and that the installation procedures are being met.

## Installation and Tightening

AISC, 5-272, para. 8 (a) - (e)

The RCSC Specification requires that fastener components be stored in suitable containers and protected from the elements. Plastic coverings are not recommended as they most likely will produce condensation that in turn will be deposited onto the fasteners. If the lubricant on the bolts is water soluble, and many bolt manufacturers use water soluble oil, it will be washed off. Also, being water soluble, this particular lubricant will eventually evaporate and invite corrosion. For these and other reasons it is of great importance to institute a sound, clean and dry storage program.

UBC, 2-449, 2227

*If bolts* are to be tightened to a snug tight condition only, the engineer must clearly state so on the drawings.

*Snug tight* can be defined as the condition that exists when the faying surfaces in the connection are in firm contact. This condition can be accomplished by making a few hits with an impact wrench or by a person using full effort with a spud wrench.

*Care should* be observed when tightening galvanized bolts. Effective lubricants have, in the past, made it possible for a person using full effort with a spud wrench, to achieve full tension.

*Tightening* may be done by turning the bolt while the nut is prevented from rotating when it is impractical to turn the nut. If impact wrenches are

*(Continued)*

Installation and Tightening *(Continued)*

used to tighten the bolts in the connection, the specification dictates they be capable of tightening the bolts in approximately 10 seconds.

*When using* the turn-of-nut method, three bolts of each diameter, length, grade, and lot number, and three nuts of each diameter, grade, and lot number, along with three washers of the grade and size to be used, shall, at the beginning of each shift, be installed in a bolt tensioning device, such as a Skidmore Wilhelm, and brought to a degree of tension no less than 5% of the required minimum bolt tension in kips, found in Table J3.7.

*When tightening* is to be performed using the calibrated wrench method, hardened washers conforming to F436 must be installed under the component being turned and installation procedures must be calibrated daily or at the start of each shift.

*When torque* wrenches are to be used, they are to set at least 5% above the minimum tension required.

*It shall* be verified during the tightening process that the calibrated torque wrench does not produce nut rotation from the snug tight condition greater than allowed by the specification.

*Galvanized A325* bolts and A490 bolts cannot be reused. Retightening of bolts that were loosened when adjacent bolts were tightened does not constitute reuse and may be retightened.

## Kips (Minimum Tension in) *(See Tables)*

AISC, 5-274, Table 4
AISC, 5-77, Table J3.7
UBC, 2-454, Table 22-IV-D
UBC, 2-591, Table J3.1

**Minimum Bolt Tension in Kips**

| Bolt Diameters in inches | A325 | A490 |
|:---:|:---:|:---:|
| 1/2 | 12 | 15 |
| 5/8 | 19 | 24 |
| 3/4 | 28 | 35 |
| 7/8 | 39 | 49 |
| 1 | 51 | 64 |
| 1-1/8 | 56 | 80 |
| 1-1/4 | 71 | 102 |
| 1-3/8 | 85 | 121 |
| 1-1/2 | 103 | 148 |

## Length (After Installation)

AISC, 5-265, para. 2 (b)

When properly installed and after final tightening, the end of the bolt shall be flush with or outside of the face of the nut.

## Load Indicator Washers

AISC, 5-306, para. 4 - 6

In modern steel construction, load indicator washers are seldom encountered. However, when they are in use, it is imperative to monitor the testing of the washers and the installation procedure. Care should be taken to ensure that the washers are installed in such a manner that the "bumps" rest against a hardened surface that is not turned during the tightening process.

UBC, 2-446, 2221.6

*Load indicator* washers are a direct tension indicator and satisfactory installation results can only be achieved by following the manufacturer's installation procedures and utilizing proper inspection techniques.

## Markings (Nut and Bolt)

AISC, 5-293, Fig. C2

High-strength nuts and bolts shall be marked as follows:

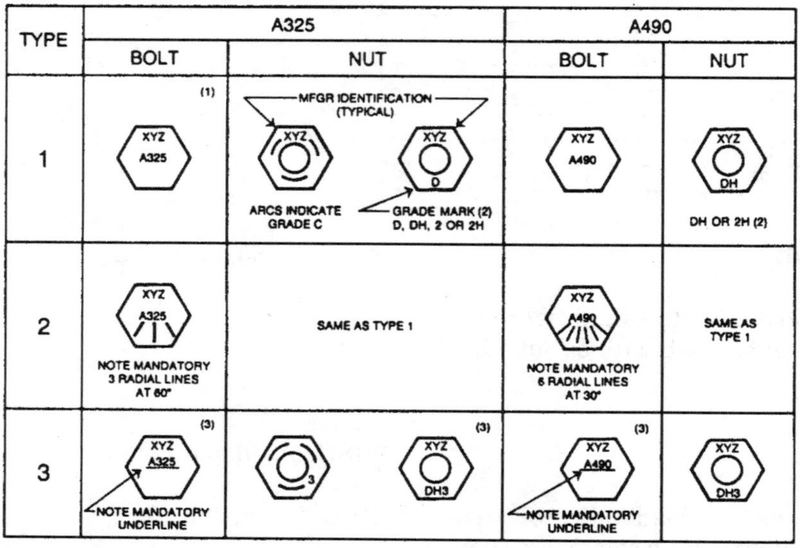

(1) ADDITIONAL OPTIONAL 3 RADIAL LINES AT 120° MAY BE ADDED.
(2) TYPE 3 ALSO ACCEPTABLE.
(3) ADDITIONAL OPTIONAL MARK INDICATING WEATHERING GRADE MAY BE ADDED.

*Fig. C2. Required marking for acceptable bolt and nut assemblies*

## Minimum Bolt Tension in kips *(See Tables)*

AISC, 5-274, Table 4
AISC, 5-77, Table J3.7
UBC, 2-454, Table 22-IV-D

### Minimum Bolt Tension in Kips

| Bolt Diameters in inches | A325 | A490 |
|---|---|---|
| 1/2 | 12 | 15 |
| 5/8 | 19 | 24 |
| 3/4 | 28 | 35 |
| 7/8 | 39 | 49 |
| 1 | 51 | 64 |
| 1-1/8 | 56 | 80 |
| 1-1/4 | 71 | 102 |
| 1-3/8 | 85 | 121 |
| 1-1/2 | 103 | 148 |

## Nominal Hole Dimensions *(See Tables)*

AISC, 5-268, Table 1

Hole sizes for different diameter bolts can be found in this specification. For convenience, these sizes are provided in the following table:

AISC, 5-71, J3.2A
AISC, Table J3.1

### Bolt Hole Dimensions in Inches

| Bolt Diameter | 1/2 in. | 5/8 in. | 3/4 in. | 7/8 in | 1 in |
|---|---|---|---|---|---|
| Standard | 9/16 | 11/16 | 13/16 | 15/16 | 1-1/16 |
| Oversize | 5/8 | 13/16 | 15/16 | 1-1/16 | 1-1/4 |
| Short Slot | 9/16 x 1 1/4 | 11/16 x 7/8 | 13/16 x 1 | 15/16 x 1-1/8 | 1-1/16 x 2-1/2 |
| Long Slot | 9/16 x 1-1/4 | 11/16 x 1-9/16 | 13/16 x 1-7/8 | 15/16 x 2-3/16 | 1-1/16 x 2-1/2 |

| Bolt Diameter | 1-1/8 | 1-1/4 | 1-3/8 | 1-1/2 |
|---|---|---|---|---|
| Standard | 1-3/16 | 1-5/16 | 1-7/16 | 1-9/16 |
| Oversize | 1-7/16 | 1-9/16 | 1-11/16 | 1-13/16 |
| Short Slot | 1-3/16 x 1-1/2 | 1-5/16 x 1-5/8 | 1-7/16 x 1-3/4 | 1-9/16 x 1-7/8 |
| Long Slot | 1-3/16 x 2-13/16 | 1-5/16 x 3-1/8 | 1-7/17 x 3-7/16 | 1-9-16 x 3-3/4 |

*Bear in* mind that base plate holes for anchor bolts may be oversized as follows (also, see Anchor Bolts in Section A):

**Bolt Size, inches (mm)**
$3/4$ (19.1)
$7/8$ (22.2)
1 < 2 (25.4 < 50.8)
>2 (> 50.8)

**Hole Size, inches (mm)**
$5/16$ (7.9) oversized
$5/16$ (7.9) oversized
$1/2$ (12.7) oversized
1 (25.4) > bolt diameter

## Nut Rotation from Snug Tight *(See Tables)*

AISC, 5-275, Table 5
UBC, 2-454, 22-IV-E

| **Nut Rotation from Snug Tight**<br>(Required amount of nut rotation when using the Turn-of-Nut Method of Tightening) | | | |
|---|---|---|---|
| **Bolt Length in Relationship to Bolt Diameter** | | | |
| Degree of slope under Bolt Head | Bolt length up to and including 4 diameters | Bolt length over 4 diameters and up to 8 diameters | Bolt length over 8 diameters and up to, but not over 12 diameters |
| Both surfaces normal | 1/3 Turn | 1/2 Turn | 2/3 Turn |
| One surface normal, one surface sloped no more than 1:20 | 1/2 Turn | 2/3 Turn | 5/6 Turn |
| Both surfaces sloped no more than 1:20 | 2/3 Turn | 5/6 Turn | 1 Turn |

The Specification allows a rotational tolerance of plus or minus 30 degrees for nuts rotated 1/2 turn and less, and plus or minus 45 degrees for nuts rotated 2/3 or more turns.

## Nuts

UBC, 2-358, 2210

UBC, . . . Nuts shall conform to the chemical and mechanical requirements of Tables 22-IV-G, 22-IV-H, and 22-IV-I. The grade and surface finish of nuts for each bolt type shall be as follows:

UBC, 2-445, 2221.3
AISC, 5-266, para. (c)

### Nuts for A325 and A490 Bolts

| A325 Bolts | Grade of Nuts | | | | | | |
|---|---|---|---|---|---|---|---|
| | A563 | | | | | A194 | |
| Type 1 Plain | C | C3 | D | DH | DH3 | 2 | 2H |
| Type 1 Galvanized | | | | DH$^g$ | | | 2H$^g$ |
| Type 2 Discontinued | | | | | | | |
| Type 3 Plain | | C3 | | | DH3 | | |

| A490 Bolts | Grade of Nuts | | | | | | |
|---|---|---|---|---|---|---|---|
| | A563 | | | | | A194 | |
| Type 1 Plain | | | | DH | DH3 | | 2H |
| Type 2 Plain | | | | DH | DH3 | | 2H |
| Type 3 Plain | | | | | DH3 | | |

and Washers

UBC, 2-448, 2226.3

> UBC, . . . Flat circular washers and square or rectangular washers shall conform to the requirements of Tables 22-IV-K, 22-IV-L, 22-IV-M, and 22-IV-N. . . .

UBC, 2-445, 2221.5
UBC, 2-358, 2210
AISC, 5-266, para. (e)

> *Design* details shall provide for washers in high-strength bolted connections as follows:

1. Where the outer face of the bolted parts has a slope greater than 1 unit vertical in 20 units horizontal (5% slope) with respect to a plane normal to the bolt axis, a hardened beveled washer shall be used to compensate for the lack of parallelism.

2. Hardened washers are not required for connections using A325 and A490 bolts, except as required in Section 2226.3, Items 3 through 7 for slip-critical connections and connections subject to direct tension or as required by Section 2227.3 for shear/bearing connections.

3. Hardened washers shall be used under the element turned in tightening when the tightening is to be performed by calibrated wrench method.

4. Irrespective of the tightening method, hardened washers shall be used under both the head and the nut when A490 bolts are to be installed and tightened to the tension specified in Table 22-IV-D in material having a specified yield point less than 40 ksi.

5. Where A325 of any diameter or A490 bolts equal to or less than 1 in. in diameter are to be installed and tightened in an oversize or short-slotted hole in an outer ply, a hardened washer conforming to Section 2221.5 shall be used.

6. When A490 bolts over 1 in. in diameter are to be installed and tightened in an oversize or short-slotted hole in an outer ply, hardened washers conforming to Section 2221.5, except washers with ⁵/₁₆ in. minimum thickness, shall be used under both the head and the nut in lieu of standard thickness washers. Multiple hardened

*(Continued)*

Nut Rotation from Snug Tight *(Continued)*

washers with a combined thickness equal to or greater than $5/16$ in. do not satisfy this requirement.

7. Where A325 bolts of any diameter or A490 bolts equal to or less than 1 in. in diameter are to be installed and tightened in a long-slotted hole in an outer ply, a plate washer or continuous bar of at least $5/16$ in. thickness with standard holes shall be provided. These washers or bars shall have a size sufficient to completely cover the slot after installation and shall be of structural grade material, but need not be hardened except as follows: when A490 bolts over 1 in. in diameter are to be used in long-slotted holes in external plies, a single hardened washer conforming to Section 2221.5, but with $5/16$ in. minimum thickness, shall be used in lieu of washers or bars of structural grade material. Multiple hardened washers with a combined thickness equal to or greater than $5/16$ in. do not satisfy this requirement.

8. Alternate design fasteners meeting the requirements of Section 2221.4 with a geometry that provides a bearing circle on the head or nut with a diameter equal to or greater than the diameter of hardened washers meeting the requirements of Section 2221.5 satisfy the requirements for washers specified in Section 2226.3, Items 4 and 5.

Washers for high-strength bolts, whether flat, beveled, circular, or rectangular, shall conform to ASTM F436.

Grades of                                      AISC, 5-266, para. (c)

**Nuts for A325 and A490 Bolts**

| A325 Bolts | Grade of Nuts | | | | | | | |
|---|---|---|---|---|---|---|---|---|
| | A563 | | | | | | A194 | |
| Type 1 Plain | C | C3 | D | DH | DH3 | | 2 | 2H |
| Type 1 Galvanized | | | | DH$^g$ | | | | 2H$^g$ |
| Type 2 Discontinued | | | | | | | | |
| Type 3 Plain | | C3 | | | DH3 | | | |

| A490 Bolts | Grade of Nuts | | | | | | |
|---|---|---|---|---|---|---|---|
| | A563 | | | | | A194 | |
| Type 1 Plain | | | | DH | DH3 | | 2H |
| Type 2 Plain | | | | DH | DH3 | | 2H |
| Type 3 Plain | | | | | DH3 | | |

## Oversized and Slotted Holes                 AISC, 5-268, Table 1

Only with the approval of the engineer does the         AISC, 5-271, para. 7 (b)
specification allow the use of oversize, short, and      AISC, 5-294, C3, para. 2
long-slotted holes.                                      AISC, 5-71, J3.2 - J3.2.e

*The use* of oversize, short, and long-slotted holes is   AISC, 5-71, Table J3.1
subject to the joint requirements found in the speci-     UBC, 2-448, 2226.2
fication.                                                 UBC, 2-594, J3.7

*This specification* allows bolt holes to be $1/16$ in.   UBC, 2-694, J3.2
greater than the diameter of the installed bolt. The      UBC, 2-698, J3.8.b
engineer shall approve all usage of oversized holes.

*Hole sizes for* different diameter bolts can be found
in this specification. For convenience, these sizes
are provided in the following table:

                                                    *(Continued)*

Oversized and Slotted Holes *(Continued)*

### Bolt Hole Dimensions in Inches

| Bolt Diameter | 1/2 in. | 5/8 in. | 3/4 in. | 7/8 in | 1 in |
|---|---|---|---|---|---|
| Standard | 9/16 | 11/16 | 13/16 | 15/16 | 1-1/16 |
| Oversize | 5/8 | 13/16 | 15/16 | 1-1/16 | 1-1/4 |
| Short Slot | 9/16 x 1 1/4 | 11/16 x 7/8 | 13/16 x 1 | 15/16 x 1-1/8 | 1-1/16 x 2-1/2 |
| Long Slot | 9/16 x 1-1/4 | 11/16 x 1-9/16 | 13/16 x 1-7/8 | 15/16 x 2-3/16 | 1-1/16 x 2-1/2 |

| Bolt Diameter | 1-1/8 | 1-1/4 | 1-3/8 | 1-1/2 |
|---|---|---|---|---|
| Standard | 1-3/16 | 1-5/16 | 1-7/16 | 1-9/16 |
| Oversize | 1-7/16 | 1-9/16 | 1-11/16 | 1-13/16 |
| Short Slot | 1-3/16 x 1-1/2 | 1-5/16 x 1-5/8 | 1-7/16 x 1-3/4 | 1-9/16 x 1-7/8 |
| Long Slot | 1-3/16 x 2-13/16 | 1-5/16 x 3-1/8 | 1-7/17 x 3-7/16 | 1-9-16 x 3-3/4 |

*Bear in* mind that base plate holes for anchor bolts may be oversized as follows (also, see Anchor Bolts in Section A):

*Standard* holes shall be provided in member-to-member connections, unless oversized, short-slotted, or long-slotted holes in bolted connections are approved by the engineer. . . .

*Oversized* holes are permitted in any or all plies of slip-critical connections, but they shall not be used in bearing-type connections. Hardened washers shall be installed over oversized holes in an outer ply. . . .

*Short*-slotted holes are permitted in any or all plies of slip-critical or bearing-type connections. . . .

*Washers* shall be installed over short-slotted holes in an outer ply; when high-strength bolts are used, such washers shall be hardened.

*Long* slotted holes are permitted in only one of the connected parts of either a slip-critical or bearing-type connection at an individual faying surface. . . .

*Where* long-slotted holes are used in an outer ply, plate washers or a continuous bar with standard holes, having a size sufficient to completely cover the slot after installation, shall be provided. In high-strength bolted connections, such plate washers or continuous bars shall not be less than $5/16$ in. thick and shall be of structural grade material, but need not be hardened. . . .

UBC, . . . When approved by the building official, oversize, short-slotted, or long-slotted holes may be used, subject to the following joint detail requirements:. . . .

*For* oversized and slotted holes, the distance required for standard holes in Subparagraph a, plus the applicable increment $C_1$ from Table J3.4, but the clear distance between holes shall not be less than one bolt diameter.

## TABLE J3.1
## Nominal Hole Dimensions

| Bolt Dia. | Hole Dimensions | | | |
|---|---|---|---|---|
| | Standard (Dia.) | Oversize (Dia.) | Short-slot (Width × length) | Long-slot (Width × length) |
| $1/2$ | $9/16$ | $5/8$ | $9/16$ × $11/16$ | $9/16$ × $1 1/4$ |
| $5/8$ | $11/16$ | $13/16$ | $11/16$ × $7/8$ | $11/16$ × $1 9/16$ |
| $3/4$ | $13/16$ | $15/16$ | $13/16$ × $1$ | $13/16$ × $1 7/8$ |
| $7/8$ | $15/16$ | $1 1/16$ | $15/16$ × $1 1/8$ | $15/16$ × $2 3/16$ |
| $1$ | $1 1/16$ | $1 1/4$ | $1 1/16$ × $1 5/16$ | $1 1/16$ × $2 1/2$ |
| $\geq 1 1/8$ | $d + 1/16$ | $d + 5/16$ | $(d + 1/16) × (d + 3/8)$ | $(d + 1/16) × (2.5 × d)$ |

## Reuse of Bolts

AISC, 5-276, para. 8 (e)

UBC, 2-449, 2227. (a)

Galvanized A325 and A490 bolts cannot be reused. Retightening of bolts that were loosened when adjacent bolts were tightened does not constitute reuse and may be retightened.

## Slope of Bolted Parts

AISC, 5-267, para. 3 (a)

*The slope* of the contact surfaces in the connection are not allowed to be greater than 1:20 without installing a hardened beveled washer to compensate for the slope.

UBC, 2-446, 2222.1

UBC, . . . All material within the grip of the bolt shall be steel. There shall be no compressible material such as gaskets or insulation within the grip. Bolted steel parts shall fit solidly together after the bolts are tightened and may be coated or uncoated. The slope of surfaces of parts in contact with the bolt head or nut shall not exceed 1:20 with respect to a plane normal to the bolt axis.

## Spacing (Minimum)

UBC, 2-698, J3.8

UBC, . . . The distance between centers of standard, oversized, or slotted fastener holes shall not be less than $2/3$ times the nominal diameter of the fastener or less than that required by the following paragraph, if applicable. . . .

The clear distance between holes shall not be less than one bolt diameter.

## Surface Conditions of Bolted Parts

AISC, 5-267, para. 3 (b)

No foreign material is permitted in the faying surface of a connection. Nothing but the steel of the connected member is permitted.

*The slope* of the contact surfaces in the connection are not allowed to be greater than 1:20 without installing a hardened beveled washer to compensate for the slope.

*While paint* is permitted on the faying surfaces of bearing-type connections, coatings are only allowed on the faying surfaces of slip-critical connections with the approval of the engineer.

## Slip-Critical Joints

AISC, 5-270, para. 5 (a)

The RCSC Specification defines slip-critical joints as joints in which any slippage would be detrimental to the structure. This includes joints:

AISC, 5-299, C5
AISC, 5-304, para. 3 & 4

1. That are subjected to fatigue loading.

2. That have bolts installed in non-standard holes.

3. That are subjected to load reversals.

4. Where both the welds and the bolts together share the load at the faying surface.

5. That the engineer has designated as such, and joints with bolts installed in slotted holes and where the force on the joint is not normal to the slot axis, unless otherwise indicated by the engineer.

*Slip-critical joints* are joints where any slippage would be undesirable. Friction type implies that the high contact pressure, or friction, between the faces of the faying surfaces, caused by the high-strength bolts, enables the joint to resist slippage.

*Bolts installed* by the turn-of-nut method of tightening provide greater slip resistance than do bolts installed by other methods. This is due to the increased clamping force.

*With* any of the four described tensioning methods, it is important to install bolts in all holes of the connection and bring them to an intermediate level of tension generally corresponding to snug tight in order to compact the joint. Even after being fully tightened, some thick parts with uneven surfaces may not be in contact over the entire faying surface. This is not detrimental to the performance of the joint. As long as the specified bolt tension is present on all bolts of the completed connection, the clamping force equal to the total of the tension in all bolts will be transferred at the locations that are in contact and be fully effective in resisting slip through friction. . . .

*(Continued)*

Slip Critical Joints *(Continued)*

*With* all methods, tightening should begin at the most rigidly fixed or stiffest point and progress toward the free edges, both in the initial snugging up and in the final tightening.

## Slip Resistance *(See Slip Critical Joints)*

AISC, 5-299, para. 1

## Snug Tight

AISC, 5-273, para. 8 (c)

In certain cases, and with engineering approval, the specification permits so designated joints to be tightened only to a snug tight condition.

AISC, 5-303, para. 2

*Snug tight* can be defined as the condition that exists when the faying surfaces in the connection are in firm contact. This condition can be accomplished by making a few hits with an impact wrench or by a person using full effort with a spud wrench.

*Care should* be observed when tightening galvanized bolts. Effective lubricants have, in the past, made it possible for a person using full effort with a spud wrench, to achieve full tension.

*It is not* practical to assume that all areas within the faying surface in all connections will be in firm contact when the bolts are at a snug tight condition.

## Tightening

AISC, 5-272, para. 8 (a) - (e)

UBC, 2-449, 2227

The RCSC Specification requires that fastener components be stored in suitable containers and protected from the elements. Plastic coverings are not recommended as they most likely will produce condensation that in turn will be deposited onto the fasteners. If the lubricant on the bolts is water soluble, and many bolt manufacturers use water soluble oil, it will be washed off. Also, being water soluble, this particular lubricant will eventually evaporate and invite corrosion. For these and other reasons it is of great importance to institute a sound, clean and dry storage program.

*If bolts* are to be tightened to a snug tight condition only, the engineer must clearly state so on the drawings.

*Snug tight* can be defined as the condition that exists when the faying surfaces in the connection are in firm contact. This condition can be accomplished by making a few hits with an impact wrench or by a person using full effort with a spud wrench.

*Care should* be observed when tightening galvanized bolts. Effective lubricants have, in the past, made it possible for a person using full effort with a spud wrench, to achieve full tension.

*Tightening* may be done by turning the bolt while the nut is prevented from rotating when it is impractical to turn the nut. If impact wrenches are used to tighten the bolts in the connection, the specification dictates they be capable of tightening the bolts in approximately 10 seconds.

*When using* the turn-of-nut method, three bolts of each diameter, length, grade, and lot number, and three nuts of each diameter, grade, and lot number, along with three washers of the grade and size to be used, shall, at the beginning of each shift, be installed in a bolt tensioning device, such as a Skidmore Wilhelm, and brought to a degree of tension no less than 5% of the required minimum bolt tension in kips, found in Table J3.7.

*When tightening* is to be performed using the calibrated wrench method, hardened washers conforming to F436 must be installed under the component being turned and installation procedures must be calibrated daily or at the start of each shift.

*When torque* wrenches are to be used, they are to be set at least 5% above the minimum tension required in **Table J3.7.**

*It shall* be verified during the tightening process that the calibrated torque wrench does not produce nut rotation from the snug tight condition greater than allowed in **Table 22-IV-E.**

*Galvanized A325* and A490 bolts cannot be reused. Retightening of bolts that were loosened when adjacent bolts were tightened does not constitute reuse and may be retightened.

# Turn-of-Nut Method

Bolts installed by the turn-of-nut method of tightening provide greater slip resistance than do bolts installed by other methods. This is due to the increased clamping force.

*The high contact* pressure achieved by the turn-of-nut method is created by the elongation of the bolt slightly into the inelastic range.

*Precautions should* be taken when tensioning galvanized bolts using the turn-of-nut method. It has been found that some lubricants encountered on galvanized bolts are so effective that the full effort of a person using an ordinary spud wrench produced the required full tension.

AISC, 5-273, para. 8 (d)

AISC, 5-275, Table 5
AISC, 5-301, para. 1
AISC, 5-304, para. 5 - 8

UBC, 2-449, 2227. (c)
UBC, 2-454, 22-IV-E

### Average Torques for A325 and A490 Bolts
(Foot-Pounds)

#### A325 Bolts

| Nominal Bolt / Nut Diameter | Average Torque | Width of Head Across Flats |
|---|---|---|
| 5/8 | 200 | 1-1/16 |
| 3/4 | 355 | 1-1/4 |
| 7/8 | 570 | 1-7/16 |
| 1 | 850 | 1-5/8 |
| 1-1/8 | 1,060 | 1-13/16 |
| 1-1/4 | 1,495 | 2 |
| 1-3/8 | 1,960 | 2-3/16 |
| 1-1/2 | 2,600 | 2-3/8 |

#### A490 Bolts

| Nominal Bolt/Nut Diameter | Average Torque | Width of Head Across Flats |
|---|---|---|
| 5/8 | 250 | 1-1/16 |
| 3/4 | 435 | 1-1/4 |
| 7/8 | 715 | 1-7/16 |
| 1 | 1,070 | 1-5/8 |
| 1-1/8 | 1,500 | 1-13/16 |
| 1-1/4 | 2,125 | 2 |
| 1-3/8 | 2,780 | 2-3/16 |
| 1-1/2 | 3,700 | 2-3/8 |

### Nut and Bolt Head Dimensions
### A325 and A490

| Bolt Diameter | Width across the Flats (Nuts and Bolts) | Nut Height |
|---|---|---|
| 1/2 | 7/8 | 31/64 |
| 5/8 | 1-1/16 | 39/64 |
| 3/4 | 1-1/4 | 47/64 |
| 7/8 | 1-7/16 | 55/64 |
| 1 | 1-5/8 | 63/64 |
| 1-1/8 | 1-13/16 | 1-7/64 |
| 1-1/4 | 2 | 1-7/32 |
| 1-3/8 | 2-3/16 | 1-11/32 |
| 1-1-2 | 2-3/8 | 1-15/32 |

This information is useful in determining proper wrench sizes.

| Nut Rotation from Snug Tight (Required amount of nut rotation when using the Turn-of-Nut Method of Tightening) | | | |
|---|---|---|---|
| Bolt Length in Relationship to Bolt Diameter | | | |
| Degree of slope under Bolt Head | Bolt length up to and including 4 diameters | Bolt length over 4 diameters and up to 8 diameters | Bolt length over 8 diameters and up to, but not over 12 diameters |
| Both surfaces normal | 1/3 Turn | 1/2 Turn | 2/3 Turn |
| One surface normal, one surface sloped no more than 1:20 | 1/2 Turn | 2/3 Turn | 5/6 Turn |
| Both surfaces sloped no more than 1:20 | 2/3 Turn | 5/6 Turn | 1 Turn |

The Specification allows a rotational tolerance of plus or minus 30 degrees for nuts rotated 1/2 turn and less, and plus or minus 45 degrees for nuts rotated 2/3 or more turns.

## Thread Run-Out

AISC, 5-288, Bolt Spec's.

Some thread run-out is permitted by this specification in the shear plane. However, extreme care should be taken to ensure that enough thread is present to prevent the nut from jamming itself on the thread run-out prior to achieving full tension. Jammed nuts can give the appearance of being fully tensioned, when in fact, they are not.

## Types of (1, 2, and 3) *(For F, N, X, See Application)*

AISC, 5-290, para. 3

The three types of A325 and A490 bolts are:

AISC, 4-9

Type 1    Medium carbon steel A325 bolts/Alloy steel A490 bolts

Type 2    Low carbon martinsitic steel for A325 and A490 bolts

Type 3    Weather resistant A325 and A490 bolts

## T. S. Bolts (F9T and F11T)

AISC, 5-306, para. 1 - 3

Alternate design fasteners such as the F9T (A325) and F11T (A490) T.S. bolts are subject to the same installation quality controls as other hardened bolts. These bolts are to be manufactured and shipped as a nut-bolt assembly.

T. S. Bolts (F9T and F11T) *(Continued)*

*As with* other fasteners, these too may provide un-
reliable results when installed improperly. The
manufacturers recommendations for installation,
along with the requirements of the specification,
must be followed to allow for satisfactory results.

*The sheared off* spline does not indicate in itself
that the bolt has been properly tensioned, it simply
indicates that at some point, someone sheared it off.
Proper inspection and monitoring of the calibration
process, bolt installation, and tensioning is the only
way to ensure proper use of these and other high-
strength bolts.

## Washer Requirements

Washers for high-strength bolts, whether flat,
beveled, circular, or rectangular, shall conform to
ASTM F436.

*The specification* provides for various situations
where hardened washers are required in connections using high-strength bolts.

1. When the slope of the face of the connected part
   exceeds 1:20, with regard to the nut and/or bolt
   face, a hardened beveled washer shall be installed.

2. When tightening is performed by the calibrated
   wrench method, hardened washers conforming
   to F436 shall be installed under the element
   turned in the tightening process.

3. When A490 bolts are being installed in material
   having less than a 40 ksi yield point (that is, A36
   steel), hardened washers must be installed under the bolt head and the nut, regardless of the
   method of tightening used.

4. If oversized or short-slotted holes are used and
   the bolts to be installed are A325 of any diameter, or A490 bolts of 1 in. or less in diameter, then
   F436 washers must be installed.

5. If oversized or short-slotted holes are used in an
   outer ply, and A490 bolts greater than 1 in. in diameter are to be installed, a $5/16$ in. minimum
   thickness F436 washer must be used under both
   the bolt head and the nut.

AISC, 5-266, para. (e)

AISC, 5-272, para. 7 (c)
AISC, 5-290, para. 3, para. 2

UBC, 2-358, 2210
UBC, 2-449, 2227. (b)

6. Where A490 bolts less than 1 in. in diameter or any diameter of A325 bolts are to be installed in long-slotted holes in an external ply, a $^5/_{16}$ in. minimum thickness bar or plate washer with standard size holes must be used.

7. When A490 bolts greater than 1 in. in diameter are to be installed in external plies of long-slotted holes, a $^5/_{16}$ in. minimum thickness single hardened F436 washer is required to be installed.

8. If round headed F9T or F11T bolts are being used and the head diameter of the bolt is at least equal to the diameter of an F436 hardened washer, the specification does not require the use of a washer under the bolt head.

**TABLE 22-IV-M—HARDENED BEVELED WASHERS[1]**

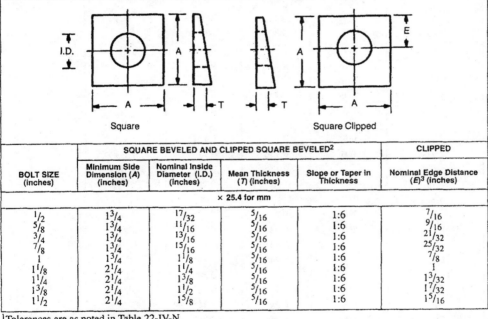

Square                                   Square Clipped

| BOLT SIZE (inches) | SQUARE BEVELED AND CLIPPED SQUARE BEVELED[2] | | | | CLIPPED |
|---|---|---|---|---|---|
| | Minimum Side Dimension (*A*) (inches) | Nominal Inside Diameter (I.D.) (inches) | Mean Thickness (*T*) (inches) | Slope or Taper in Thickness | Nominal Edge Distance (*E*)[3] (inches) |
| | | | × 25.4 for mm | | |
| $^1/_2$ | $1^3/_4$ | $^{17}/_{32}$ | $^5/_{16}$ | 1:6 | $^7/_{16}$ |
| $^5/_8$ | $1^3/_4$ | $^{11}/_{16}$ | $^5/_{16}$ | 1:6 | $^9/_{16}$ |
| $^3/_4$ | $1^3/_4$ | $^{13}/_{16}$ | $^5/_{16}$ | 1:6 | $^{21}/_{32}$ |
| $^7/_8$ | $1^3/_4$ | $^{15}/_{16}$ | $^5/_{16}$ | 1:6 | $^{25}/_{32}$ |
| 1 | $1^3/_4$ | $1^1/_8$ | $^5/_{16}$ | 1:6 | $^7/_8$ |
| $1^1/_8$ | $2^1/_4$ | $1^1/_4$ | $^5/_{16}$ | 1:6 | 1 |
| $1^1/_4$ | $2^1/_4$ | $1^3/_8$ | $^5/_{16}$ | 1:6 | $1^3/_{32}$ |
| $1^3/_8$ | $2^1/_4$ | $1^1/_2$ | $^5/_{16}$ | 1:6 | $1^7/_{32}$ |
| $1^1/_2$ | $2^1/_4$ | $1^5/_8$ | $^5/_{16}$ | 1:6 | $1^5/_{16}$ |

[1]Tolerances are as noted in Table 22-IV-N.
[2]Rectangular beveled washers shall conform to the dimensions shown above, except that one side may be longer than that shown for the *A* dimension.
[3]Clipped edge *E* shall not be closer than $^7/_8$ of the bolt diameter from the center of the washer.

*(Continued)*

Washer Requirements *(Continued)*

### TABLE 22-IV-L—HARDENED CIRCULAR AND CLIPPED CIRCULAR WASHERS[1]

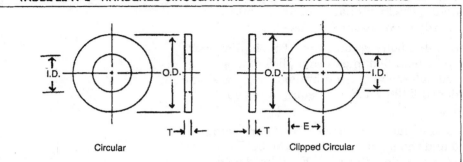

Circular                                Clipped Circular

| BOLT SIZE (inches) | CIRCULAR AND CLIPPED CIRCULAR | | | | CLIPPED |
| | Nominal Outside Diameter (O.D.) (inches) | Normal Inside Diameter (I.D.) (inches) | Thickness (T) (inches) | | Minimum Edge Distance (E)[2] (inches) |
| | | | min. | max. | |
| | | × 25.4 for mm | | | |
| $1/4$ | $5/8$ | $9/32$ | 0.051 | 0.080 | $7/32$ |
| $5/16$ | $11/16$ | $11/32$ | 0.051 | 0.080 | $9/32$ |
| $3/8$ | $13/16$ | $13/32$ | 0.051 | 0.080 | $11/32$ |
| $7/16$ | $59/64$ | $15/32$ | 0.051 | 0.080 | $13/32$ |
| $1/2$ | $1\,1/16$ | $17/32$ | 0.097 | 0.177 | $7/16$ |
| $5/8$ | $1\,5/16$ | $11/16$ | 0.122 | 0.177 | $9/16$ |
| $3/4$ | $1\,15/32$ | $13/16$ | 0.122 | 0.177 | $21/32$ |
| $7/8$ | $1\,3/4$ | $15/16$ | 0.136 | 0.177 | $25/32$ |
| 1 | 2 | $1\,1/8$ | 0.136 | 0.177 | $7/8$ |
| $1\,1/8$ | $2\,1/4$ | $1\,1/4$ | 0.136 | 0.177 | 1 |
| $1\,1/4$ | $2\,1/2$ | $1\,3/8$ | 0.136 | 0.177 | $1\,3/32$ |
| $1\,3/8$ | $2\,3/4$ | $1\,1/2$ | 0.136 | 0.177 | $1\,7/32$ |
| $1\,1/2$ | 3 | $1\,5/8$ | 0.136 | 0.177 | $1\,5/16$ |
| $1\,3/4$ | $3\,3/8$ | $1\,7/8$ | 0.178[3] | 0.28[3] | $1\,17/32$ |
| 2 | $3\,3/4$ | $2\,1/8$ | 0.178[3] | 0.28[3] | $1\,3/4$ |
| $2\,1/4$ | 4 | $2\,3/8$ | 0.24[4] | 0.34[4] | 2 |
| $2\,1/2$ | $4\,1/2$ | $2\,5/8$ | 0.24[4] | 0.34[4] | $2\,3/16$ |
| $2\,3/4$ | 5 | $2\,7/8$ | 0.24[4] | 0.34[4] | $2\,13/32$ |
| 3 | $5\,1/2$ | $3\,1/8$ | 0.24[4] | 0.34[4] | $2\,5/8$ |
| $3\,1/4$ | 6 | $3\,3/8$ | 0.24[4] | 0.34[4] | $2\,7/8$ |
| $3\,1/2$ | $6\,1/2$ | $3\,5/8$ | 0.24[4] | 0.34[4] | $3\,1/16$ |
| $3\,3/4$ | 7 | $3\,7/8$ | 0.24[4] | 0.34[4] | $3\,5/16$ |
| 4 | $7\,1/2$ | $4\,1/8$ | 0.24[4] | 0.34[4] | $3\,1/2$ |

[1] Tolerances are as noted in Table 22-IV-N.
[2] Clipped edge, $E$, shall not be closer than $7/8$ of the bolt diameter from the center of the washer.
[3] $3/16$ inch (4.8 mm) nominal.
[4] $1/4$ inch (6.4 mm) nominal.

# E   REINFORCING STEEL

## Allowable Gaps

D1.4, . . . For indirect butt joints with splice plates, the maximum joint clearance between the bars shall be not more than $^3/_4$ in. (See Figure 3.3(A).)

*For indirect* lap joints (see Figure #.4 (B)), the maximum separation between the bar and the splice plate shall be no more than $^1/_4$ of the bar diameter, but not more than $^3/_{16}$ in.

*For direct lap* joints, if the bar deviates by more than $^1/_2$ of the bar diameter, or by no more than $^1/_4$ in. from each other while the bars remain in approximately the same plane, the joint shall be made through a splice bar or plate, and the requirements for an indirect lap joint shall apply (see 3.7.2 and 4.2.6).

## Anchorages, Base Plates, and Inserts          D1.4, Fig. 3.5

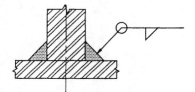

**A - EXTERNAL FILLET WELD**

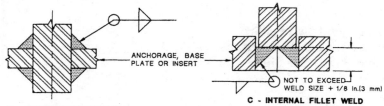

ANCHORAGE, BASE
PLATE OR INSERT

NOT TO EXCEED
WELD SIZE + 1/8 In.(3 mm)

**B - EXTERNAL FILLET WELD**

**C - INTERNAL FILLET WELD**

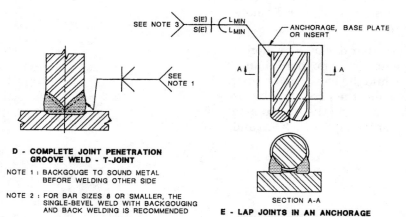

SEE NOTE 3

$S(E)$ | $L_{MIN}$
$S(E)$ | $L_{MIN}$

ANCHORAGE, BASE PLATE
OR INSERT

SEE
NOTE 1

**D - COMPLETE JOINT PENETRATION
GROOVE WELD - T-JOINT**

NOTE 1 : BACKGOUGE TO SOUND METAL
BEFORE WELDING OTHER SIDE

NOTE 2 : FOR BAR SIZES 8 OR SMALLER, THE
SINGLE-BEVEL WELD WITH BACKGOUGING
AND BACK WELDING IS RECOMMENDED

NOTE 3 : $L_{MIN} = 2$ X BAR DIAMETER

SECTION A-A

**E - LAP JOINTS IN AN ANCHORAGE
USING FLARE-BEVEL-GROOVE WELDS**

## Arc Strikes          D1.4, 5.3

D1.4, . . . Arc strikes outside the area of permanent welds shall be avoided, especially on reinforcing bars. Cracks or blemishes resulting from arc
strikes shall be ground to a smooth contour and
checked to ensure soundness.

## Bars in Tension

UBC, 2-198, 1912.14.3.3

UBC, . . . A full-welded splice shall have the bars butted and welded to develop in tension at least 125% of specified yield strength of the bar.

## Bar Size

For Direct Butt Splice

D1.4, 3.3

D1.4, . . . With the exception of 3.2.2, reinforcing bars may be welded with direct or indirect butt joints, lap joints, or T-joints. . . .

*However,* direct butt joints are preferable for bars greater than No. 6.

For Lap Splice

D1.4, 3.2.2

D1, . . . Welded lap joints shall be limited to bar sizes No. 6 and smaller.

Transition in

D1.4, 3.1

D1.4, . . . Direct butt joints in tension in axially aligned bars of different size shall be made as shown in Figure 3.1. . . .

D1.4, Fig. 3.1

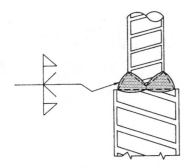

Figure 3.1 — Direct Butt Joint  Showing Transition Between Bars of Different Sizes (See 3.1)

## Carbon Equivalent

D1.4, 1.3.2 - 1.3.4.3

D1.4, . . . The carbon equivalent of reinforcing bars shall be calculated as shown in 1.3.4.1 or 1.3.4.2, as applicable. . . .

D1.4, Table 5.2

*(Continued)*

Carbon Equivalent *(Continued)*

*For* all steel bars, except those designated as ASTM A706, the carbon equivalent shall be calculated suing the chemical composition, as shown in the mill test report, by the following method:
C.E. = %C + %Mn/6. . . .

*For* steel bars designated ASTM A706, the carbon equivalent shall be calculated using the chemical composition, as shown in the mill test report, by the following formula: C.E. = %C + %Mn/6 + %Cu/40 + %Ni/20 + %CR/10 − %Mo/50 − %V/10. The carbon equivalent shall not exceed 0.55%. . . .

*If* mill test reports are not available, chemical analysis may be made on bars representative of the bars to be welded. If the chemical composition is not known or obtained:

1. For bars No. 6 or less, use a minimum preheat of 300 deg. F.

2. For bars No. 7 or larger, use a minimum preheat of 400 deg. F.

3. For all ASTM A706 bar sizes, use Table 5.2 C.E. values of "over 0.45% to 0.55% inclusive".

### Table 5.2
### Minimum Preheat and Interpass Temperature[1,2]
### (see 5.2.1)

| Carbon Equivalent[3,4] (C.E.) Range, % | Size of Reinforcing Bar | Shielded Metal Arc Welding with Low Hydrogen Electrodes, Gas Metal Arc Welding, or Flux Cored Arc Welding Minimum Temperature | |
|---|---|---|---|
| | | °F | °C |
| Up to 0.40 | up to 11 inclusive | none[5] | none[5] |
| | 14 and 18 | 50 | 10 |
| Over 0.40 to 0.45 inclusive | up to 11 inclusive | none[5] | none[5] |
| | 14 and 18 | 100 | 40 |
| Over 0.45 to 0.55 inclusive | up to 6 inclusive | none[5] | none[5] |
| | 7 to 11 inclusive | 50 | 10 |
| | 14 to 18 | 200 | 90 |
| Over 0.55 to 0.65 inclusive | up to 6 inclusive | 100 | 40 |
| | 7 to 11 inclusive | 200 | 90 |
| | 14 to 18 | 300 | 150 |
| Over 0.65 to 0.75 | up to 6 inclusive | 300 | 150 |
| | 7 to 18 inclusive | 400 | 200 |
| Over 0.75 | 7 to 18 inclusive | 500 | 260 |

1. When reinforcing steel is to be welded to main structural steel, the preheat requirements of the structural steel shall also be considered (see ANSI/AWS D1.1, table titled "Minimum Preheat and Interpass Temperature." The minimum preheat requirement to apply in this situation shall be the higher requirement of the two tables. However, extreme caution shall be exercised in the case of welding reinforcing steel to quenched and tempered steels, and such measures shall be taken as to satisfy the preheat requirements for both. If not possible, welding shall not be used to join the two base metals.

2. Welding shall not be done when the ambient temperature is lower than 0°F (-18°C). When the base metal is below the temperature listed for the welding process being used and the size and carbon equivalent range of the bar being welded, it shall be preheated (except as otherwise provided) in such a manner that the cross section of the bar for not less than 6 in. (150 mm) on each side of the joint shall be at or above the specified minimum temperature. Preheat and interpass temperatures shall be sufficient to prevent crack formation.

3. After welding is complete, bars shall be allowed to cool naturally to ambient temperature. Accelerated cooling is prohibited.

4. Where it is impractical to obtain chemical analysis, the carbon equivalent shall be assumed to be above 0.75%. See also 1.3.4.

5. When the base metal is below 32°F (0°C), the base metal shall be preheated to at least 70°F (20°C), or above, and maintained at this minimum temperature during welding.

## Contractor Obligations

D1.4, 7.6 - 7.6.5

D1.4, . . . The contractor shall be responsible for visual inspection and necessary correction of all deficiencies in materials and workmanship in accordance with the requirements of Sections 3 and 4, and 4.4 or other parts of the code, as applicable. . . .

*The* contractor shall comply with all requests of the inspector(s) to correct deficiencies in materials and workmanship as provided in the contract documents. . . .

*In* the event that faulty welding or its removal for rewelding damages the base metal so that in the judgment of the engineer its retention is not in accordance with the intent of the contract documents, the contractor shall remove and replace the damaged base metal or shall compensate for the deficiency in a manner approved by the engineer.

## Cracks

D1.4, 4.4.2

D1.4, . . . Welds shall have no cracks in either the weld metal or heat-affected zone.

## Craters

D1.4, 4.4.4

D1.4, . . . All craters shall be filled to the full cross section of the weld.

## Crossing Bars

D1.4, 4.2.7

D1.4, . . . Welding of bars that cross shall not be permitted unless authorized by the engineer.

UBC, 2-148,1907.5.4
SBCCI, 1606.4.4

# Direct Butt Splice

D1.4, Fig. 3.2

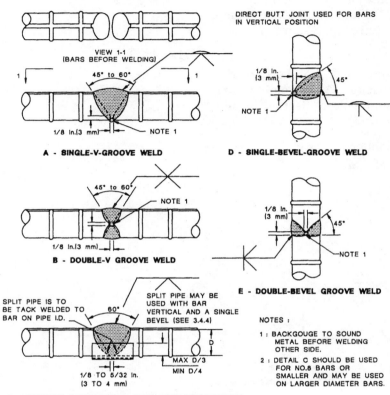

DIRECT BUTT JOINT USED FOR BARS
WHOSE AXES ARE HORIZONTAL

VIEW 1-1
(BARS BEFORE WELDING)

45° to 60°

1/8 In.(3 mm)   NOTE 1

A - SINGLE-V-GROOVE WELD

45° to 60°   NOTE 1

1/8 In.(3 mm)

B - DOUBLE-V GROOVE WELD

SPLIT PIPE IS TO
BE TACK WELDED TO
BAR ON PIPE I.D.

60°

SPLIT PIPE MAY BE
USED WITH BAR
VERTICAL AND A SINGLE
BEVEL (SEE 3.4.4)

D

MAX D/3
MIN D/4

1/8 TO 5/32 In.
(3 TO 4 mm)

C - SINGLE-V GROOVE WELD WITH SPLIT PIPE BACKING
(NOTE 2)

DIRECT BUTT JOINT USED FOR BARS
IN VERTICAL POSITION

1/8 In.
(3 mm)   45°

NOTE 1

D - SINGLE-BEVEL-GROOVE WELD

1/8 In.
(3 mm)   45°

NOTE 1

E - DOUBLE-BEVEL GROOVE WELD

NOTES :

1 : BACKGOUGE TO SOUND
METAL BEFORE WELDING
OTHER SIDE.

2 : DETAIL C SHOULD BE USED
FOR NO.8 BARS OR
SMALLER AND MAY BE USED
ON LARGER DIAMETER BARS.

Area of

D1.4, 2.3.1

D1.4, . . . The effective weld area shall be the
nominal cross sectional area of the bar being
welded (See Figure 3.2). If different size bars are
being welded, the weld area shall be bases on the
smaller bar.

Bar Size for

D1.4, 3.3

D1.4, . . . With the exception of 3.2.2, reinforcing bars may be welded with direct or indirect
butt joints, lap joints, or T-joints (Figures 3.2, 3.3,
3.4, and 3.5); however, direct butt joints are
preferable for bars greater than No. 6.

*(Continued)*

Direct Butt Splice *(Continued)*

Details of                                                           D1.4, 3.4 - 3.4.4

> D1.4, . . . A direct butt joint shall be single            D1.4, Fig. 3.2
> welded or double welded and shall have complete
> joint penetration and complete fusion. . . .

Offsets in                                                          D1.4, 4.2.3

> D1.4, . . . The joint members shall be aligned so
> as to minimize eccentricities. After welding, bars
> in direct butt joints shall not be offset at the joint
> by more than the following:

Bar sizes No. 10 or smaller          $1/8$ in.

Bar sizes No. 11 and No. 14          $3/16$ in.

Bar sizes No. 18                     $1/4$ in.

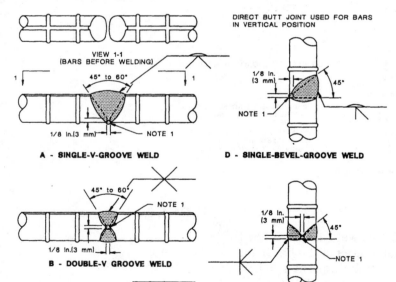

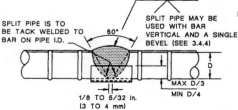

## Effective Weld Areas

D1.4, 2.3 - 2.3.3.3

### Direct Butt Splice

D1.4, 2.3.1

D1.4, . . . The effective weld area shall be the nominal cross sectional area of the bar being welded (see Figure 3.2). If different bar sizes are being welded, the weld area shall be bases on the smaller bar.

### Fillet Welds

D1.4, 2.3.3

D1.4, . . . The effective weld area shall be the effective weld length multiplied by the effective throat. . . .

### Flare-Bevel and Flare-V-Groove Welds

D1.4, 2.3.2

The effective weld area shall be the effective weld length multiplied by the effective weld size (see Figure 2.1).

## Electrodes for

| | |
|---|---|
| S.M.A.W. | D1.4, 5.7 - 5.7.2 |
| F.C.A.W. | D1.4, 5.8 - 5.8.1.2 |
| G.M.A.W. | D1.4, 5.8 - 5.8.1.2. |

## Filler Metal Requirements

D1.4, 5.1 - 5.1.2<br>D1.4, Table 5.1

## Table 5.1
## Matching Filler Metals Requirements
### (See 5.1)

| Group | Steel Specification | | Min Yield Point/Strength ksi | Min Yield Point/Strength MPa | Min Tensile Strength ksi | Min Tensile Strength MPa | Electrode Specification[4] | Yield Point/Strength[1] ksi | Yield Point/Strength[1] MPa | Tensile Strength[1] ksi | Tensile Strength[1] MPa |
|---|---|---|---|---|---|---|---|---|---|---|---|
| I | ASTM A615 | Grade 40 | 40 | — | 70 | — | SMAW AWS A5.1 and A5.5 E7015, E7016, E7018, E7028, E7015-X, E7016-X, E7018-X | 60 | 415 | 72 | 495 |
| | ASTM A615M | Grade 300 | — | 300 | — | 500 | | 57 | 390 | 70 | 480 |
| | ASTM A617 | Grade 40 | 40 | — | 70 | — | GMAW AWS A5.18 ER70S-X | 60 | 415 | 72 | 495 |
| | ASTM A617M | Grade 300 | — | 300 | — | 500 | FCAW AWS A5.20 E7XT-X (Except -2, -3, -10, -GS) | 60 | 415 | 72 | 495 |
| II | ASTM A616 | Grade 50 | 50 | — | 80 | — | SMAW AWS A5.5 E8015-X, E8016-X, E8018-X | 67 | 460 | 80 | 550 |
| | ASTM A616M | Grade 350 | — | 350 | — | 550 | | | | | |
| | ASTM A706 | Grade 60 | 60 | — | 80 | — | GMAW AWS A5.28 ER80S-X | 68 | 470 | 80 | 550 |
| | ASTM A706M | Grade 400 | — | 400 | — | 550 | FCAW AWS A5.29 E8XTX-X | 68 | 470 | 80-100 | 550-690 |
| III | ASTM A615 | Grade 60 | 60 | — | 90 | — | SMAW AWS A5.5 E9015-X, E9016-X, E9018-X | 77 | 530 | 90 | 620 |
| | ASTM A615M | Grade 400 | — | 400 | — | 600 | | | | | |
| | ASTM A616 | Grade 60 | 60 | — | 90 | — | GMAW AWS A5.28 ER90S-X | 78 | 540 | 90 | 620 |
| | ASTM A616M | Grade 400 | — | 400 | — | 600 | | | | | |
| | ASTM A617 | Grade 60 | 60 | — | 90 | — | FCAW AWS A5.29 E9XTX-X | 78 | 540 | 90-110 | 620-760 |
| | ASTM A617M | Grade 400 | — | 400 | — | 600 | | | | | |

(Continued)

**Table 5.1**
**(Continued)**

| | Steel Specification Requirements | | | | Filler Metal Requirements | | | | |
|---|---|---|---|---|---|---|---|---|---|
| Group | Steel Specification | Minimum Yield Point/Strength ksi | MPa | Minimum Tensile Strength ksi | MPa | Electrode Specification[4] | Yield Point/Strength[1] ksi | MPa | Tensile Strength[1] ksi | MPa |
| IV | ASTM A615 Grade 75[2] | 75 | — | 100 | — | SMAW AWS A5.5 E10015-X, E10016-X, E10018-X | 87 | 600 | 100 | 690 |
| | ASTM A615M Grade 500[3] | — | 500 | — | 700 | E10018-M | 88-100 | 610-690 | 100 | 690 |
| | | | | | | GMAW AWS A5.28 ER100S-X | 88-102 | 610-700 | 100 | 690 |
| | | | | | | FCAW AWS A5.29 E10XTX-X | 88 | 610 | 100-120 | 690-830 |

Notes:

1. This table is based on filler metal as-welded properties. Single values are minimums. Hyphenated values indicated minimum and maximum.

2. Applicable to bar sizes Nos. 11, 14, and 18.

3. Applicable to bar sizes Nos. 35, 45, and 55.

4. Filler metals classified in the postweld heat treated (PWHT) condition by the AWS filler metal specification may be used when given prior approval by the Engineer. Consideration shall be made of the differences in tensile strength, ductility and hardness between the PWHT versus as-welded condition.

Filler Metal Requirements *(Continued)*

Atmospheric Exposure Time Limits                           D1.4, Table 5.3

### Table 5.3
### Permissible Atmospheric Exposure of Low-Hydrogen Electrodes
### (See 5.7.2.1)

| Electrode | Column A (hours) | Column B (hours) |
|---|---|---|
| A5.1 | | |
| E70XX | 4 max | over 4 to 10 max |
| A5.5 | | |
| E70XX-X | 4 max | over 4 to 10 max |
| E80XX-X | 2 max | over 2 to 10 max |
| E90XX-X | 1 max | over 1 to 5 max |
| E100XX-X | 1/2 max | over 1/2 to 4 max |
| E110XX-X | 1/2 max | over 1/2 to 4 max |

Notes:

1. Column A: Electrodes exposed to the atmosphere for longer periods than shown shall be redried before use.

2. Column B: Electrode exposure to the atmosphere for longer periods than those established by test shall be redried before use.

3. Entire Table: Electrodes shall be issued and held in quivers, or other small containers which may be open. Heated containers are not mandatory.

## Fillet Welds                                          D1.4, 2.3.3 - 2.3.3.3

### Effective Throat                                      D1.4, 2.3.3.3

D1.4, . . . The effective throat shall be the minimum distance minus any convexity between the weld root and face of the fillet weld.

### Effective Weld Area                                   D1.4, 2.3.3

D1.4, . . . The effective weld area shall be the effective weld length multiplied by the effective throat. . . .

### Effective Weld Length                                 D1.4, 2.3.3.1

The effective weld length shall be the overall length of the full size fillet. No reduction in effective weld length shall be made for either the start or finish, if the weld is the specified size at those locations.

Curved Fillet                                                           D1.4, 2.3.3.2

D1.4, . . . The effective weld length of a curved
fillet shall be measured along the weld axis.

Minimum Effective Length                                               D1.4, 2.3.3.1

D1.4, . . . The effective weld length shall be the
overall length of the full size fillet. No reduction
in effective weld length shall be made for either
the start or finish, if the weld is the specified size
at those locations.

## Flare-Bevel and Flare-V Groove Welds

D1.4, 2.3.2 - 2.3.2.3
D1.4, Fig. 2.1

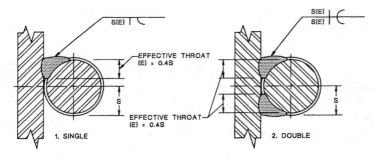

**A - FLARE-BEVEL-GROOVE WELD**

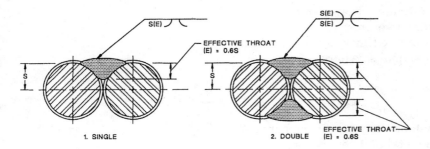

**B - FLARE-V-GROOVE WELD**

NOTE:   1.   RADIUS OF REINFORCING BAR = S

2.   THESE ARE SECTIONAL VIEWS.  BAR
DEFORMATIONS ARE SHOWN ONLY
FOR ILLUSTRATIVE PURPOSES.

Figure 2.1 — Effective Weld Sizes for Flare-Groove Welds (See 2.3.2.3)

*(Continued)*

Flare-Bevel and Flare-V Groove Welds *(Continued)*

| | |
|---|---|
| Effective Throat | D1.4, 2.3.2.3 |

D1.4, . . . The effective weld size, when filled flush to the solid section of the reinforcing steel bar, shall be 0.4 of the bar radius for flare-bevel-groove welds and 0.6 of the bar radius for flare-groove-welds. Larger effective weld sizes may be used to determine allowable stresses provided the welding procedure qualifies the larger weld size. When bars of unequal diameter are being joined, the effective weld size shall be based on the radius of the smaller bar.

| | |
|---|---|
| Effective Weld Area | D1.4, 2.3.2 |

D1.4, . . . The effective weld area shall be the effective weld length multiplied by the effective weld size (see Figure 2.1)

| | |
|---|---|
| Effective Weld Length | D1.4, 2.3.2.1 |

D1.4, . . . The effective weld length shall be the weld length of the specified weld size. No reduction in effective length shall be made for either the start or finish if the weld is the specified size at these locations.

| | |
|---|---|
| Minimum Effective Length | D1.4, 2.3.2.2 |

D1.4, . . . The minimum effective weld length shall not be less than two times the bar diameter for equal size bars or two times the smaller bar diameter for two unequal size bars.

## Fusion                                                          D1.4, 4.4.3

D1.4, . . . There shall be thorough fusion between weld metal and base metal and between successive passes in the weld.

## Galvanized Bar                                              D1.4, 5.6 - 5.6.3

D1.4, . . . When welding galvanized base metal, one of the following options shall be met:

1. Welding of galvanized base metal, without prior removal of the coating, shall be performed in accordance with welding procedure qualified to the requirements of this code. Note that the welding procedure will normally involve larger root openings in joint, electrodes with lower silicon content, and slower welding speeds.

2. Welding of galvanized base metal may be done after removing all coating from within 2 in. of the weld joint. In this option, the welding shall be performed using a welding procedure for uncoated reinforcing bar qualified in accordance with this code. The galvanized coating may be removed with oxyfuel gas flame, abrasive shot blasting, or other suitable means. . . .

*When* welding galvanized surfaces, suitable ventilation shall be provided to prevent the concentration of fumes. . . .

## Gaps in

Direct Lap Splices                                                   D1.4, 4.2.5

D1.4, . . . For direct lap joints, if the bar deviates by more than $1/2$ of the bar diameter, or by no more than $1/4$ in. from each other while the bars remain in approximately the same plane, the joint shall be made through a splice bar or plate, and the requirements for an indirect lap joint shall apply. . . .                     D1.4, Fig. 3.4

*(Continued)*

## Gaps in *(Continued)*

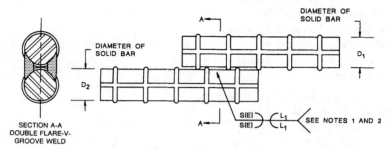

DIAMETER OF
SOLID BAR

DIAMETER OF
SOLID BAR

$D_1$

$D_2$

SECTION A-A
DOUBLE FLARE-V-
GROOVE WELD

A -

S(E)

S(E)

$L_1$

$L_1$

SEE NOTES 1 AND 2

**A - DIRECT LAP JOINT
WITH BARS IN CONTACT**

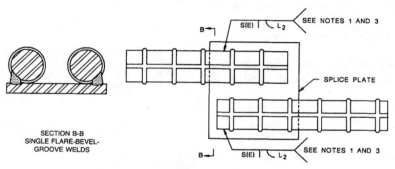

S(E)

$L_2$

SEE NOTES 1 AND 3

SPLICE PLATE

SECTION B-B
SINGLE FLARE-BEVEL-
GROOVE WELDS

S(E)

$L_2$

SEE NOTES 1 AND 3

**B - INDIRECT LAP JOINTS
WITH BARS SEPARATED**

NOTES :

1. THE EFFECTS OF ECCENTRICITY SHALL BE
   CONSIDERED OR RESTRAINT PROVIDED IN
   THE DESIGN OF THE JOINT

2. $L_1$    =  2 $D_1$ (MIN.) ; $D_1 \leq D_2$

3. $L_2$ MIN =  2 X DIAMETER OF BAR

4. GAPS BETWEEN BARS AND PLATES WILL VARY BASED
   ON HEIGHT OF DEFORMATIONS.

**Figure 3.4 — Lap Joints (See 3.2.1)**

Indirect Butt Splices

D1.4, 4.2.4

D1.4, . . . For indirect butt joints with splice
plates, the maximum joint clearance between the
bars shall not be more than ¾ in. . . .

D1.4, Fig. 3.3

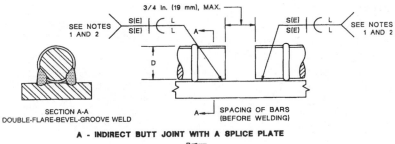

SECTION A-A
DOUBLE-FLARE-BEVEL-GROOVE WELD

A - INDIRECT BUTT JOINT WITH A SPLICE PLATE

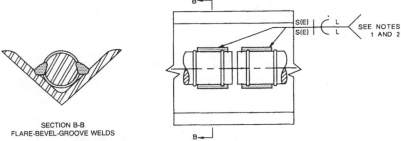

SECTION B-B
FLARE-BEVEL-GROOVE WELDS

B - INDIRECT BUTT JOINT WITH A SPLICE ANGLE

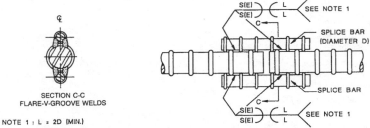

SECTION C-C
FLARE-V-GROOVE WELDS

C - INDIRECT BUTT JOINT WITH TWO SPLICE BARS

NOTE 1 : L = 2D (MIN.)

NOTE 2 : VARIATION OF THIS WELD
USING SINGLE FLARE-V
WELDS IS PERMITTED PROVIDED
ECCENTRICITY IS CONSIDERED IN DESIGN

NOTE 3 : GAPS BETWEEN BARS
OR BARS AND PLATES WILL VARY DEPENDING
ON HEIGHT OF DEFORMATIONS

NOTE 4   DEFORMATIONS SHOWN ON SECTIONAL VIEWS
ARE FOR ILLUSTRATIVE PURPOSES ONLY.

**Figure 3.3 — Indirect Butt Joints (See 3.5)**

Indirect Lap Splices                               D1.4, 4.2.6

D1.4, . . . For indirect lap joints (see Figure
3.4(B)), the maximum separation between the bar
and the splice plate shall be no more than $1/4$ of
the bar diameter, but not more than $3/16$ in. . . .

## Groove Weld Reinforcement

D1.4, 4.4.1

D1.4, . . . The fillet weld faces shall be slightly convex or slightly concave as shown in Figures 4.1(A) and 4.1(B) or flat, and with none of the unacceptable profiles exhibited in Figure 4.1(C).

D1.4, Fig. 4.1

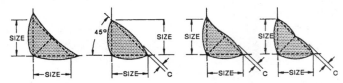

**(A) DESIRABLE FILLET WELD PROFILES**    **(B) ACCEPTABLE FILLET WELD PROFILES**

NOTE : CONVEXITY, C, OF A WELD OR INDIVIDUAL SURFACE BEAD SHALL NOT EXCEED THE VALUE OF THE FOLLOWING TABLE:

| MEASURED LEG SIZE OR WIDTH OF INDIVIDUAL SURFACE BEAD, L | MAX. CONVEXITY |
|---|---|
| L ≤ 5/16 In. (8mm) | 1/16 In. (1.6mm) |
| 5/16 In. < L < 1 In. (25mm) | 1/8 In. (3mm) |
| L ≥ 1 In. | 3/16 In. (5mm) |

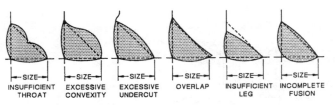

**(C) UNACCEPTABLE FILLET WELD PROFILES**

Figure 4.1 — Acceptable and Unacceptable Weld Profiles (See 4.4)

## Indirect Butt Splices

D1.4, 3.5

D1.4, . . . An indirect butt splice shall be made with either single or double-flare-groove welds between the bars and the splice member (see Figure 3.3).

D1.4, Fig. 3.3

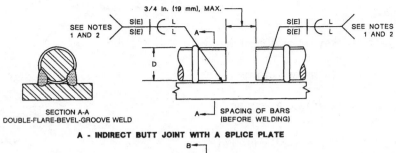

SECTION A-A
DOUBLE-FLARE-BEVEL-GROOVE WELD

SPACING OF BARS
(BEFORE WELDING)

**A - INDIRECT BUTT JOINT WITH A SPLICE PLATE**

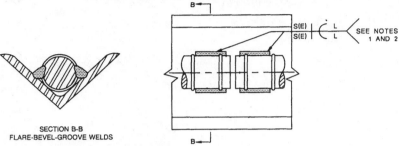

SECTION B-B
FLARE-BEVEL-GROOVE WELDS

**B - INDIRECT BUTT JOINT WITH A SPLICE ANGLE**

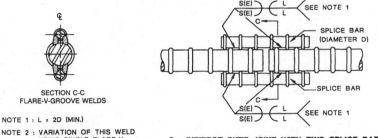

SECTION C-C
FLARE-V-GROOVE WELDS

SPLICE BAR
(DIAMETER D)

SPLICE BAR

NOTE 1 : L = 2D (MIN.)

NOTE 2 : VARIATION OF THIS WELD
USING SINGLE FLARE-V
WELDS IS PERMITTED PROVIDED
ECCENTRICITY IS CONSIDERED IN DESIGN

NOTE 3 : GAPS BETWEEN BARS
OR BARS AND PLATES WILL VARY DEPENDING
ON HEIGHT OF DEFORMATIONS

NOTE 4 : DEFORMATIONS SHOWN ON SECTIONAL VIEWS
ARE FOR ILLUSTRATIVE PURPOSES ONLY.

**C - INDIRECT BUTT JOINT WITH TWO SPLICE BARS**

**Figure 3.3 — Indirect Butt Joints (See 3.5)**

Gaps in

D1.4, 4.2.4

D1.4, . . . For indirect butt joints with splice plates, the maximum joint clearance between the bars shall not be more than ³/₄ in. (see Figure 3.3(A)).

## Insert Plate Connections

D1.4, 3.7 - 3.7.5

D1.4, . . . Precast members may be interconnected by welding reinforcing bars that project through the ends of the precast members or by welding together insert plates which have been cast into the precast member. The heat of welding may cause localized damage to the concrete.

D1.4, Fig. 3.5

*Joints* of projecting reinforcing bars shall conform to 3.3 through 3.6, as applicable.

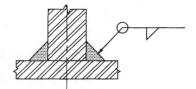

A - EXTERNAL FILLET WELD

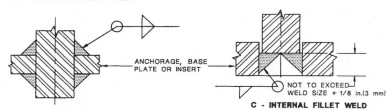

ANCHORAGE, BASE PLATE OR INSERT

NOT TO EXCEED WELD SIZE + 1/8 in.(3 mm)

B - EXTERNAL FILLET WELD

C - INTERNAL FILLET WELD

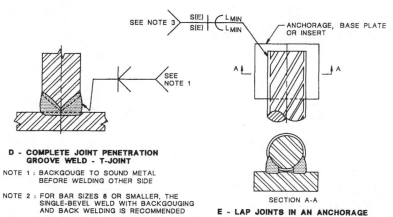

SEE NOTE 3    S(E)    L MIN
              S(E)    L MIN

ANCHORAGE, BASE PLATE OR INSERT

SEE NOTE 1

D - COMPLETE JOINT PENETRATION GROOVE WELD - T-JOINT

NOTE 1 : BACKGOUGE TO SOUND METAL BEFORE WELDING OTHER SIDE

NOTE 2 : FOR BAR SIZES 8 OR SMALLER, THE SINGLE-BEVEL WELD WITH BACKGOUGING AND BACK WELDING IS RECOMMENDED

NOTE 3 : L MIN = 2 X BAR DIAMETER

SECTION A-A

E - LAP JOINTS IN AN ANCHORAGE USING FLARE-BEVEL-GROOVE WELDS

# Inspection

D1.4, Chapter 7

### Materials

D1.4, 7.2

D1.4, . . . The inspector shall make certain that only base metals are used that conform to the requirements of this code. . . .

*The inspector* shall be furnished with complete detail drawings showing the size, length, type, and location of all welds to be made.

*The inspector* shall verify the welding and subsequent testing of any procedure qualification specimens that are required. The inspector shall inspect the welding equipment to be used for the work to make certain that it is in such condition as to enable qualified welders to apply qualified procedures and attain results prescribed elsewhere in this code.

*When* the quality of a welder's work appears to be below the requirements of this code, the inspector may require testing of a welder's qualification by means of a partial or complete requalification in accordance with 6.3.

### Welder, Operator

D1.4, 7.4 - 7.4.2

D1.4, . . . The inspector shall permit welding to be performed only by welders who are qualified in accordance with the requirements of this code. . . .

*When* the quality of a welder's work appears to be below the requirements of this code, the inspector may require testing of a welder's qualification by means of a partial or complete requalification in accordance with 6.3.

### Work and Records

D1.4, 7.5 - 7.5.7

D1.4, . . . The inspector shall verify that the size, length, and location of all welds conform to the requirements of this code and to the detail drawings, that no specified welds are omitted, and that no unspecified welds have been added without approval. . . .

*(Continued)*

Inspection, Work and Records *(Continued)*

> *The inspector* shall verify that the welding procedures used meet the requirements of this code. . . .

> *The inspector* shall verify that electrodes are used only in the position and with the type of welding current and polarity for which they are classified. . . .

> *The inspector* shall, at suitable intervals, observe the technique and performance of each welder to verify that the applicable requirements of this code are met. . . .

> *Inspectors* shall identify with a distinguishing mark or other recording methods all parts or joints that they have inspected and accepted. Any recording method that is mutually agreeable may be used. Die stamping of dynamically loaded structures is not permitted without the approval of the engineer. . . .

> *The inspector* shall keep a record of qualifications of all welders, all procedure qualifications or other tests that are made, and other such information as may be required.

Obligations of Contractor                         D1.4, 7.6

> D1.4, . . . The contractor shall be responsible for visual inspection and necessary correction of all deficiencies in materials and workmanship in accordance of Sections 3 and 4 and 4.4. . . .

> *The contractor* shall comply with all requests of the inspector(s) to correct deficiencies in materials and workmanship as provided in the contract documents. . . .

> *In* the event that faulty welding or its removal for rewelding damages the base so that in the judgment of the engineer its retention is not in accordance with the intent of the contract documents, the contractor shall remove and replace the damaged base metal or shall compensate for the deficiency in a manner approved by the engineer. . . .

*When* nondestructive testing other than visual inspection is specified in the information furnished to the bidders, it shall be the contractor's responsibility to ensure that all specified welds meet the quality requirements of 4.4.

Nondestructive Testing                                              D1.4, 7.7

D1.4, . . . When nondestructive testing other than visual is required, it shall be so stated in the information furnished to the bidders. This information shall designate the welds to be examined, the extent of the examination of each weld, and method of testing. . . .

*Welds* tested nondestructively that do not meet the requirements of this code shall be repaired using the applicable provisions of this code.

## Lap Splice Limitations                                          D1.4, 3.2.2

D1.4, . . . Welded lap joints shall be limited to bar size No. 6 and smaller.

## Lap Welded Splice Details                                        D1.4, 3.6

D1.4, . . . A lap weld shall be made with double-flare-V-groove welds (see Figure 3.4(A)), except that single-flare-V-groove welds may be used when the joint is accessible from only one side, and approved by the engineer. . . .                                                   D1.4, Fig. 3.4

*An* indirect lap joint shall be made with single-flare-bevel groove welds between the bars and the splice plate, with the bars being separated (see Figure 3.4(B)).

Direct Lap Splices                                                 D1.4, 4.2.5

D1.4, . . . For direct lap joints, if the bar deviates by more than $1/2$ of the bar diameter, or by no more than $1/4$ in. from each other while the bars remain in approximately the same plane, the joint shall be made through a splice bar or plate, and the requirements for an indirect lap joint shall apply (see 3.7.2 and 4.2.6).                       D1.4, Fig. 3.4

*(Continued)*

Lap Welded Splice Details *(Continued)*

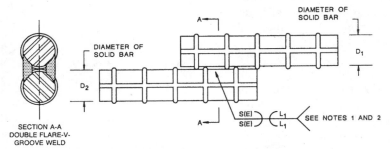

**A - DIRECT LAP JOINT WITH BARS IN CONTACT**

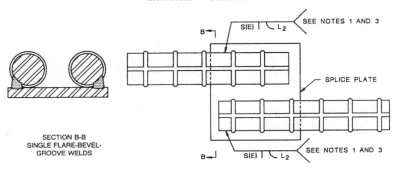

**B - INDIRECT LAP JOINTS WITH BARS SEPARATED**

NOTES :

1. THE EFFECTS OF ECCENTRICITY SHALL BE CONSIDERED OR RESTRAINT PROVIDED IN THE DESIGN OF THE JOINT

2. $L_1$     =   2 $D_1$ (MIN.) ; $D_1 \leq D_2$

3. $L_2$ MIN =  2 X DIAMETER OF BAR

4. GAPS BETWEEN BARS AND PLATES WILL VARY BASED ON HEIGHT OF DEFORMATIONS.

**Figure 3.4 — Lap Joints (See 3.2.1)**

Indirect Lap Splices

D1.4, . . . For indirect lap joints (see Figure 3.4(B)), the maximum separation between the bar and the splice plate shall be no more than $1/4$ of the bar diameter, but not more than $3/16$ in.

D1.4, 4.2.6
D1.4, Fig. 3.3

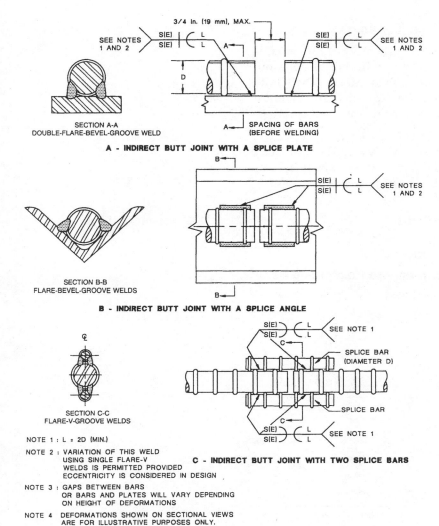

3/4 in. (19 mm), MAX.

SEE NOTES 1 AND 2

S(E)
S(E)

L
L

A

SEE NOTES 1 AND 2

S(E)
S(E)

L
L

D

SECTION A-A
DOUBLE-FLARE-BEVEL-GROOVE WELD

A

SPACING OF BARS
(BEFORE WELDING)

A - INDIRECT BUTT JOINT WITH A SPLICE PLATE

B

SECTION B-B
FLARE-BEVEL-GROOVE WELDS

S(E)
S(E)

L
L

SEE NOTES 1 AND 2

B

B - INDIRECT BUTT JOINT WITH A SPLICE ANGLE

S(E)
S(E)

L
L

SEE NOTE 1

C

SPLICE BAR
(DIAMETER D)

SECTION C-C
FLARE-V-GROOVE WELDS

SPLICE BAR

C

NOTE 1 : L = 2D (MIN.)

NOTE 2 : VARIATION OF THIS WELD
USING SINGLE FLARE-V
WELDS IS PERMITTED PROVIDED
ECCENTRICITY IS CONSIDERED IN DESIGN

NOTE 3 : GAPS BETWEEN BARS
OR BARS AND PLATES WILL VARY DEPENDING
ON HEIGHT OF DEFORMATIONS

NOTE 4   DEFORMATIONS SHOWN ON SECTIONAL VIEWS
ARE FOR ILLUSTRATIVE PURPOSES ONLY.

S(E)
S(E)

L
L

SEE NOTE 1

C - INDIRECT BUTT JOINT WITH TWO SPLICE BARS

**Figure 3.3 — Indirect Butt Joints (See 3.5)**

## Nondestructive Testing

D1.4, 7.7 - 7.7.5

D1.4, . . . When nondestructive testing other than visual inspection is specified in the information furnished to the bidders, it shall be the contractor's responsibility to ensure that all specified welds meet the quality requirements of 4.4. . . .

*(Continued)*

Nondestructive Testing *(Continuing)*

*When* nondestructive testing other than visual is required, it shall be so stated in the information furnished to the bidders. This information shall designate the welds to be examined, the extent of the examination of each weld, and method of testing. . . .

*Welds* tested nondestructively that do not meet the requirements of this code shall be repaired using the applicable provisions of this code.

# Nominal Dimensions of Bar                              D1.4, Appendix B

## Appendix B

## Nominal Dimensions of Standard Reinforcing Bars

(This Appendix is not a part of ANSI/AWS D1.4-92, *Structural Welding Code — Reinforcing Steel*, but is included for information purposes only.)

| Bar size† | Unit Weight | | Diameter | | Cross-Sectional Area | |
|---|---|---|---|---|---|---|
|  | lb/ft | kg/m | in. | mm | in.² | mm² |
| 3 | 0.376 | 0.56 | 0.375 | 9.52 | 0.11 | 71.0 |
| 4 | 0.668 | 1.00 | 0.500 | 12.70 | 0.20 | 129.0 |
| 5 | 1.043 | 1.55 | 0.625 | 15.88 | 0.31 | 200.0 |
| 6 | 1.502 | 2.26 | 0.750 | 19.05 | 0.44 | 283.9 |
| 7 | 2.044 | 3.64 | 0.875 | 22.22 | 0.60 | 387.1 |
| 8 | 2.670 | 3.98 | 1.000 | 25.40 | 0.79 | 509.7 |
| 9 | 3.400 | 5.07 | 1.128 | 28.65 | 1.00 | 645.2 |
| 10 | 4.303 | 6.41 | 1.270 | 32.26 | 1.27 | 819.4 |
| 11 | 5.313 | 7.92 | 1.410 | 35.81 | 1.56 | 1006.5 |
| 14 | 7.65 | 11.40 | 1.693 | 43.00 | 2.25 | 1451.7 |
| 18 | 13.60 | 20.26 | 2.257 | 57.33 | 4.00 | 2580.8 |

*The nominal dimensions of a deformed bar are equivalent to those of a plain round bar having the same weight per foot as the deformed bar.
†The bar size number is based on the number of eighths of an inch in the nominal diameter of the bar.

| Metric Bar Size | Diameter (mm.) | Inch-Pound Bar Size | Diameter (in.) |
|---|---|---|---|
| #10 | 9.5 | #3 | 0.375 |
| #13 | 12.7 | #4 | 0.500 |
| #16 | 15.9 | #5 | 0.625 |
| #19 | 19.1 | #6 | 0.750 |
| #22 | 22.2 | #7 | 0.875 |
| #25 | 25.4 | #8 | 1.000 |
| #29 | 28.7 | #9 | 1.128 |
| #32 | 32.3 | #10 | 1.270 |
| #36 | 35.8 | #11 | 1.410 |
| #43 | 43.0 | #14 | 1.693 |
| #57 | 57.3 | #18 | 2.257 |

## ASTM STANDARD REINFORCING BARS

| BAR SIZE DESIGNATION | NOMINAL AREA SQ. INCHES | NOMINAL WEIGHT POUNDS PER FT. | NOMINAL DIAMETER INCHES |
|---|---|---|---|
| #3 | 0.11 | 0.376 | 0.375 |
| #4 | 0.20 | 0.668 | 0.500 |
| #5 | 0.31 | 1.043 | 0.625 |
| #6 | 0.44 | 1.502 | 0.750 |
| #7 | 0.60 | 2.044 | 0.875 |
| #8 | 0.79 | 2.670 | 1.000 |
| #9 | 1.00 | 3.400 | 1.128 |
| #10 | 1.27 | 4.303 | 1.270 |
| #11 | 1.56 | 5.313 | 1.410 |
| #14 | 2.25 | 7.65 | 1.693 |
| #18 | 4.00 | 13.60 | 2.257 |

## Obligations of Contractor                    D1.4, 7.6 - 7.6.5

D1.4, . . . The contractor shall be responsible for visual inspection and necessary correction of all deficiencies in materials and workmanship in accordance of Sections 3 and 4 and 4.4. . . .

*The contractor* shall comply with all requests of the inspector(s) to correct deficiencies in materials and workmanship as provided in the contract documents. . . .

*In* the event that faulty welding or its removal for rewelding damages the base so that in the judgment of the engineer its retention is not in accor-

*(Continued)*

Obligations of Contractor *(Continuing)*

dance with the intent of the contract documents, the contractor shall remove and replace the damaged base metal or shall compensate for the deficiency in a manner approved by the engineer. . . .

*When* nondestructive testing other than visual inspection is specified in the information furnished to the bidders, it shall be the contractor's responsibility to ensure that all specified welds meet the quality requirements of 4.4.

## Offsets (Direct Butt Splice)                    D1.4, 4.2.3

D1.4, . . . The joint members shall be aligned so as to minimize eccentricities. After welding, bars in direct butt joints shall not be offset at the joint by more than the following:

| | |
|---|---|
| Bar sizes No. 10 or smaller | $1/8$ in. |
| Bar sizes No. 11 and No. 14 | $3/16$ in. |
| Bar size No. 18 | $1/4$ in. |

## Overlap                                         D1.4, 4.4.5

D1.4, . . . Welds shall be free from overlap.

## Piping Porosity                                 D1.4, 4.4.7

D1.4, . . . The sum of diameters of piping porosity in flare-groove and fillet welds shall not exceed $3/8$ in. in any linear inch of weld and shall not exceed $9/16$ in. in any 6 in. length of weld.

## Precast Connections                             D1.4, 3.7 - 3.7.5

D1.4, . . . Precast members may be interconnected by welding reinforcing bars that project through the ends of precast members or by welding together insert plates which have been cast into the precast members. The heat of welding may cause localized damage to the concrete. . . .

*Joints* of projecting reinforcing bars shall conform to 3.3 through 3.6, as applicable.

## Preheat and Interpass Temperatures

D1.4, Table 5.2

### Table 5.2
### Minimum Preheat and Interpass Temperature[1,2]
### (see 5.2.1)

| Carbon Equivalent[3,4] (C.E.) Range, % | Size of Reinforcing Bar | Shielded Metal Arc Welding with Low Hydrogen Electrodes, Gas Metal Arc Welding, or Flux Cored Arc Welding | |
|---|---|---|---|
| | | Minimum Temperature | |
| | | °F | °C |
| Up to 0.40 | up to 11 inclusive | none[5] | none[5] |
| | 14 and 18 | 50 | 10 |
| Over 0.40 to 0.45 inclusive | up to 11 inclusive | none[5] | none[5] |
| | 14 and 18 | 100 | 40 |
| Over 0.45 to 0.55 inclusive | up to 6 inclusive | none[5] | none[5] |
| | 7 to 11 inclusive | 50 | 10 |
| | 14 to 18 | 200 | 90 |
| Over 0.55 to 0.65 inclusive | up to 6 inclusive | 100 | 40 |
| | 7 to 11 inclusive | 200 | 90 |
| | 14 to 18 | 300 | 150 |
| Over 0.65 to 0.75 | up to 6 inclusive | 300 | 150 |
| | 7 to 18 inclusive | 400 | 200 |
| Over 0.75 | 7 to 18 inclusive | 500 | 260 |

1. When reinforcing steel is to be welded to main structural steel, the preheat requirements of the structural steel shall also be considered (see ANSI/AWS D1.1, table titled "Minimum Preheat and Interpass Temperature." The minimum preheat requirement to apply in this situation shall be the higher requirement of the two tables. However, extreme caution shall be exercised in the case of welding reinforcing steel to quenched and tempered steels, and such measures shall be taken as to satisfy the preheat requirements for both. If not possible, welding shall not be used to join the two base metals.

2. Welding shall not be done when the ambient temperature is lower than 0°F (-18°C). When the base metal is below the temperature listed for the welding process being used and the size and carbon equivalent range of the bar being welded, it shall be preheated (except as otherwise provided) in such a manner that the cross section of the bar for not less than 6 in. (150 mm) on each side of the joint shall be at or above the specified minimum temperature. Preheat and interpass temperatures shall be sufficient to prevent crack formation.

3. After welding is complete, bars shall be allowed to cool naturally to ambient temperature. Accelerated cooling is prohibited.

4. Where it is impractical to obtain chemical analysis, the carbon equivalent shall be assumed to be above 0.75%. See also 1.3.4.

5. When the base metal is below 32°F (0°C), the base metal shall be preheated to at least 70°F (20°C), or above, and maintained at this minimum temperature during welding.

## Preparation of Materials                          D1.4, 4.1 - 4.1.2

D1.4, . . . Surfaces to be welded shall be free from
fins, tears, cracks, or other defects that would adversely affect the quality or strength of the weld.
Surfaces to be welded, and surfaces adjacent to a
weld, shall also be free from loose or thick scale,
slag, rust, moisture, grease, epoxy coating, or other
foreign material that would prevent proper welding
or produce objectionable fumes. Mill scale that
withstands vigorous wire brushing, a thin inhibi-
tive coating, or antispatter compound may remain.

Notches and Gouges                                   D1.4, 4.1.2

D1.4, . . . Roughness exceeding this value, and
occasional notches or gouges not more than
$3/16$ in. deep on otherwise satisfactory surfaces,
shall be removed by machining or grinding. Bars
for direct butt joints that have sheared ends
shall be trimmed back beyond the area deformed
by shearing.

## Progression of Welding                            D1.4, 5.5

D1.4, . . . Welds made in the vertical position (see
Figure 6.1(C), position 3G or Figure 6.2(C), position
3G), shall use vertical-up progression.

## Qualification Procedures                          D1.4, Chapter 6

D1.4, . . . Welding procedures for reinforcing bar     D1.4, Fig 6.1 and 6.2
joints, anchorage, base plate, and insert connections
that are to be employed in executing contract work
under this section shall be established in a procedure specification and shall be qualified prior to use
by tests as prescribed in 6.2 to the satisfaction of
the engineer. . . .

*At* the engineer's discretion, evidence of previous
qualification of the welding procedures and welders
to be employed may be accepted. . . .

*Welding* procedure qualification shall be required
for complete joint penetration groove welds and

flare-groove welds for each welding position (see Figures 6.1 and 6.2).

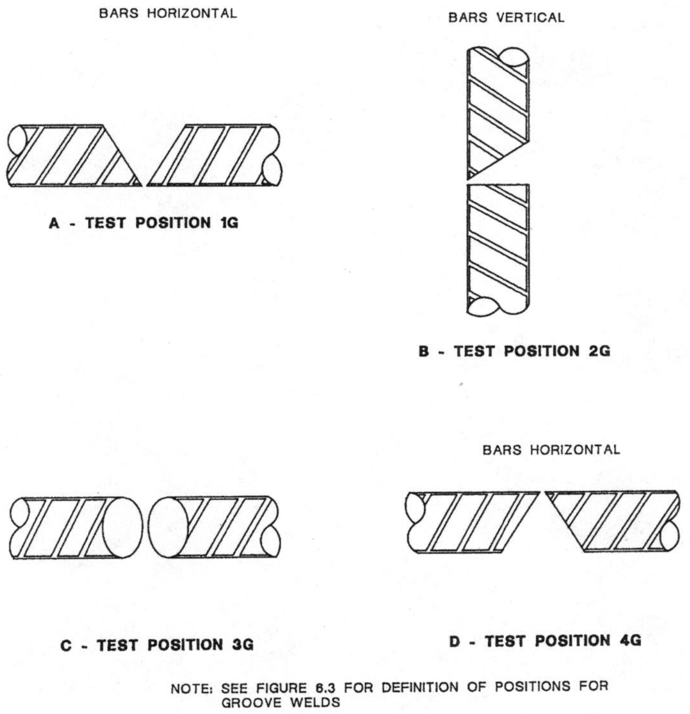

BARS HORIZONTAL

A - TEST POSITION 1G

BARS VERTICAL

B - TEST POSITION 2G

C - TEST POSITION 3G

BARS HORIZONTAL

D - TEST POSITION 4G

NOTE: SEE FIGURE 6.3 FOR DEFINITION OF POSITIONS FOR GROOVE WELDS

**Figure 6.1 — Direct Butt Joint Test Positions for Groove Welds (See 6.2.4.1)**

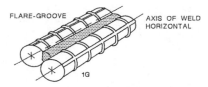

A - TEST POSITION : FLAT

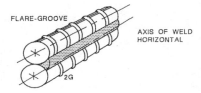

B - TEST POSITION : HORIZONTAL

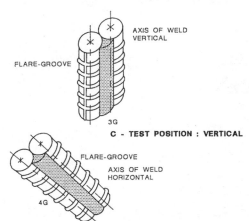

C - TEST POSITION : VERTICAL

D - TEST POSITION : OVERHEAD

NOTE: SEE FIGURES 6.3 AND 6.4 FOR DEFINITIONS OF
POSITIONS FOR FLARE-GROOVE AND FILLET WELDS.

Figure 6.2 — Indirect Butt Joint Test Positions for Flare-Groove Welds
or Fillet Welds (See 6.2.4.2)

## Requalification Variables

D1.4, 6.2.1 - 6.2.1.7

D1.4, . . . Changes in 6.2.1.5 through 6.2.1.7 shall
be considered in essential variables. Joint welding
procedures including such changes shall require requalification. . . .

Shielded Metal Arc Welding:

1. A change increasing filler metal strength level, for example, a change from E70XX to E80XX-X, but not vice versa.

2. An increase in the diameter of the electrode used over that required by the welding procedure specification.

3. A change of electrode amperage and voltage that are not within the ranges recommended by the electrode manufacturer.

4. A change in position in which welding is performed as defined in 6.2.4.1 and 6.2.4.2.

5. A change in the type of groove, for example, a change from a flare-V-groove to a flare-bevel-groove.

6. A change in the shape in any one type of groove involving:

    a. A decrease in the included angle of the groove exceeding 5 deg.

    b. A decrease in the root opening of the groove exceeding $1/16$ in.

    c. An increase in the root face of the groove exceeding $1/16$ in.

    d. The omission, but not inclusion, of backing material. . . .

Welder Qualifications                                           D1.4, 6.3 - 6.3.2.4

D1.4, . . . A welder shall be qualified for each process to be used in fabrication. . . .

A *change* in the position of welding to one for which the welder is not qualified requires qualification in that position.

# Reinforcing Base Metal                                        D1.4, 1.3

D1.4, . . . Reinforcing steel base metals in this               SBCCI, 1602.5.3
code shall conform to the requirements of the latest
edition of one of the ASTM specifications listed below. Combinations of any of these reinforcing steel
base metals may be welded together. . . .

## Requalification Variables *(see Qualification Procedures)*

D1.4, 6.,2.1 - 6.2.1.7

## Shielding Gas

D1.4, 5.8.2

D1.4, . . . When a gas or gas mixture is used for shielding in gas metal arc or flux cored arc welding, it shall be of a welding grade having a dew point of −40 deg. F or lower. When requested by the engineer, the gas manufacturer shall furnish certification that the gas or gas mixture will meet the dew point requirement.

## Size of

ASTM Bar Size Table

### ASTM STANDARD REINFORCING BARS

| BAR SIZE DESIGNATION | NOMINAL AREA SQ. INCHES | NOMINAL WEIGHT POUNDS PER FT. | NOMINAL DIAMETER INCHES |
|---|---|---|---|
| #3 | 0.11 | 0.376 | 0.375 |
| #4 | 0.20 | 0.668 | 0.500 |
| #5 | 0.31 | 1.043 | 0.625 |
| #6 | 0.44 | 1.502 | 0.750 |
| #7 | 0.60 | 2.044 | 0.875 |
| #8 | 0.79 | 2.670 | 1.000 |
| #9 | 1.00 | 3.400 | 1.128 |
| #10 | 1.27 | 4.303 | 1.270 |
| #11 | 1.56 | 5.313 | 1.410 |
| #14 | 2.25 | 7.65 | 1.693 |
| #18 | 4.00 | 13.60 | 2.257 |

## Tack Welds

D1.4, 5.4

D1.4, . . . Tack welds that do not become a part of permanent welds are prohibited unless authorized by the engineer. Tack welds shall be made using preheat and welded with electrodes meeting the requirements of final welds. They shall be thoroughly cleaned and subjected to the same quality requirements as the final welds.

## Transition in Bar Size

D1.4, 3.1

D1.4, . . . Direct butt joints in tension in axially aligned bars of different size shall be made as shown in Figure 3.1 (see Appendix B for bar sizes).

D1.4, Fig. 3.1

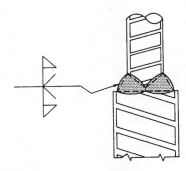

Figure 3.1 — Direct Butt Joint Showing Transition Between Bars of Different Sizes (See 3.1)

## Undercut

D1.4, 4.4.6

D1.4, . . . Undercut depth greater than $1/32$ in. in the solid section of the bar or structural member shall not be allowed.

## Welder Qualifications *(See Qualification Procedures)*

D1.4, 1.4, 6.3 - 6.3.2.4

# Weld Profiles

D1.4, 4.4.1

D1.4, . . . The fillet weld faces shall be slightly concave, as shown in Figures 4.1(A) and 4.1(B) or flat, and with none of the unacceptable profiles exhibited in Figure 4.1(C).

D1.4, Fig 4.1

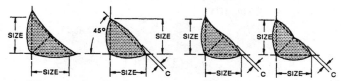

**(A) DESIRABLE FILLET WELD PROFILES**     **(B) ACCEPTABLE FILLET WELD PROFILES**

NOTE : CONVEXITY, C, OF A WELD OR INDIVIDUAL SURFACE BEAD SHALL NOT EXCEED THE VALUE OF THE FOLLOWING TABLE:

| MEASURED LEG SIZE OR WIDTH OF INDIVIDUAL SURFACE BEAD, L | MAX. CONVEXITY |
|---|---|
| L ≤ 5/16 In. (8mm) | 1/16 In. (1.6mm) |
| 5/16 In. < L < 1 In. (25mm) | 1/8 In. (3mm) |
| L ≥ 1 In. | 3/16 In. (5mm) |

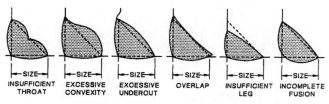

| INSUFFICIENT THROAT | EXCESSIVE CONVEXITY | EXCESSIVE UNDERCUT | OVERLAP | INSUFFICIENT LEG | INCOMPLETE FUSION |

**(C) UNACCEPTABLE FILLET WELD PROFILES**

Figure 4.1 — Acceptable and Unacceptable Weld Profiles (See 4.4)

# Wind Velocity

D1.4, 5.2.2

D1.4, . . . Welding shall not be done when ambient temperature is lower that 0 deg. F, when surfaces to be welded are exposed to rain, snow, or wind velocities greater than 5 miles per hour, or when welders are exposed to inclement conditions.

# F  SHEET STEEL (DECKING)

## Approval Needed

D1.3, . . . All references to the need for approval shall be interpreted to mean approval by the building commissioner, the engineer, or a duly-designated person who acts for and on behalf of the owners on all matters within the scope of this specification.

D1.3, 1.1.6

## Ambient Temperature

D1.3, . . . Welding shall not be done when the ambient temperature is lower than 0 deg. F; when surfaces are wet or exposed to rain, sleet, snow, or high wind; or when welders are exposed to inclement conditions.

D1.3, 4.1.3

## Base Metals

D1.3, 1.2

## Butt Joints

D1.3, . . . As covered by this specification, this type of weld is restricted to the welding of sheet steel to sheet steel in all positions of welding.

D1.3, 1.1.2

## Cracks

D1.3, . . . A weld shall be acceptable by visual inspection, provided all four of the following criteria are met:

*The* weld has no cracks. . . .

D1.3, 4.5.2.1

## Diameter of Spot and Puddle Welds

D1.3, 2.2.2.1

D1.3, . . . The minimum allowable effective diameter is ³⁄₈ in.

## Flare-Groove Welds

D1.3, 1.1.5

D1.3, . . . As covered by this specification, these types of welds may be used (see Table 1.1) in any position of welding involving:

1. Two sheet steels for flare-V-grooves.

2. Two sheet steels, as well as a sheet and a thicker steel member for flare-bevel-groove welds and fillet welds (see Table 1.1).

Bevel Welds

D1.3, 3.6

D1.3, . . . Single-flare-bevel groove welds shall preferably be made in the horizontal or flat position. The minimum length shall be ³⁄₄ in. (see Figure 3.5).

V-Groove Welds

D1.3, 3.7

D1.3, . . . Single-flare-V-groove welds shall preferably be made in the flat position. The minimum length shall be ³⁄₄ in. (see Figure 3.6).

## Fillet Welds

D1.3, 3.5

D1.3, . . . Fillet welds in lap and T-joints shall preferably be made in the horizontal position. The minimum length shall be ³⁄₄ in. (see Figures 3.4A and 3.4B).

D1.3, C2.2.4

Minimum Length

D1.3, 3.5

D1.3, . . . The minimum length shall be ³⁄₄ in.

## Filler Metal Requirements

D1.3, 5.1 - 5.5

D1.3, . . . For groove welds in butt joints, the electrodes or electrode and gas combinations shall be in accordance with Table 5.1.

D1.3, Table 5.1

## Table 5.1
## Matching Filler Metal Requirements
## (see 5.1.1)

| Steel Specification | Minimum Yield Point | | Minimum Tensile Strength | | Filler Metal Requirements |
|---|---|---|---|---|---|
| | ksi | MPa | ksi | MPa | |
| A446 Gr A | 33 | 230 | 45 | 310 | SMAW AWS A5.1 or A5.5 |
| Gr B | 37 | 255 | 52 | 360 | E60XX, E70XX or E70XX-X |
| Gr C | 40 | 275 | 55 | 380 | |
| A570 Gr 30 | 30 | 205 | 49 | 340 | SAW AWS A5.17 or A5.23 |
| Gr 33 | 33 | 230 | 52 | 360 | F6X-EXXX or F7AX-EXXX |
| Gr 36 | 36 | 250 | 53 | 365 | |
| Gr 40 | 40 | 275 | 55 | 380 | GMAW AWS A5.18 |
| Gr 45 | 45 | 310 | 60 | 415 | ER70S-X |
| Gr 50 | 50 | 345 | 65 | 450 | |
| A606 | 45 | 310 | 65 | 450 | |
| A607 Gr 45 | 45 | 310 | 60 | 415 | FCAW AWS A5.20 |
| Gr 50 | 50 | 345 | 65 | 450 | E60T-X or E70T-X (except 2 and 3) |
| A611 Gr A | 25 | 175 | 42 | 290 | |
| Gr B | 30 | 205 | 45 | 310 | |
| Gr C | 33 | 230 | 48 | 335 | |
| Gr D | 40 | 275 | ·52 | 360 | |
| A607 Gr 55 | 55 | 380 | 70 | 485 | SMAW AWS A5.1 or A5.5 E70XX or E70XX-X SAW AWS A5.17 or A5.23 F7AX-EXXX GMAW AWS A5.18 ER70S-X FCAW AWS A5.20 E70T-X (except 2 and 3) |
| A446 Gr E | 80 | 550 | 82 | 565 | SMAW AWS A5.5 E80XX-X |
| A607 Gr 60 | 60 | 415 | 75 | 515 | SAW AWS A5.23 F8AX-EXXX |
| Gr 70 | 70 | 485 | 85 | 585 | GMAW AWS A5.28 ER80S |
| A611 Gr E | 80 | 550 | 82 | 565 | FCAW AWS A5.29 E8XT |

Note: Low hydrogen electrodes must be used when required by ANSI/AWS D1.1. See 1.1.1.

## Formed Steel Decking (FSD) *(See also,* Studs, *Section A)*

AISC, 5-60, I5.1 - I5.3

Composite construction shall, where applicable, be limited to decking having a 3 in. maximum rib height and an average rib height of not less than 2 in.

AISC, 2-246
AISC, 2-255
UBC, 2-686, I5 - I5.3

*Welded shear* connectors shall be ¾ in. or less in diameter.

*(Continued)*

Formed Steel Decking (FSD) *(Continued)*

*After installation* by welding, shear connectors shall extend a minimum of 1 ½ in. above the top of the steel deck.

*The top* of the composite slab shall be a minimum of 2 in. above the top of the decking.

*The maximum* center-to-center spacing of shear connectors on a beam or girder shall not be greater than 36 in.

*Decking shall* be anchored, by welds or shear connectors, at a maximum center-to-center spacing of 16 in. Welds and studs are used as deck anchors to help resist uplifting of the deck.

*Studs must* be placed in the deck ribs; therefore, stud spacing must be in 6 in. multiples. Additional studs or "puddle welds" must be used, as required, so that the space between deck attachments (welds or studs) does not exceed 16 in. . . .

*Unless* located directly over the web, the diameter of studs shall not be greater than 2 ½ times the thickness of the flange to which they are welded. The minimum center-to-center spacing of stud shear connectors shall be 6 diameters along the longitudinal axis of the supporting composite beam.

*Eight times* the slab thickness is the maximum allowable center-to-center spacing of shear connectors

## Groove Welds (Butt Joints)                          D1.3, 1.1.2

D1.3, . . . As covered by this specification, this type of weld is restricted to the welding of sheet steel to sheet steel in all positions of welding.

# Gage Numbers and Equivalent Thicknesses

D1.3, Appendix D

**Table D1**
**Gage Numbers and Equivalent Thicknesses**
**Hot-Rolled and Cold-Rolled Sheet**

| Manufacturers' Standard Gage Number | Thickness Equivalent in. | mm |
|---|---|---|
| 3 | 0.2391 | 6.073 |
| 4 | 0.2242 | 5.695 |
| 5 | 0.2092 | 5.314 |
| 6 | 0.1943 | 4.935 |
| 7 | 0.1793 | 4.554 |
| 8 | 0.1644 | 4.176 |
| 9 | 0.1495 | 3.800 |
| 10 | 0.1345 | 3.416 |
| 11 | 0.1196 | 3.038 |
| 12 | 0.1046 | 2.657 |
| 13 | 0.0897 | 2.278 |
| 14 | 0.0747 | 1.900 |
| 15 | 0.0673 | 1.709 |
| 16 | 0.0598 | 1.519 |
| 17 | 0.0538 | 1.366 |
| 18 | 0.0478 | 1.214 |
| 19 | 0.0418 | 1.062 |
| 20 | 0.0359 | 0.912 |
| 21 | 0.0329 | 0.836 |
| 22 | 0.0299 | 0.759 |
| 23 | 0.0269 | 0.660 |
| 24 | 0.0239 | 0.607 |
| 25 | 0.0209 | 0.531 |
| 26 | 0.0179 | 0.455 |
| 27 | 0.0164 | 0.417 |
| 28 | 0.0149 | 0.378 |

Note: Table D1 is for information only. This product is commonly specified to decimal thickness, not to gage number.

**Table D2**
**Gage Numbers and Equivalent Thicknesses**
**Galvanized Sheet**

| Galvanized Sheet Gage Number | Thickness Equivalent in. | mm |
|---|---|---|
| 8 | 0.1681 | 4.270 |
| 9 | 0.1532 | 3.891 |
| 10 | 0.1382 | 3.510 |
| 11 | 0.1233 | 3.132 |
| 12 | 0.1084 | 2.753 |
| 13 | 0.0934 | 2.372 |
| 14 | 0.0785 | 1.993 |
| 15 | 0.0710 | 1.803 |
| 16 | 0.0635 | 1.613 |
| 17 | 0.0575 | 1.460 |
| 18 | 0.0516 | 1.311 |
| 19 | 0.0456 | 1.158 |
| 20 | 0.0396 | 1.006 |
| 21 | 0.0366 | 0.930 |
| 22 | 0.0336 | 0.853 |
| 23 | 0.0306 | 0.777 |
| 24 | 0.0276 | 0.701 |
| 25 | 0.0247 | 0.627 |
| 26 | 0.0217 | 0.551 |
| 27 | 0.0202 | 0.513 |
| 28 | 0.0187 | 0.475 |
| 29 | 0.0172 | 0.437 |
| 30 | 0.0157 | 0.399 |
| 31 | 0.0142 | 0.361 |
| 32 | 0.0134 | 0.340 |

Note: Table D2 is for information only. This product is commonly specified to decimal thickness, not to gage number.

# Inspection

D1.3, 7.1 - 7.3.2

D1.3, . . . Welds shall be inspected visually and shall meet the requirements of 4.5.

D1.3, 4.5 - 4.5.2.4
D1.3, 4.5.1 - 4.5.2.4

*The inspector* shall make certain, prior to welding, that qualified and valid welding procedure specifications applicable to the contract are available, as required by this specification, and that all welders are qualified and are thoroughly familiar with these procedures. . . .

Inspection *(Continued)*

*At* any time, and specifically while arc spot welds are being made, the inspector may request that the melting rate of the electrodes, wire feed speed, or welding current be compared with that established in the welding procedure qualification test. If these melting rates are 5% or more below those specified, new welds using the correct current shall be made adjacent to those welds made with inadequate current.

*When* the quality of a welder's work is judged by the inspector to be below the requirements of this specification, requalification of the welder may be required.

*A weld* shall be acceptable by visual inspection, provided all four of the following criteria are met:

1. The weld has no cracks.

2. The weld has a minimum reinforcement of $1/32$ in. for all square groove, arc spot, and arc seam welds.

3. The cumulative length of undercut is no longer than L/8, where L is the specified length of the weld or, in the case of arc spot welds, the circumference, provided fusion exists between the weld metal and the base metal. Depth of undercut is not a subject of inspection and need not be measured. Melt-through that results in a hole is unacceptable.

4. Faces of fillet welds shall be flat or slightly convex.

# Melting Rate                                D1.3, C5.2.4

D1.3, . . . A good measure of welding current can be provided by the melting rate (M) of the electrode.

$$M = \frac{\text{inches of electrode melted}}{\text{time in minutes}}$$

*Melting* rate is a method of measuring welding current that has long been in use. Many published shielded arc welding procedures still include the melting rate (M) along with the welding current and other data needed.

*Once* the welding current for a given size and classification of shielded metal arc electrode has been established, the welder should place a new electrode into the holder. The welder should proceed to weld at this current level for 1 minute (60 seconds) and then measure the length of the electrode melted during this time interval. This can easily be done by placing a steel measuring tape along the electrode stub, as shown in Figure C9. . . .

*For* most welding procedures on sheet steel, it will take less than 60 seconds to melt off the electrode. In such a case, a shorter welding time should be used. For example, the electrode may be melted for 30 seconds, and the melted length of electrode should be multiplied by two. A 20 second period may be used with a multiplier of three, or a 15 second period, with a multiplication factor of four.

## Minimum Length of

Flare-Bevel Groove Weld                                    D1.3, 3.6

  D1.3, . . . The minimum length shall be
  $^3/_4$ in. . . .

Flare-V Groove Weld                                         D1.3, 3.7

  D1.3, . . . The minimum length shall be
  $^3/_4$ in. . . .

Fillet Weld                                                 D1.3, 3.5

  D1.3, . . . The minimum length shall be
  $^3/_4$ in. . . .

## Processes                                                 D1.3, 1.3 - 1.3.3

  D1.3, . . . This specification provides for welding with the shielded metal arc (S.M.A.W.), gas metal arc (G.M.A.W.), flux cored arc (F.C.A.W.), or submerged arc (S.A.W.) welding processes. (NOTE: Any

  *(Continued)*

Processes *(Continued)*

variation of gas metal arc welding, short-circuiting transfer, is acceptable.)

*When* stud welding through the flat portion of decking or roofing onto supporting structural members, the procedure shall conform to Section 7, Stud Welding, of ANSI/AWS D1.1, *Structural Welding Code-Steel.*

*Other* welding processes may be used when approved by the engineer. In such cases, the engineer shall specify any additional qualification requirements necessary to assure satisfactory joints for the intended service.

## Preparation of Material

D1.3, 4.1 - 4.1.3

D1.3, . . . Surfaces to be welded shall be smooth, uniform, and free from fins, tears, cracks, or other imperfections that would adversely affect the quality or strength of the weld. Surfaces to be welded and surfaces adjacent to a weld shall also be free from loose or thick scale, slag, rust, moisture, grease, or other foreign material that would prevent proper welding or produce objectionable fumes. Mill scale that can withstand vigorous wire brushing, a thin rust-inhibitive coating, a galvanized coating, or an antispatter compound may remain.

## Puddle Welds

D1.3, 1.1.3

D1.3, . . . As covered by this specification, this type of weld is restricted to the welding of sheet steel to thicker supporting members in the flat position (see Table 1.1). Neither the thickness of a single sheet nor the combined thickness of sheets welded to supporting members shall exceed 0.15 in. . . .

D1.3, 2.2.2.1

*The* maximum allowable effective diameter is ³⁄₈ in. . . .

## Qualification Procedures and Tests

D1.3, 6.1 - 6.8.2.12

## Quality of Welds

D1.3, 4.4

D1.3, . . . Welds shall be visually inspected for their location, size, and length in conformance to the drawings and specifications, as well as for their qualities such as bead shape, reinforcement, and undercut.

## Requalification

D1.3, 7.3.2

D1.3, . . . When the quality of a welder's work is judged by the inspector to be below the requirements of this specification, requalification of the welder may be required.

Variables

D1.3, C4.3

## Seam Welds

D1.3, 1.1.4

D1.3, . . . As covered by this specification, this type of weld is restricted (see Table 1.1) to the welding of joints involving:

1. Sheet to thicker supporting member in the flat position.
2. Sheet to sheet in the flat or horizontal position (see Table 1.1).

Arc Seam Welds

D1.3, C2.2.3

D1.3, . . . Some decks are made with narrow flutes that may restrict the size of the arc spot weld that can be made. In this case, an oblong arc spot weld, called an arc seam weld, is made. Its additional length makes up for the smaller diameter. The minimum length of the weld is 1 1/2 in. and is measured between the centers of the circular portions of the weld (see Figure C4).

*(Continued)*

Seam Welds *(Continued)*

Specification of                                                      D1.3, 3.4

> D1.3, . . . The minimum width of weld metal at
> the faying surface of arc seam welds shall be
> 3/8 in. The maximum distance from the longitudi-
> nal axis of an arc seam weld or from the end of
> an arc seam weld to the edge of the sheet steel
> shall not be less than that obtained when using
> the formula in 3.3.2, but not less than 1.5d (see
> Figure 3.3A). . . .
>
> *Arc* seam welds between the sheet and the sup-
> porting member shall be made in the flat posi-
> tion. . . .
>
> *Arc* seam welds between steel sheets may be
> made in the horizontal position, provided the fit-
> up is exceptionally good.

## Shielding Gas                                                      D1.3, 5.4.3

> D1.3, . . . A gas or gas mixture used for shielding
> in gas metal arc welding or flux cored arc welding,
> shall be of a welding grade having a dew point of
> −40 deg. F or lower. When requested by the engi-
> neer, the gas manufacturer shall furnish certifica-
> tion that the gas or gas mixture meets the procure-
> ment specification and will meet the dew point
> requirement.

## Spot Welds                                                        D1.3, 1.1.3

> D1.3, . . . Neither the thickness of a single sheet          UBC, 2-519, E2
> nor the combined thickness of sheets welded to sup-
> porting members shall exceed 0.15 in. . . .
>
> UBC, . . . Arc spot welds permitted by this specifi-
> cation are for welding sheet steel to thicker sup-
> porting members in the flat position. Arc spot welds
> (puddle welds) shall not be made on steel where the
> thinnest connected part is over 0.15 in. thick, nor
> through a combination of steel sheets having a total
> thickness over 0.15 in.

Minimum Diameter of                           D1.3, 2.2.2.1

D1.3, . . . The minimum allowable effective diameter is ⅜ in.

Minimum Reinforcement of                      D1.3, 4.5.2.2

D1.3, . . . The weld has a minimum reinforcement of ¹⁄₃₂ in. for all square groove, arc spot, and arc seam welds.

On Supporting Members                         D1.3, 3.3

D1.3, . . . Arc spot welds made through one or two thicknesses of sheet steel onto a supporting member shall be performed in the flat position, as shown in Figure 3.2A. For sheets thinner than 0.028 in., a steel washer, as shown in Figure 3.2B, shall be used to prevent burn back. The weld metal shall have a diameter of at least ⅜ in. at the fusion zone.

*The* minimum distance from the center of an arc spot weld to any edge of the sheet material shall not be less than . . . but not less than 1.5 d, where d is the face diameter and where *P* is the force transferred by the arc spot weld (see Figures 3.2A and 3.2C).

## Studs *(See also, Section A)*            D1.3, 6.7.6

D1.3, . . . When qualifying studs to be fused through sheet steels onto structural members, the sheet steel must be placed tightly against the structural member. The quality control requirements of ANSI/AWS D1.1, *Structural Welding Code-Steel,* shall apply.

## Square Groove Weld                        D1.3, 3.2

D1.3, . . . Square grooves shall be used in butt   D1.3, Fig. 3.1
joints with the welding done preferably in the flat position. Root opening (gap) shall be qualified during the joint welding procedure qualification test and shall be subject to the limitations of 6.6 (see Figure 3.1).

*(Continued)*

Square Groove Weld *(Continued)*

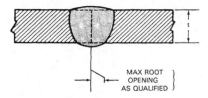

MAX ROOT
OPENING
AS QUALIFIED

Figure 3.1 — Square-Groove Welds in Butt Joints (See 3.2)

## Thickness Limitations

D1.3, 1.1.1

D1.3, . . . This specification is applicable to the arc welding of sheet steels or strip steels, or both, including cold-formed members, 0.18 in. or less in thickness.

## Welding Processes

D1.3, 1.3 - 1.3.3

D1.3, . . . This specification provides for welding with the shielded metal arc (S.M.A.W.), gas metal arc (G.M.A.W.), flux cored arc (F.C.A.W.), or submerged arc (S.A.W.) welding processes. (NOTE: Any variation of gas metal arc welding, short-circuiting transfer, is acceptable.)

*When* stud welding through the flat portion of decking or roofing onto supporting structural members, the procedure shall conform to Section 7, Stud Welding, of ANSI/AWS D1.1, *Structural Welding Code-Steel.*

*Other* welding processes may be used when approved by the engineer. In such cases, the engineer shall specify any additional qualification requirements necessary to assure satisfactory joints for the intended service.

## Welding Sheet Steel

UBC, 2-518, E2

UBC, . . . Arc welds on steel where each connected part is over 0.18 in. in thickness shall be made in accordance with AISC Specification (Section A6). . . .

UBC, 2-784, E2

*Except* as modified herein, arc spot welds on steel where at least one of the connected parts is 0.18 in. or less in thickness shall be made in accordance with the AWS D-1.3 (Section 6). . . .

## Weld Washers

D1.3, . . . Weld washers are to be used in containing the arc spot welds in sheet steel thinner than 0.028 in. (see Figure 2.1).

*Weld* washers shall be made of one of the sheet steels listed in 1.2.1 and shall have a thickness between 0.05 and 0.08 in., with a minimum prepunched hole diameter of ³⁄₈ in.

UBC, . . . Weld washers shall be used when the thickness of the sheet is less than 0.028 in. Weld washers shall have a thickness between 0.05 and 0.08 in. with a minimum prepunched hole of ³⁄₈ inch diameter.

D1.3, 2.2.2.2

D1.3, Fig. 2.1
UBC, 2-519, E2.2
UBC, 2-785, E2.2

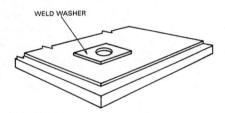

WELD WASHER

Figure 2.1 — Typical Weld Washer
(See 2.2.2.2)

# Index